高等学校 Java 课程系列教材

Android 手机程序设计实用教程

耿祥义　张跃平　编著

清华大学出版社
北　京

内 容 简 介

手机已经进入智能手机时代，基于 Android 操作系统的智能手机正在受到广泛的关注，市场对 Android 手机应用程序的需求正在迅速增长，因此，学习开发 Android 手机程序是非常有意义的。本书注重 Android 系统的特点，特别是手机程序设计的特点，使用 Android 4.2，重点讲解 Android 手机应用开发的核心内容。本书注重教材的可读性和实用性，许多例题都经过精心的考虑，既能帮助读者理解知识，同时又具有启发性和实用性。全书共分 12 章，分别是 Android 简介与开发环境、Android 应用程序的结构、常用 View 视图、常用 ViewGroup 视图、常用专用视图、菜单、动作栏与对话框、2D 绘图、Intent 对象、常用后台对象、使用 SD 卡、文件的读写、使用 SQLite 数据库等内容。

本书适合高等院校相关专业作为 Android 手机程序设计的教材，以及自学者和 Android 手机软件开发人员参考使用。

本书封面贴有清华大学出版社防伪标签，无标签者不得销售。
版权所有，侵权必究。侵权举报电话：010-62782989　13701121933

图书在版编目（CIP）数据

Android 手机程序设计实用教程/耿祥义等编著.--北京：清华大学出版社，2013(2018.7 重印)
高等学校 Java 课程系列教材
ISBN 978-7-302-32100-2

Ⅰ.①A…　Ⅱ.①耿…　Ⅲ.①移动电话机－应用软件－程序设计－教材　Ⅳ.①TN929.53
中国版本图书馆 CIP 数据核字(2013)第 082728 号

责任编辑：魏江江　薛　阳
封面设计：杨　兮
责任校对：时翠兰
责任印制：刘海龙

出版发行：清华大学出版社
　　网　　址：http://www.tup.com.cn, http://www.wqbook.com
　　地　　址：北京清华大学学研大厦 A 座　　邮　编：100084
　　社 总 机：010-62770175　　　　　　　　邮　购：010-62786544
　　投稿与读者服务：010-62776969, c-service@tup.tsinghua.edu.cn
　　质 量 反 馈：010-62772015, zhiliang@tup.tsinghua.edu.cn
　　课 件 下 载：http://www.tup.com.cn, 010-62795954

印 装 者：清华大学印刷厂
经　　销：全国新华书店
开　　本：185mm×260mm　　印　张：21.5　　字　数：524 千字
版　　次：2013 年 8 月第 1 版　　　　　　　印　次：2018 年 7 月第 6 次印刷
印　　数：9001～10500
定　　价：39.00 元

产品编号：052856-01

前　言

　　本书注重 Android 系统的特点，特别是手机程序设计的特点，使用 Android 4.2，重点讲解 Android 手机应用开发的核心内容。本书注重教材的可读性和实用性，许多例题都经过精心的考虑，既能帮助读者理解知识，同时又具有启发性和实用性。全书共分 12 章，分别是 Android 简介与开发环境、Android 应用程序的结构、常用 View 视图、常用 ViewGroup 视图、常用专用视图、菜单、动作栏与对话框、2D 绘图、Intent 对象、常用后台对象、使用 SD 卡、文件的读写、使用 SQLite 数据库等内容。

　　第 1 章介绍 Android 简介与开发环境，对 Android 开发平台给予了详细讲解。第 2 章讲解 Android 应用程序的结构，使读者能快速了解 Android 应用程序的基本结构以及开发过程需要的一些基本知识。第 3 章讲解常用 View 视图，这些 View 视图不仅在开发 Android 程序中有较高的使用频率，而且也体现了程序设计的一些重要的思想。第 4 章讲解常用 ViewGroup 视图，这些 ViewGroup 视图对于美化程序的界面是非常重要的，本章选择的例子都充分体现了 ViewGroup 视图的重要性。第 5 章讲解常用专用视图，这些专用视图对于开发具有某些特定功能的程序是非常重要的，因此，本章的例子非常注重实用性，读者可以举一反三开发一些类似的应用程序。第 6 章讲解菜单、动作栏与对话框，本章的内容在手机程序设计中占有非常重要的地位，使用方式也很有特色，为此，本章例子都充分考虑手机程序设计的特点，讲解如何在手机程序设计中合理地使用各种菜单以及对话框和动作栏。第 7 章讲解 2D 绘图，特别讲解了在游戏开发中经常使用的 SurfaceView 类，掌握本章的内容对于开发手机游戏设计是非常重要的。第 8 章讲解 Intent 对象，是 Android 开发应用程序中最重要的核心内容，因此所选内容和例子都充分体现了 Intent 对象在 Android 应用开发中的重要地位和实用价值。第 9 章讲解常用后台对象，掌握这些常用的后台对象，对于提高程序的运行效率是非常重要的，本章不仅讲解了重要的 Service 后台对象，也讲解了怎样让前台和后台更好地交互数据的相关类，特别讲解了 AsyncTask 类，该类对于处理程序前台和后台之间的数据交互是非常重要的。第 10 章讲解使用 SD 卡，合理有效地使用 SD 卡对于手机程序设计是至关重要的，因此本章所给出的例子能充分体现使用 SD 卡的好处。第 11 章讲解文件的读写，在讲解上特别注重体现 Android 系统读写文件的特点，由于程序通过文件读写能体现更强大的功能，为此本章几乎覆盖了 Android 系统的全部读写文件的知识内容，例子都非常具有实用价值。第 12 章讲解怎样使用 SQLite 数据库，不仅讲解了怎样在应用程序中创建数据库，而且本章的创新点是讲解了怎样在程序中外挂数据库，这对于充分利用数据库是非常重要的。

本书的例题全部在 Android 4.2 环境下编译通过。登录清华大学出版社网站 http://tup.tsinghua.edu.cn 可下载本书的全部源代码。根据 Android 开发环境的特点，本书的源代码全部按项目格式提供，而且这些项目都是编译通过的，读者可以方便地阅读和调试这些项目。

希望本教材能对读者学习 Android 应用开发有所帮助，并请读者批评指正(xygeng0629@sina.com)。

编者
2013 年 6 月

目 录

第 1 章 Android 简介与开发环境 ·· 1
    1.1 Android 简介 ·· 1
    1.2 搭建 Android 开发环境 ··· 2
    1.3 创建虚拟设备 ·· 7
    1.4 开发 Android 手机程序 ··· 9
    1.5 安装与卸载 Android 程序 ··· 11
    1.6 工程中一些重要的文件 ·· 13
    1.7 Android 的帮助文档 ··· 15
    1.8 Android SDK＋Eclipse 环境 ······································ 15
  习题 1 ·· 15

第 2 章 Android 程序的结构 ·· 17
    2.1 Activity 对象与程序的基本结构 ·································· 17
    2.2 Android 应用程序的配置文件 ···································· 18
    2.3 设置主要的 Activity 对象 ··· 21
    2.4 Activity 对象的外观及状态 ······································· 23
    2.5 视图资源 ··· 26
    2.6 值资源 ·· 30
    2.7 图像资源 ··· 33
    2.8 获取资源 ··· 35
  习题 2 ·· 36

第 3 章 常用 View 视图 ··· 38
    3.1 View 视图的常用属性与度量值 ··································· 39
    3.2 TextView 视图 ·· 41
    3.3 EditText 视图 ··· 44
    3.4 Button 视图 ··· 48
    3.5 ToggleButton 视图 ·· 52
    3.6 CheckBox 视图 ·· 55
    3.7 RadioButton 视图 ·· 58

3.8　Spinner 视图 …………………………………………………………… 61
3.9　ListView 视图 …………………………………………………………… 67
3.10　动态创建 Spinner 视图和 ListView 视图 …………………………… 71
3.11　GridView 视图 ………………………………………………………… 78
3.12　ScrollView 视图 ……………………………………………………… 81
3.13　HorizontalScrollView 视图 …………………………………………… 83
3.14　使用样式资源简化视图文件 ………………………………………… 84
习题 3 ……………………………………………………………………………… 85

第 4 章　常用的 ViewGroup 视图 ……………………………………………… 87

4.1　LinearLayout 视图 …………………………………………………… 87
4.2　RelativeLayout 视图 …………………………………………………… 91
4.3　TableLayout 视图 ……………………………………………………… 95
4.4　TabHost 视图 …………………………………………………………… 97
4.5　GridLayout 视图 ……………………………………………………… 103
4.6　FrameLayout 视图 …………………………………………………… 106
4.7　AbsoluteLayout 视图 ………………………………………………… 109
习题 4 ……………………………………………………………………………… 113

第 5 章　常用的专用 View 视图 ……………………………………………… 114

5.1　DigitalClock 视图、AnalogClock 视图与 CalendarView 视图 …… 114
5.2　DatePicker 视图与 TimePicker 视图 ………………………………… 117
5.3　ImageView 视图与 ImageButton 视图 ……………………………… 120
5.4　Chronometer 视图 …………………………………………………… 123
5.5　Toast 视图 ……………………………………………………………… 125
5.6　ProgressBar 视图 ……………………………………………………… 127
5.7　VideoView 视图 ……………………………………………………… 129
5.8　WebView 视图 ………………………………………………………… 132
习题 5 ……………………………………………………………………………… 135

第 6 章　菜单、动作栏与对话框 ……………………………………………… 136

6.1　菜单资源 ………………………………………………………………… 136
6.2　选项菜单 ………………………………………………………………… 137
6.3　上下文菜单 ……………………………………………………………… 142
6.4　弹出式菜单 ……………………………………………………………… 146
6.5　动作栏 …………………………………………………………………… 149
6.6　动作栏与选项菜单 ……………………………………………………… 153
6.7　AlertDialog 对话框 …………………………………………………… 155
6.8　DatePickerDialog 对话框与 TimePickerDialog 对话框 …………… 162

6.9　ProgressDialog 对话框 …………………………………………………… 166
6.10　使用 Dialog 创建对话框 ………………………………………………… 170
6.11　长按事件与对话框 ………………………………………………………… 173
习题 6 ……………………………………………………………………………… 175

第 7 章　2D 绘图 ……………………………………………………………… 177
7.1　Drawable 类 ………………………………………………………………… 177
7.2　Canvas 类 …………………………………………………………………… 178
7.3　SurfaceView 类 ……………………………………………………………… 183
7.4　使用画布绘制位图 ………………………………………………………… 187
习题 7 ……………………………………………………………………………… 189

第 8 章　Intent 对象与 Activity 对象 ……………………………………… 190
8.1　Intent 对象及使用步骤 …………………………………………………… 190
8.2　Intent 对象与 AndroidManifest 配置文件 ……………………………… 194
8.3　内置范畴与自定义范畴 …………………………………………………… 200
8.4　内置动作与自定义动作 …………………………………………………… 204
8.5　Intent 对象的附加数据 …………………………………………………… 208
8.6　启动拨号的 Activity 对象 ………………………………………………… 212
8.7　启动发送短信的 Activity 对象 …………………………………………… 214
8.8　启动播放视频的 Activity 对象 …………………………………………… 216
8.9　启动使用 Google 地图的 Activity 对象 ………………………………… 219
8.10　启动使用浏览器的 Activity 对象 ……………………………………… 222
8.11　启动发送 E-mail 的 Activity 对象 ……………………………………… 224
8.12　具有多个 Activity 对象的程序 ………………………………………… 228
8.13　让 Activity 对象返回数据 ……………………………………………… 231
8.14　启动使用照相机的 Activity 对象 ……………………………………… 237
习题 8 ……………………………………………………………………………… 239

第 9 章　常用后台对象 ………………………………………………………… 240
9.1　Activity 对象与 Service 对象、BroadcastReceiver 对象 ……………… 240
9.2　Service 对象及生命周期 …………………………………………………… 241
9.3　使用多个 Service 对象 …………………………………………………… 246
9.4　IntentService 类 …………………………………………………………… 249
9.5　AsyncTask 类 ……………………………………………………………… 253
9.6　广播及接收 ………………………………………………………………… 258
9.7　PendingIntent 类 …………………………………………………………… 261
习题 9 ……………………………………………………………………………… 267

第 10 章　使用 SD 卡 ··· 268

 10.1　设置 SD 卡的大小 ··· 268
 10.2　上传文件到 SD 卡 ··· 269
 10.3　查看 SD 卡中的内容 ··· 269
 10.4　显示 SD 卡中的图像 ··· 270
 10.5　播放 SD 卡中的视频或 MP3 ··· 272
 习题 10 ··· 274

第 11 章　文件的读写 ·· 275

 11.1　使用输入/输出流在数据区读写文件 ··· 275
 11.2　使用 SharedPreferences 对象在数据区读写文件 ······································· 279
 11.3　在 SD 卡中读写文件 ··· 286
 11.4　读取 assets(资产)中的文件 ··· 289
 11.5　读取\res\raw(原始资源)中的文件 ··· 291
 11.6　解析 XML 文件 ··· 294
 11.7　基于文本文件的电话簿 ·· 297
 11.8　基于 XML 数据库的英-汉字典 ·· 304
 习题 11 ··· 309

第 12 章　使用 SQLite 数据库 ··· 310

 12.1　连接 SQLite 数据库 ·· 310
 12.2　外挂 SQLite 数据库 ·· 315
 12.3　SQLiteDatabase 类的两个重要方法 ·· 317
 12.4　事务 ··· 321
 12.5　基于数据库的消费记载 ·· 325
 习题 12 ··· 332

第1章　Android 简介与开发环境

主要内容：
- Android 简介；
- 搭建 Android 开发环境；
- 创建虚拟设备；
- Android 程序的开发与安装；
- 工程中一些重要的文件；
- Android SDK＋Eclipse 环境。

1.1　Android 简介

1. Android 与智能手机

手机和个人计算机相比，尽管历史不长，但也和个人计算机一样迅速地成为人们的重要工具之一。目前，随着手机硬件功能的提高和完善，手机已经进入到智能手机阶段。智能手机（Smartphone）的主要特点是：具有独立的操作系统，可以通过移动通信网络连接到 Internet，可以由用户自行安装软件、不断地完善提高其智能水平。比如，人们可以用智能手机与计算机、PDA 之间交换图片、铃声、游戏、程序、流媒体等数字格式的数据，可以用智能手机控制计算机、家电设备等。

Android 手机就是智能手机发展阶段的一个重要标志。Android 是 Google 公司推出的手机操作系统的名称，使用 Android 操作系统的手机就会具有智能手机的性能。Android 一词最早出现于 19 世纪的一部科幻小说《未来夏娃》，小说的作者将外表像人的机器起名为 Android，因此，Android 一词的本义指"机器人"。Android 的 Logo 是一个绿色的机器人（图 1.1），绿色也是 Android 的标志。

图 1.1　Android 的 Logo

2. Android 操作系统及应用开发

Android 操作系统是 Google 公司在 2007 年 11 月 5 日公布的手机操作系统，其本质是基于 Linux 内核的操作系统。Android 操作系统最初是"Android"公司开发的一款手机操作系统，2005 年，Google 公司收购"Android"公司后，继续完善 Android 操作系统，并在 2008 年 9 月正式发布了 Android 1.0 手机操作系统（版本为 Android 1.0），标志着 Android 手机的诞生。

手机的智能性不仅需要优良的硬件和操作系统,更需要开发出优秀的应用软件。Android 操作系统支持 Java 语言,即可以使用 Java 语言编写运行于 Android 1.0 操作系统上的应用程序,一款 Android 手机可以通过安装应用软件不断地提高自己的应用性能、提高智能水平。Java 语言不仅是很优秀的语言、适合进行网络应用的开发,而且具有广泛的开发团队,这非常有利于 Android 手机技术的发展和 Android 手机用户群的壮大。

3. Android 的发展史

2008 年 9 月发布 Android 1.0 操作系统及相应的 Android 1.0 SDK Tools。
2009 年 4 月发布 Android 1.5 操作系统及相应的 Android 1.5 SDK Tools。
2009 年 9 月发布 Android 1.6 操作系统及相应的 Android 1.6 SDK Tools。
2010 年 1 月发布 Android 2.1 操作系统及相应的 Android 2.1 SDK Tools。
2010 年 5 月发布 Android 2.2 操作系统及相应的 Android 2.2 SDK Tools。
2010 年 12 月发布 Android 2.3 操作系统及相应的 Android 2.3 SDK Tools。
2011 年 2 月发布 Android 2.3.3 操作系统及相应的 Android 2.3.3 SDK Tools。
2011 年 2 月发布 Android 3.0 操作系统及相应的 Android 3.0 SDK Tools。
2012 年 8 月发布 Android 4.0 操作系统及相应的 Android 4.0 SDK Tools。
2012 年 10 月发布 Android 4.1 操作系统及相应的 Android 4.1 SDK Tools。
2012 年 12 月发布 Android 4.2 操作系统及相应的 Android 4.2 SDK Tools。

1.2 搭建 Android 开发环境

搭建开发 Android 程序的开发环境需要下列工具。
- Java SE 提供的 JDK。
- Android SDK。
- SDK platform。

1. 安装 Java SE 提供的 JDK

Java SE(曾称为 J2SE)称为 Java 标准版或 Java 标准平台。Java SE 提供了标准的 Java Development Kit(JDK)。

可以登录官方网址 http://java.sun.com 或 http://www.oracle.com/technetwork/java/javase/downloads/index.html 免费下载 Java SE 提供的 JDK。当前最新的 JDK 版本为 JDK 7,Sun 公司把这一最新的版本命名为 JDK 7,但人们仍然习惯地称作 JDK 1.7。开发 Android 手机程序,建议下载的 JDK 版本不低于 1.5,即 JDK 版本至少是 JDK 1.5 或以上版本。

本书使用 Windows 操作系统,本书将使用针对 Windows 操作系统平台的 JDK,因此下载的版本为 jdk-7-windows-i586.exe,如果读者使用其他的操作系统,可以下载相应的 JDK。

双击下载后的 jdk-7-windows-i586-.exe 文件图标出现安装向导界面,选择接受软件安装协议。可以使用默认的安装路径 C:\program Files\Java\JDK1.7.0,但也可以修改安装路径为 C:\JDK7 或 D:\JDK7。需要注意的是,安装 JDK 过程中,JDK 还额外提供一个 Java 运行环境——JRE(Java Runtime Environment),并提示是否修改 JRE 默认的安装路径 C:\program Files\Java\JRE7。

建议采用该默认的安装路径,如果修改该默认安装路径,修改后的安装路径不可以与 JDK 的安装路径相同。安装完毕后,比如 JDK 安装到 D:\jdk7,D:\jdk7 目录形成了如图 1.2 所示的目录结构。

图 1.2　JDK 的目录结构

2. 安装 Android SDK

Android SDK(Android Software Development Kit)是 Android 专属的软件开发工具包。

1) 下载并安装 Android SDK

可以登录 Android 的官方网址 http://developer.android.com 下载适合于相应操作系统的 Android SDK。如果无法登录(大陆地区),可以在常用的搜索引擎中搜索关键字 Android SDK,找到提供下载 Android SDK 的网址。

由于安装 Android SDK 之后,可以随时使用其中的 SDK 管理器在线升级 Android SDK 的版本,比如教材所用计算机上的 Android SDK 最初版本是 Android 2.2.3 SDK,其 SDK 的压缩文件是 android-sdk_r18-windows.zip。

将 android-sdk_r18-windows.zip 解压到某个目录,这里,解压缩到 D:\。解压缩后会自动产生一个名字为 android-sdk-windows 的文件夹,形成的目录结构如图 1.3 所示。

注:如果下载的是 android-sdk_r20-windows.exe,其安装过程也是解压缩,可选择解压缩到 D:\android-sdk-windows。

2) 设置 path 的值

Android SDK 安装目录(这里是 D:\android-sdk-windows)下的 tools 子目录中提供了用于 SDK 自身组件安装、卸载管理,提供模拟器工具以及其他开发所需的第三方工具。为了能在命令行随时使用 tools 下的命令,需要将 D:\android-sdk-windows\tools 设置成系统环境变量 path 的值中的一个值。

对于 Windows 2000/2003/XP,右键单击"我的电脑"(对于 Windows 7,右键单击"计算机"),在弹出的快捷菜单中选择"属性",弹出"系统特性"对话框,再单击该对话框中的"高级选项",然后单击按钮"环境变量",添加系统环境变量。一般来说系统已经设置过环境变量 path,并且 path 已经有了一些值,因此可单击该变量进行编辑操作,将需要的值 D:\android-sdk-windows\tools 加入即可,如图 1.4 所示(将新加入的值与其他已有的值用分号分隔,如果新加入的值是最后一项,不需要末尾加分号)。

图 1.3　Android SDK 的目录结构

图 1.4　编辑环境变量 path 的值

注:在给环境变量 path 增加新的值后,需重新打开 MS-DOS 命令行窗口才可以使 path 新增加的值有效。

3. 安装 SDK Platform

下载的 Android SDK 仅仅提供了最基本的 SDK 的"管理"工具和基本的调试工具，为了开发 Android 应用程序还需要下载一个 SDK Platform（开发平台）以及相应的 Platform-tools（平台工具）。开发 Android 应用程序的 SDK Platform，由 Android 的专用包（专用类库）和虚拟设备构成，即 SDK Platform 提供了编写、调试 Android 应用程序所需要的专用类和虚拟设备。安装了 Android SDK 之后，会发现 Android SDK 安装目录（这里是 D:\android-sdk-windows）下的 platform 子目录下没有任何文件，因此需要下载 Android 的专用包和虚拟设备，比如下载 Android 4.1.2（API 16）或 Android 4.2（API 17），还需要下载相应的平台开发工具（Platform-tools）。

1）使用 SDK 管理器（SDK Manager.exe）

进入 Android SDK 安装目录运行（双击 SDK Manager.exe）打开 SDK 管理器，稍等片刻将出现如图 1.5 所示的 SDK 管理器界面。

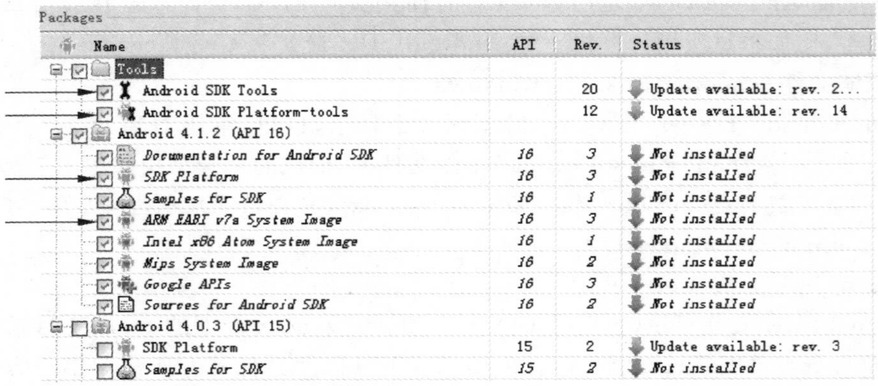

图 1.5 SDK 管理器

（1）更新 Android SDK

SDK 管理器会让读者选择是否更新已经安装的 Android SDK 的版本，如果需要更新，就选择 SDK 管理器界面上 Tools 下的 Android SDK Tools 进行更新。本教材选择了更新，并选择使用 Android 4.1.2（API 16）（作者在更新时，Android 4.2（API 17）还没有发布）。如果读者决定使用 Android 4.2（API 17），选择相应的 Android 4.2（API 17）即可。

（2）Platform-tools

SDK 管理器提供下载和更新程序设计需要的 Platform-tools（平台工具），因此必须把 Tools 目录下的 Android SDK Platform-tools 选项选中（见图 1.5 中第 2 个"→"所指的选项），以便获得相应的 Android Platform-tools。SDK 管理器将平台工具存放到 Android SDK 安装目录的\platform-tools 的子目录中（platform-tools 子目录由 SDK 管理器负责建立）。

（3）Platform

开发 Android 应用程序需要 SDK Platform（开发平台），即需要 Android 的专用类库和虚拟设备。因此必须下载至少一个 Android API 到 Android SDK 安装目录的\platforms 子目录中。SDK 管理器自动选择一组推荐的软件包（packages），可以简单地选择推荐的软件包,SDK 管理器就会安装选定的包到 Android SDK 安装目录下的\platforms 子目录中。如果考虑到下载速度,可以放弃选择一些帮助文档和暂时不使用的软件包(如果暂时不编写 Google 地图相关的应用,可暂时放弃下载 Google API),只选择图 1.5 中箭头(→)所指的选项即可,其中选项 SDK Platform 和 ARM EABI v7a System Image 是必须要下载的,否则无法进行 Android 应用程序的开发或调试。

作者下载的 Android 4.1.2(API 16)作为其中的一个 SDK Platform(开发平台),那么\platforms 子目录中就会有名字为 android-16 的文件夹(如果读者下载的 Android 4.2(API 17)作为其中的一个 SDK Platform,那么\platforms 子目录中就会有名字为 android-17 的文件夹),表示 Android 可以使用 Android 4.1.2(API 16)作为一个开发平台。不同版本的 Android API 代表不同的 SDK Platform 版本,允许下载多个 Android API,使得可以开发旧版本的 Android 程序。本教材所用的计算机已累计下载了 3 个版本的 SDK Platform。

在图 1.5 中选择好需要下载的选项后,单击 Install Package 按钮,在弹出的接受协议对话框中选择 Accept 后,SDK 管理器会将 Android API,比如版本是 Android 4.1.2(API 16)的 SDK Platform 安装到 Android SDK 安装目录下的\platforms 子目录中,\platforms 子目录中会有名字为 android-16 的文件夹。

注:由于 Android 开发平台比较庞大,如果下载 Android 开发平台的全部组件需要一定的时间。

下载成功后,Android SDK 安装目录将发生一定的变化(如图 1.6 所示)。

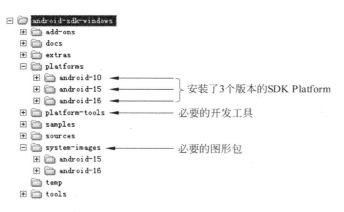

图 1.6 安装 Platform 和 Platform-tools 之后的 Android SDK

- 原有\platforms 子目录中会出现一个或多个存放专用类库和虚拟设备的文件夹,即会有一个版本或多个版本的 SDK Platform(见图 1.6 中箭头←所指的目录)。
- Android SDK 安装目录会增加一个名字为 platform-tools 的子目录(见图 1.6 中箭头←所指的目录)。
- 对于 Android 4.0 版本之后的版本,SDK 管理器会在 Android SDK 安装目录下增加一个名字为 system-images 的文件夹,用于提供虚拟设备所需要的图形包(见图 1.6 中箭头←所指的目录)。

Android SDK 安装目录发生怎样的变化依赖于下载的内容的多少以及更新情况,但 platforms 目录下至少要有一个版本的 SDK Platform。Android 4.0 版本后,system-images

目录也是必需的,本教材所用的计算机,Android SDK 安装目录发生的变化如图 1.6 所示。

2) 开发平台的序号

SDK 管理器将 platforms 子目录下的开发平台(SDK Platform),按其发布时间顺序地编号,比如,platforms 子目录中有 android-10,android-15 和 android-16 三个子目录,即有 Android 2.2.3(API 10),Android 4.0.3(API 15)和 Android 4.1.2(API 16)版本的 SDK Platform,那么将来开发应用程序时,可以使用 target id 指定所使用的 SDK Platform,例如:

target 1 意思是版本为 Android 2.2.3(API 10)的 SDK Platform。
target 2 意思是版本为 Android 4.0.3(API 15)的 SDK Platform。
target 3 意思是版本为 Android 4.1.2(API 16)的 SDK Platform。

可以用命令行方式进入 Android SDK 安装目录的 tools 子目录,执行 android list target:

```
D:\android-sdk-windows\tools> android list target
```

查看 SDK Platform 的版本情况。

本教材运行的效果如图 1.7 所示。

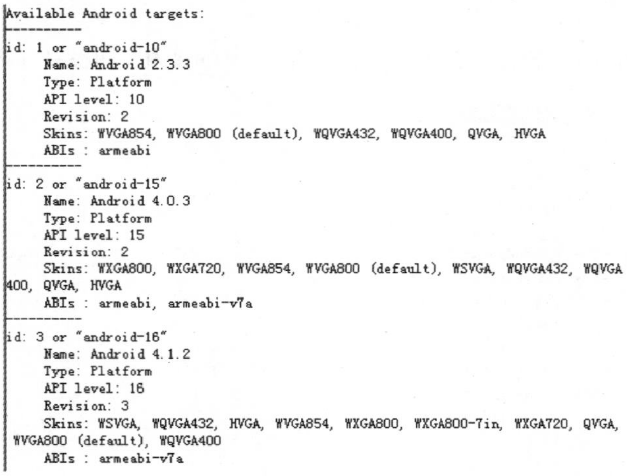

```
Available Android targets:
----------
id: 1 or "android-10"
     Name: Android 2.3.3
     Type: Platform
     API level: 10
     Revision: 2
     Skins: WVGA854, WVGA800 (default), WQVGA432, WQVGA400, QVGA, HVGA
     ABIs : armeabi
----------
id: 2 or "android-15"
     Name: Android 4.0.3
     Type: Platform
     API level: 15
     Revision: 2
     Skins: WXGA800, WXGA720, WVGA854, WVGA800 (default), WSVGA, WQVGA432, WQVGA
400, QVGA, HVGA
     ABIs : armeabi, armeabi-v7a
----------
id: 3 or "android-16"
     Name: Android 4.1.2
     Type: Platform
     API level: 16
     Revision: 3
     Skins: WSVGA, WQVGA432, HVGA, WVGA854, WXGA800, WXGA800-7in, WXGA720, QVGA,
 WVGA800 (default), WQVGA400
     ABIs : armeabi-v7a
```

图 1.7　SDK 可使用的开发平台

注:需要特别注意的是,如果下载了 Google API 或安装了其他厂商的支持 Android 的开发平台,那么这些平台将被存放在 Android SDK 安装目录的 add-ons 文件中(通常情况下,add-ons 文件中无任何子目录),并和 platforms 下的开发平台一同排序编号。

3) 离线安装

如果读者的计算机无法连接到 Google 提供的资源,那就需要将别人安装的 Android Platform、Android Platform-tools 和 system-images 复制到你的 Android SDK 安装目录下。当然,也可以把别人安装好的整个目录 android-sdk-windows(大约 1GB 左右)复制到你的计算机上即可,在这种情形下,你唯一需要安装的就是 Java SE 提供的 JDK(见 1.2 节一开始介绍的"安装 Java SE 提供的 JDK")。

4）设置 path 的值

Android SDK 安装目录下 platform-tools 子目录提供开发程序所需的专用工具。为了能在命令行随时使用 platform-tools 下的命令，需要将 D:\android-sdk-windows\platform-tools 设置成系统环境变量 path 的值中的一个值（编辑 path 的知识点如图 1.4 所示）。

1.3　创建虚拟设备

搭建好 Android 开发环境后，还需要创建一个虚拟设备，即 AVD。AVD 的全称为 Android Virtual Device，就是运行 Android 程序的虚拟设备（比如手机模拟器）。创建 AVD 的方法有两种，一是通过 AVD 管理器，二是通过命令行创建。

1. AVD 管理器

进入 Android SDK 安装目录找到 AVD 管理器 AVD Manager.exe，并运行 AVD 管理器（双击 AVD Manager.exe），稍等片刻将出现 AVD 管理器界面，单击该界面上的 New... 按钮，出现创建 AVD 的对话框，如图 1.8 所示。

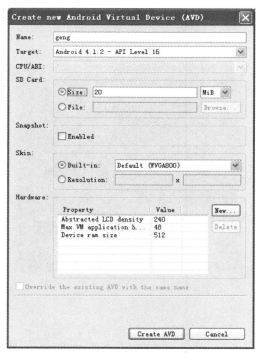

图 1.8　创建虚拟设备的对话框

在图 1.8 所示的 AVD 的对话框上进行如下设置：
- 在 Name 文本框输入要创建的 AVD 的名字，比如：geng。
- 单击 Target 下拉列表，将列出 Android SDK 已安装的 SDK Platform，在下拉表的选项中选择一个 SDK Platform，例如，选择基于 Android 4.1.2（API 16）的 SDK Platform。那么，所创建的 AVD 就可以模拟运行使用 Android 4.1.2（API 16）开发

平台设计的 Android 程序。
- SD Card(Secure Digital Memory Card)的 Size 设置为 20 即可(最小为 9)。
- 其他选项取默认设置即可。

单击 AVD 的对话框下面的 Create AVD 按钮,创建名字为 geng 的 AVD。如果成功地创建了 AVD,就可以在 AVD 管理器的界面上选中所创建的名字为 geng 的 AVD,单击 AVD 管理器的界面右侧的 Start 按钮(如图 1.9 所示),启动名字为 geng 的 AVD,要耐心多等待一会,就会出现如图 1.10 所示的 AVD。

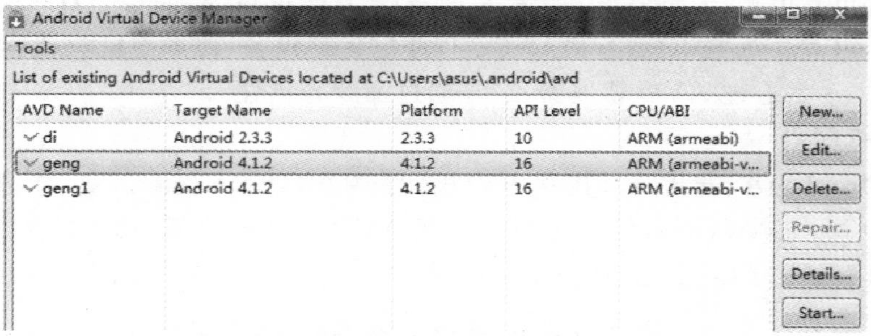

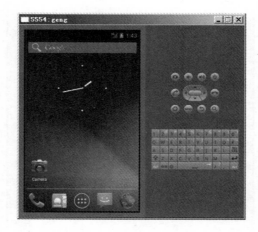

图 1.9　AVD 管理器

图 1.10　名字为 geng 的 AVD

2. 命令行方式

使用命令行进入 Android 安装目录下的 tools 子目录,执行如下命令:

```
D:\android-sdk-windows\tools> android create avd -- target 3 -- name zhang
```

其中 android 是命令,后面是参数。
- 参数 create avd 的作用是创建 AVD。
- 参数 target 用来指定 AVD 适用的平台,如果 platforms 下有多个开发平台,这些平台按 Google 公司发布的时间顺序从 1 到 n 排序。本书这里取顺序是 3(本书安装了 3 个平台,如 1.2 节的图 1.6 所示)。
- 参数 name 用来指定 AVD 的名字。

执行上述命令创建名字为 zhang 的 AVD 的过程如图 1.11 所示。

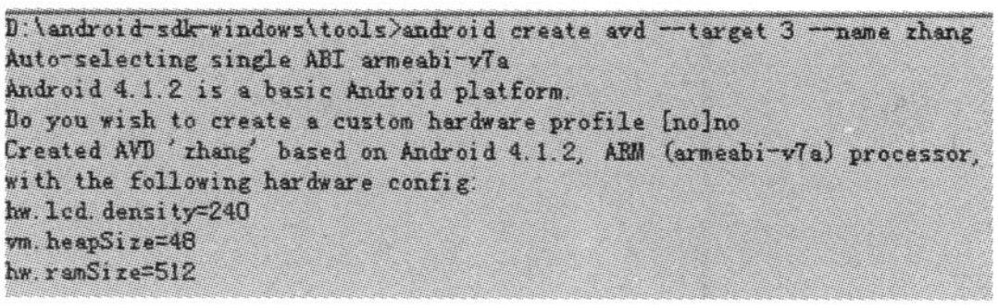

```
D:\android-sdk-windows\tools>android create avd --target 3 --name zhang
Auto-selecting single ABI armeabi-v7a
Android 4.1.2 is a basic Android platform.
Do you wish to create a custom hardware profile [no]no
Created AVD 'zhang' based on Android 4.1.2, ARM (armeabi-v7a) processor,
with the following hardware config:
hw.lcd.density=240
vm.heapSize=48
hw.ramSize=512
```

图 1.11　创建名字为 zhang 的 AVD

也可以用简写方式创建 AVD：

D:\android-sdk-windows\tools>android create avd -t 3 -n zhang

启动 AVD 管理器，在 AVD 管理器界面上选中名字为 zhang 的 AVD，单击 AVD 管理器的界面右侧的 Start 按钮，启动名字为 zhang 的 AVD(要耐心多等待一会)。

1.4　开发 Android 手机程序

本教材使用命令行方式，因此，读者需要熟悉几个简单的命令行操作。当需要使用命令行命令时，需打开 MS-DOS 命令行窗口。需要熟悉几个简单的 DOS 操作命令，例如，从逻辑分区 C 转到逻辑分区 D，需在命令行依次输入 D 和冒号并回车确定。进入某个子目录(文件夹)的命令是："cd　目录名"；退出某个子目录的命令是："cd .."。例如，从目录 example 退到目录 boy 的操作是："c:\boy>example>cd .."，直接退到磁盘根目录的命令是："cd\"。

另外，你需要一个文本编辑器，如 Windows 平台提供的"记事本(NotePad)"，使用文本编辑器编辑程序中需要的 Java 源文件以及 XML 文件。

开发 Android 手机程序的步骤如下。
- 创建(Create)工程；
- 编译(Debug)工程。

1. 创建(Create)工程

使用 Android SDK 提供的 android 命令建立一个项目工程(android.exe 在 Android SDK 安装目录下的 tools 子目录中，为了能在命令行随时使用 tools 下的可执行文件，需要将 D:\android-sdk-windows\tools 设置成系统环境变量 path 的值中的一个值)。

1) 查看平台，确定工程的目标(targets)

开发的项目需选择一个 SDK Platform，当前开发环境可能安装了多个 SDK Platform，开发环境为这些 SDK Platform 分别指定了一个 id 号，因此需要知道当前的开发环境可以选择哪些 SDK Platform(可参见 1.2 节)。

用命令行进入 Android SDK 安装目录的 tools 子目录中，然后执行如下命令：

android list target

例如：

D:\android-sdk-windows\tools>android list target 将列出全部的 SDK Platform 及相

应的 id 号。本教材一共安装了 3 个 SDK Platform(如 1.2 节的图 1.7 所示)。

2) 创建工程

使用 android.exe 建立工程的语法如下：

```
android create project -- target <target-id>
                      -- name <工程名称>
                      -- path <path-to-workspace>/<工程名称>
                      -- activity <Activity 的子类名称>
                      -- package <包名>
```

不可以直接在磁盘的根目录下建立工程,需要在磁盘根目录下建立一个子目录,即通常所说的工作空间(Workspace),该目录的名字可任意给定,比如在磁盘 D 下建立名字为 2000 的子目录(工作空间)。

用命令行进入 D:\2000 目录中,建立名字为 Hello 的工程：

```
D:\2000> android create project -- target 3 -- name Hello -- path ./Hello -- activity MainActivity -- package tom.jiafei
```

- 参数 target 指定工程的目标 id,即工程要使用的 SDK Platform(这里取 id 是 3)。
- 参数 name 指定工程的名称(这里为 Hello)。
- 参数 path 指定工程的路径,一般取当前路径即可(这里是./)。
- 参数 activity 指定 Activity 的子类名称(这里为 MainActivity)。
- 参数 package 指定工程中类的包名(这里为 tom.jiafei)。

使用 android.exe 建立工程的简写语法如下：

```
android create project -t <target-id>
                       -n <工程名称>
                       -p <path-to-workspace>/<工程名称>
                       -a <Activity 的子类名称>
                       -k <包名>
```

例如：

```
D:\2000> android create project -t 3 -n Hello -p ./Hello -a MainActivity -k tom.jiafei
```

如果创建工程成功,命令行将输出创建的工程的有关信息。

注：两个不同的工程,需要有不同的包名,否则虚拟设备或 Android 手机将混淆包名相同的工程。

3) 查看工程

工程 Hello 对应的 Hello 目录的结构如图 1.12 所示。

其中\src\tom\jiafei 下存放着工程的源文件。\res 下存放着项目需要的资源。有关工程里的文件的细节后续章节会详细讲解。

图 1.12 工程的目录结构

2. 编译(Debug)工程

使用 ant 工具编译(Debug)工程,因此需要安装 ant 工具(在网络上很容易得到 ant 工具)。本教材使用的是 apache-ant-1.8.4-bin.zip,将 apache-ant-1.8.4-bin.zip 解压缩到

D:\，如图 1.13 所示。

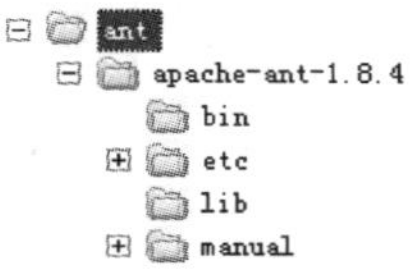

bin 子目录中提供的 ant 命令可用于编译(Debug)工程,需要将 D:\ant\apache-ant-1.8.4\bin 设置成系统环境变量 path 的值中的一个值。

进入到 Hello 工程的根目录,即 D:\2000\Hello 中,执行 ant debug,如下所示：

图 1.13　ant 工具

D:\2000\Hello> ant debug

编译(Debug)成功后,项目的 bin 目录下面会出现"Hello-debug.apk"文件,该文件就是可用 Android 手机执行的应用程序文件。

1.5　安装与卸载 Android 程序

需要将工程下产生的可用 Android 手机执行的 apk 文件(即 Android 应用程序)安装到 AVD 或真实的手机上,以便查看程序的运行效果。

1. 安装到 AVD(Install)

1) 启动 AVD

首先要用 AVD 管理器启动一个 AVD(比如,启动 1.3 节创建的名字为 geng 的 AVD),并解开 AVD 模拟器上的屏幕锁。

2) 安装 apk 到启动的 AVD 中

进入 Hello 工程的根目录,即 D:\2000\Hello 中,使用 adb 命令将 bin 目录下的 apk 文件 Hello-debug.apk 部署到 AVD:

D:\2000\Hello> adb install bin/Hello-debug.apk

adb 命令在 Android SDK 安装目录的 platform-tools 文件夹中,要确保 Android SDK 安装目录\platform-tools 是环境变量 path 的一个值。

如果安装成功,命令行将输出如下信息：

```
* daemon not running. starting it now on port 5037 *
* daemon started successfully *
81 KB/s (37865 bytes in 0.453s)
        pkg: /data/local/tmp/Hello-debug.apk
Success
```

如果安装失败,则出现下列错误提示：

```
* daemon not running. starting i
* daemon started successfully *
error: device not found
```

请首先执行下列命令：

```
adb kill-server
adb start-server
adb remount
```

然后再进行安装：

D:\2000\Hello> adb install bin/Hello-debug.apk

3）在模拟器中运行程序

打开 AVD 并解开屏幕锁。首先单击模拟器主界面上的 图标，打开应用程序所在的界面，如图 1.14 所示（有些模拟器是向右拖动鼠标拉开 screen）。

图 1.14　主界面

在模拟器应用界面找到名字为 MainActivity（创建工程时的主 Activity 类的名称）的应用程序的图标（可以在应用界面上向左或向上拖动鼠标，以便看到下一屏幕上的应用图标），双击这个应用程序的图标（图标上默认的图像是小机器人，如图 1.15 所示），就可以看到程序的运行效果了（如图 1.16 所示）。

图 1.15　apps 界面上的工程　　　　图 1.16　工程运行效果

2. 安装到手机

首先用 USB 数据线连接程序所在的计算机与 Android 手机，连接成功后，在 Android 手机上激活 Android 4.1。

在计算机上进入 Hello 工程的根目录，即 D:\2000\Hello 中，使用 adb 命令将 apk 文件安装到 Android 手机：

D:\2000\Hello> adb install bin/Hello-debug.apk

如果安装失败，则出现下列错误提示：

```
* daemon not running. starting i
* daemon started successfully *
   error: device not found
```

请首先执行下列命令：

```
adb kill-server
adb start-server
adb remount
```

然后再进行安装：

D:\2000\Hello > adb install bin/Hello - debug.apk

3. 卸载（uninstall）

如果对源工程进行了修改，那么必须从模拟器或手机上卸载曾安装过的当前工程，才能使用 adb 命令将修改后的工程的 apk 文件安装到模拟器或手机上。

进入工程的根目录：D:\2000\Hello，执行 ant uninstall 命令，就可以卸载当前模拟器或手机中的 Hello 工程的 apk 文件。如下所示：

D:\2000\Hello > ant uninstall

4. 快捷方式

如果使用 adb 命令安装一个曾安装过的程序，就需要从模拟器或手机上卸载曾安装过的程序，才能使用 adb 命令再次将新的 apk 程序文件安装到模拟器或手机上，这对调试者来说显得有点繁琐。

ant 工具提供了一个将 debug 和 install 合二为一的命令：

ant debug install

该命令编译（Debug）工程，并同时将编译工程得到的程序文件（apk）安装到虚拟设备或手机中，而且会自动替换曾安装过的程序文件（如果未曾安装，就安装该程序文件）。

进入工程的根目录，这里是 Hello 目录，执行 ant debug install 命令，如下所示：

D:\2000\Hello > ant debug install

本书在后续章节将一直使用快捷方式调试程序。

注：使用 ant debug install 命令时必须保证已经启动一个 AVD 虚拟设备或某个 Android 手机已经连接到当前计算机，并且手机要打开 USB 调试模式。如果单独使用 ant debug 命令，不要求启动一个虚拟设备 AVD 或某个 Android 手机已经连接到当前计算机。

1.6 工程中一些重要的文件

尽管还没有正式开始学习 Android 程序设计，但已经熟悉了程序开发以及部署的基本步骤。

- 工程目录下的\res 中存放应用程序的资源文件。
- 工程目录下的\src 中存放应用程序的 Java 源文件。
- 在编译（Debug）Hello 工程后，工程目录下多出的\gen\tom\jiafei 目录中存放着应用程序所需要的系统类，用户不能修改它，以后在学习具体应用程序时会给予介绍。

现在去查看 Hello 工程，用文本编辑器打开工程\res\values 下的 strings.xml 文件，内容如下：

strings.xml

< ?xml version = "1.0" encoding = "utf - 8"? >
< resources >
    < string name = "app_name" > MainActivity </string >

</resources>

在模拟器中运行程序时,读者会发现 strings.xml 文件中的具有 name 属性值为"app_name"的<string>标记(XML 语法也称元素)所标记的文本内容刚好是程序的图标的名字,也是显示在模拟器(手机屏幕)上方的标题(其原因在第 2 章的 2.1 节讲解)。该文本的默认内容是 Activity 子类的名字,将<string…>…</string>标记的内容修改为:你好,我喜欢! 修改后的文件如下:

strings.xml

```
<?xml version = "1.0" encoding = "utf-8"?>
<resources>
    <string name = "app_name">你好,我喜欢!</string>
</resources>
```

在保存修改后的文件时,需用另存方式替换原来的文件,并将编码选择为 UTF-8,文件类型选择为"所有文件",如图 1.17 所示。

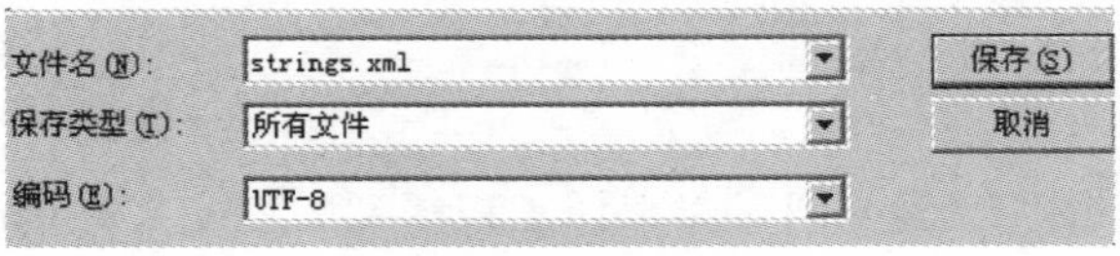

图 1.17 保存修改的 XML 文件

注:有关 XML 语言的知识建议读者参看作者在清华大学出版社出版的《XML 程序设计》或《XML 基础教程》。

用文本编辑器打开工程\res\layout 下的 main.xml 文件,内容如下:

main.xml

```
<?xml version = "1.0" encoding = "utf-8"?>
<LinearLayout xmlns:android = "http://schemas.android.com/apk/res/android"
    android:orientation = "vertical"
    android:layout_width = "match_parent"
    android:layout_height = "match_parent"   >
<TextView
    android:layout_width = "match_parent"
    android:layout_height = "wrap_content"
    android:text = "Hello World,MainActivity"     />
</LinearLayout>
```

将其中的<TextView…/>空标记中的 android:text 属性的值"Hello World,MainActivity",修改为"你好,Android,我来了"。在保存修改后的文件时,需用另存方式,并将编码选择为 UTF-8,文件类型选择为"所有文件"。

现在,用快捷方式编译工程,安装应用程序到 AVD(有关知识点参见 1.5 节)。用命令行进入 D:\2000\Hello,执行如下命令:

```
D:\2000>Hello> ant debug install
```

程序运行效果如图 1.18 和图 1.19 所示。

图 1.18　图标的文字是：你好，我喜欢！　　　　图 1.19　运行效果

1.7　Android 的帮助文档

可以登录 Android 的官方网址 http://developer.android.com 在线查看帮助文档。登录后选择界面上的 Reference 就可查看 Android 的专用类库的帮助文档。也可使用 SDK 管理器下载帮助文档（docs），启动 SDK 管理器选择下载 docs 后，SDK 管理器将帮助文档下载到 Android SDK 安装目录的 docs 的子目录之中，\docs\Reference\下是 Android 的专用类库的全部帮助文档（HTML 格式）。

1.8　Android SDK＋Eclipse 环境

目前大多数介绍 Android 手机程序开发环境的文章或书籍都是基于 Eclipse＋Android SDK 的开发环境（或 MyEclipse＋Android SDK 环境）。尽管集成（IDE）环境有利于快速地开发应用程序，但是，对于开发 Android 手机程序，优势并不明显，其原因是，Android 支持 ant 工具，使用 ant 工具可以方便地 debug 一个工程，因此，如果读者喜欢命令行方式，就完全没必要再使用 Eclipse。另外，如果读者的计算机配置不是很高，Eclipse 的运行速度会受到一定的影响，反而影响了开发速度。读者可以首先熟悉本书介绍的命令行方式，在掌握了使用命令行方式开发 Android 程序之后，再去熟悉、掌握一个流行的集成开发环境（IDE）就是一件比较容易的事情了（毕竟 IDE 环境可以免去我们记忆类库中类的方法的名称）。

本书不打算介绍 Android SDK＋Eclipse 环境，因为可以在网络上很容易查找到相关的文章，这些文章详细介绍了怎样搭建和使用 Android SDK＋Eclipse 环境。除了 1.2 节所需要的环境外，搭建 Eclipse＋Android SDK 环境只需额外下载一个 Android SDK 和 Eclipse 之间的插件（ADT），如果读者比较熟悉 Eclipse，并已经熟悉了使用命令行方式开发 Android 程序，搭建和使用 Android SDK＋Eclipse 环境不应当有什么困难。

习　题　1

1. 说出智能手机的三个主要特点。
2. Android 开发环境需要的 Java SDK 的最低的版本是多少？
3. Android SDK 是否包含有 SDK Platform（开发平台）？
4. SDK 管理器可以下载或更新 Android 系统提供的 SDK Platform，请问 SDK 管理器将下载或更新 SDK Platform 存放到 Android SDK 安装目录的哪个子目录中？

5. 如果装备使用其他手机厂商提供的 SDK Platform（开发平台），该平台应该存放到 Android SDK 安装目录的哪个子目录中？

6. Android SDK 安装目录的\platforms 目录中是否可以存放多个 Android 系统提供的 SDK Platform？

7. 用命令行方式进入 Android SDK 安装目录的 tools 子目录，执行 android list target 的目的是什么？

8. 编译工程后得到的应用程序的扩展名是什么，该应用程序文件被保存在工程的哪个目录中？

9. 创建名字是 Boy 的工程，主要的 Activity 子类的名字是 World，使用的包名是 moon.flower。用"ant debug install"命令编译、安装程序到虚拟设备 AVD。

10. 如果 XML 文件的编码是 UTF-8，保存 XML 文件时，编码应当选择为什么编码？

11. 如果创建工程时，使用的包名是 sun.water，那么程序代码的 Java 源文件（这些源文件中使用的包名都是 sun.water）需要保存在工程的哪个子目录中？

第 2 章　Android 程序的结构

主要内容：
- Activity 对象与程序的基本结构；
- Android 应用程序的配置文件；
- 设置主要的 Activity 对象；
- Activity 对象的外观及状态；
- 视图(layout)资源；
- 值(values)资源；
- 图像(drawable)资源。

2.1　Activity 对象与程序的基本结构

Activity 类是设计 Android 程序中最重要的类。Activity 类在 android.app 包中。编写 Android 程序需要扩展 Activity 类，即编写 Activity 类的子类。

在 Android 应用程序设计中，人们习惯称 Activity 类的子类的对象为一个 Activity 对象，或简称一个 activity(将 Activity 单词的首字母小写)。一个 Android 应用程序(*.pak 格式的文件，见 1.4 节)可以包含一个或多个 Activity 对象，这些 Activity 对象是应用程序中最重要的一部分，如图 2.1 所示(Android 应用程序还可以包含有 Service 等后台对象，有关细节在第 9 章讨论)。一个 Android 应用程序必须有一个 Activity 对象被设置为主要的 Activity 对象，Android 运行环境(ADV 模拟器或 Android 手机)通过加载主要的 Activity 对象开始运行一个 Android 应用程序。Android 应用程序中的一个 Activity 对象可以请求加载它包含的其他的 Activity 对象，Android 应用程序正是通过其中的这些 Activity 对象来体现自身的功能(见第 8 章)。如果应用程序包含若干个 Activity 对象，就必须在应用程序的配置文件中为每个 Activity 对象配置一个＜activity＞标记，即在应用程序中注册 Activity 对象，只有这样应用程序才可以使用这个 Activity 对象(见 2.2 节)。

图 2.1　Android 应用程序的基本结构

注：在第8章之前，我们的应用程序中只有一个主要的Activity对象。

2.2 Android应用程序的配置文件

AndroidManifest.xml配置文件位于工程的根目录中，其内容都是一些和当前应用程序密切相关的信息，编译（Debug）工程时，将根据工程根目录下的AndroidManifest.xml文件生成pak文件（一个Android应用程序）。

在后续的学习中会不断地学习怎样编辑AndroidManifest.xml，目前的应用程序比较简单，因此AndroidManifest.xml的内容也相对比较简单。

需要再次强调的是，XML文件和Java文件不同，默认的是UTF-8编码，因此在保存XML文件时必须将编码选择为"UTF-8"、保存类型选择为"所有文件"。

1. 创建工程获得AndroidManifest.xml配置文件

Android是通过创建工程、debug（编译）工程得到Android应用程序的pak文件，在这里，通过分析工程中提供的AndroidManifest.xml配置文件来熟悉Android程序的基本结构。

用命令行方式进入D:\2000，创建名字为Example2_1的工程：

```
D:\2000 > android create project – t 3 – n Example2_1
– p ./Example2_1 – a FirstActivity – k tom.jiafei
```

创建Example2_1工程之后，工程的结构如图2.2所示。

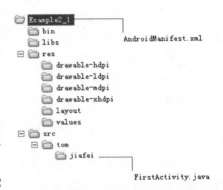

图2.2 Example2_1工程的结构

请注意以下两点：
- 工程根目录（Example2_1）下有一个名字为AndroidManifest.xml的配置文件。
- 工程根目录下的\src\tom\jiafei目录中有一个名字为FirstActivity.java的Java源文件。

AndroidManifest.xml文件的内容如下（箭头是我们添加的解释）。

AndroidManifest.xml

```
<?xml version = "1.0" encoding = "utf – 8"?>
< manifest xmlns:android = "http://schemas.android.com/apk/res/android"
        package = "tom.jiafei"                          ←用包名作为应用程序的标识
        android:versionCode = "1"                       ←应用程序的版本号
        android:versionName = "1.0">                    ←应用程序的版本名称
    < application android:label = "@string/app_name"    ←应用程序的标题（见图2.3）
                android:icon = "@drawable/ic_launcher"> ←应用程序的图标（见图2.3）
        < activity android:name = "FirstActivity"       ←创建Activity对象的类
                android:label = "@string/app_name">     ←Activity对象的标题（见图2.4）
            < intent – filter >
                < action android:name = "android.intent.action.MAIN" />  ←是主要的Activity
                < category android:name = "android.intent.category.LAUNCHER" /
        </ intent – filter >
```

```
        </activity>
    </application>
</manifest>
```

图标 →
标题 →

标题 →

图 2.3　应用程序的图标与标题　　　　　图 2.4　Activity 对象的标题

2. AndroidManifest.xml 文件的重要细节

1）确定应用程序的唯一标识

AndroidManifest.xml 文件的根标记＜manifest…＞中的 package 属性的值为包名，例如：

```
package = " tom.jiafei "
```

包名是在创建工程时给出的，指定了应用程序的唯一标识，因此在创建不同的工程时，一定保证不同的工程使用不同的包名（参见 1.4 节）。

2）给出应用程序的版本号和名称

AndroidManifest.xml 文件的根标记＜manifest…＞中的 versionCode 属性的值为应用程序的版本号，值是正整数。该值的作用是让其他程序判断我们的程序目前是第几版。versionCode 属性的值默认是 1，对于学习阶段，可以不必修改这个默认值。versionName 属性的值为应用程序的版本的名称，值可以是一个字符串，比如"计算器高级版"，versionName 的默认取值是 1.0，在学习阶段可不必修改 versionName 的值。

3）确定应用程序图标上的文字和图像

＜application…＞…＜/application＞标记中的 label 属性值给出应用程序图标上的文字（双击该图标运行程序），label 属性值可以取工程\res\values 目录下的 strings.xml 文件中给出的值，例如：android:label = "@string/app_name"表示 label 属性值是 strings.xml 文件中某个＜string＞标记中的文本，该＜string＞标记的 name 属性值是"app_name"，例如，如果 strings.xml 文件中有如下的＜string＞标记：

```
<?xml version = "1.0" encoding = "utf-8"?>
<resources>
    <string name = "app_name">清华大学</string>    <!-- name 属性值是"app_name"的标记 -->
    <string name = "bird">飞翔吧</string>            <!-- name 属性值是"bird"的标记 -->
</resources>
```

那么应用程序图标上的文字就是"清华大学"。

如果将＜application…＞…＜/application＞标记中的 label 属性值改为：android:label = "@string/bird"，那么应用程序图标上的文字就是"飞翔吧"。

label 属性值也可以直接给定，例如：

```
android:label = "你好,我是应用程序图标上的文字"
```

注：如果不修改 strings.xml 文件,具有 name 属性值是"app_name"的＜string＞标记所包含的文本是创建工程时给出的 Activity 子类的名字(参见1.5节)。

＜application...＞…＜/application＞标记中的 icon 属性值指定应用程序图标上使用的图像。创建工程后,工程的\res 目录下依次有名字分别为 drawable-hdpi、drawable-ldpi、drawable-mdpi 和 drawable-xhdpi 的子目录,分别放着大小不同的 4 幅机器人图像(名字都是 ic_launcher.png)。android:icon 默认取值如下：

android:icon = "@drawable/ic_launcher"

编译器生成的应用程序中(apk 文件中)会包含这 4 幅图像,Android 运行环境,比如 ADV 或手机,在运行应用程序时会根据屏幕的分辨率选择 4 幅图像中的一幅。

4) 确定应用程序中的 Activity 对象

一个应用程序可以包含若干个 Activity 对象,当包含多个 Activity 对象时,需要在配置文件中为每个 Activity 对象配置一个＜activity＞标记,即在程序中注册 Activity 对象,只有这样程序才可以使用这个 Activity 对象。

AndroidManifest.xml 文件中的标记＜application...＞…＜/application＞中的＜activty＞…＜/activity＞子标记是说明应用程序有几个 Activity 对象,并且哪个是主要的 Activity 对象。如果某个＜activity＞…＜/activity＞标记包含有＜intent-filter＞子标记,并使用＜action.../＞空标记的 name 属性说明自己是 MAIN,使用＜category.../＞空标记的 name 属性说明自己是 LAUNCHER。那么,Android 运行环境将首先加载这个 Activity 对象,即这个 Activity 对象是应用程序中的主要的 Activity 对象。

＜activity＞…＜/activity＞标记中的 name 属性值是创建 Activity 对象的类名,例如：

android:name = "FirstActivity"

因此,工程的"\src\包名\"目录中要提供相应类的 Java 源文件,比如\src\tom\jiafei 目录中的 FirstActivity.java 源文件。

＜activty＞…＜/activity＞标记中的 label 属性值是 Activity 对象对外显示的标题,label 属性值可以是 strings.xml 文件中的某个＜string＞标记中的文本内容,例如：

android:label = "@string/app_name"

那么,Activity 对象对外显示的标题就是 strings.xml 文件中具有 name 属性值为"app_name"的＜string＞标记中的文本内容。

label 属性值也可以直接给定,例如：

android:label = "This is a main activity"

关于应用程序图标上的名字和主要的 activity 的标题,请务必注意以下两点：
- 如果主要的 activity 使用 label 属性设置了标题,那么应用程序事先设置的图标上的名字将被更改为主要的 activity 的标题。
- 如果主要的 activity 没有使用 label 属性设置标题,那么 activity 的标题就是应用程序设置的图标上的名字。

2.3　设置主要的 Activity 对象

一个应用程序可以包含多个 Activity 对象,但其中必须有一个 Activity 对象被设置为主要的 Activity 对象。当创建工程时,其中的 Activity 的子类的对象被设置成应用程序的主要的 Activity 对象。现在让我们修改 AndroidManifest.xml 文件中的内容,以便让应用程序包含两个 Activity 对象,然后重新指定主要的 Activity 对象。

首先将下列 Java 源文件(其内容不必深究,在 2.5 节还要详细介绍)保存到 2.2 节中创建的 Example2_1 工程的 \src\tom\jiafei 目录中。

SecondActivity.java

```java
package tom.jiafei;
import android.app.Activity;
import android.os.Bundle;
public class SecondActivity extends Activity {        //子类名为 SecondActivity
    public void onCreate(Bundle savedInstanceState) {
        super.onCreate(savedInstanceState);
        setContentView(R.layout.ok); //和 FirstActivity.java 不同,这里为 R.layout.ok
    }
}
```

目前,Example2_1 工程的 src 目录下的 tom\jiafei 目录中已经分别有了两个 Activity 对象的 Java 源文件,一个为 FirstActivity.java(创建工程时自动诞生的),另一个为 SecondActivity.java(创建工程后,我们又向工程新添加的)。

AndroidManifest.xml 文件中,默认地将 FirstActivity.java 对应的 Activity 对象作为主要的 Activity 对象。以下我们修改 AndroidManifest.xml 文件,将 SecondActivity.java 对应的 Activity 对象作为主要的 Activity 对象。用文本编辑器打开工程根目录下的 AndroidManifest.xml 文件,修改后的 AndroidManifest.xml 如下(请注意加重字体注释的内容):

AndroidManifest.xml

```xml
<?xml version = "1.0" encoding = "utf-8"?>
<manifest xmlns:android = "http://schemas.android.com/apk/res/android"
    package = "tom.jiafei"
    android:versionCode = "1"
    android:versionName = "1.0">
  <application android:label = "@string/bird"
               android:icon = "@drawable/ic_launcher">
    <!-- 新的 SecondActivity 类的 Activity 对象是 MAIN: -->
    <activity android:name = "SecondActivity"
              android:label = "@string/bird">
      <intent-filter>
        <action android:name = "android.intent.action.MAIN" />
        <category android:name = "android.intent.category.LAUNCHER" />
      </intent-filter>
    </activity>
```

```xml
            <!-- 原有的activity,不再是MAIN: -->
            <activity android:name = "FirstActivity"
                      android:label = "@string/app_name">
                <intent-filter>
                    <category android:name = "android.intent.category.LAUNCHER" />
                </intent-filter>
            </activity>
        </application>
</manifest>
```

在保存修改后的文件时,需用另存方式替换原来的文件,并将编码选择为UTF-8,文件类型选择为"所有文件"(这一点务必切记)。

以下为我们新增加的Activity对象配置strings.xml文件中的内容,即配置Activity的标题,并新增加一个用于配置和Activity对象视图相关的ok.xml文件,有关它们的作用将在2.5节和2.6节详细讲解。

1. 用文本编辑器修改 strings.xml

打开工程\res\values下的strings.xml文件,进行修改,多增加一个<string>标记,并设置该标记的name属性的值为"bird",该<string>标记中的文本将作为SecondActivity类的Activity对象的标题(参见前面刚刚修改过的AndroidManifest.xml),修改后的strings.xml如下:

strings.xml

```xml
<?xml version = "1.0" encoding = "utf-8"?>
<resources>
    <string name = "app_name">FirstActivity</string>
    <string name = "bird">飞翔吧</string> <!-- 新增添的标记,包含的文本:飞翔吧 -->
</resources>
```

在保存修改后的文件时,需用另存方式替换原来的strings.xml文件,并将编码选择为UTF-8,文件类型选择为"所有文件"。

2. 增加一个和视图相关的XML文件

SecondActivity.java对应的Activity对象将用到下列XML文件ok.xml,将ok.xml(内容模仿\res\layout目录中已有的main.xml)保存到工程\res\layout目录中(保存文件时,需将编码选择为UTF-8,文件类型选择为"所有文件")。

ok.xml

```xml
<?xml version = "1.0" encoding = "utf-8"?>
<LinearLayout xmlns:android = "http://schemas.android.com/apk/res/android"
    android:orientation = "vertical"
    android:layout_width = "match_parent"
    android:layout_height = "match_parent"
    >
<TextView
    android:layout_width = "match_parent"
    android:layout_height = "wrap_content"
    android:text = "大家好,我是主要的activity了"
```

```
            />
</LinearLayout>
```

启动 ADV 模拟器,用命令行进入工程的根目录,即 D:\ 2000\Example2_1,编译(Debug)工程,并安装应用程序:

```
D:\2000\Example2_1> ant debug install
```

在 ADV 中运行程序,就会发现 ADV 加载了主要的 activity,效果如图 2.5 所示。

图 2.5　主要 activity 的效果

2.4　Activity 对象的外观及状态

1. Activity 对象的外观

Activity 对象中含有一个 Window 类型的成员对象,习惯称为 Activity 对象的窗口。Activity 对象通过自己的窗口和用户进行交互,当 Activity 对象被加载时,它的窗口将占据整个设备的屏幕。

Activity 对象的窗口分为两部分,顶部用来显示 Activity 对象的标题,标题下方的全部空间用于放置视图(View)。图 2.6 是 Activity 在手机中的外观示意图,图 2.7 是 Activity 在 ADV 中的外观。

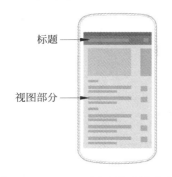

图 2.6　Activity 对象的标题与视图　　　　图 2.7　ADV 中 Activity 对象的外观

2. Activity 对象的状态

Activity 对象由运行环境负责创建和加载,用户只需在 AndroidManifest.xml 文件中给出创建 Activity 对象的类即可。

一个 Activity 对象在它的生命周期中会涉及如下 4 个状态。

1) 活动状态(active)

当 Activity 对象的窗口呈现在设备屏幕的前面(foreground),并处于有焦点(focus)的状态,此时的 Activity 对象处于活动状态。处于活动状态的 Activity 对象的窗口可以和用户进行交互式操作。

2) 暂停状态(pause)

当 Activity 对象的窗口失去焦点,但在设备屏幕上没有被完全遮挡,比如其他的处于活动状态的 Activity 对象的窗口或其他应用程序的窗口部分遮挡了当前 Activity 对象的窗口,那么当前 Activity 对象处于暂停状态。处于暂停状态的 Activity 对象不能和用户进行交互式操作,但仍然驻留在内存中,可以恢复到活动状态。需要注意的是,当运行环境发现内存空间处于极度消耗状态时,运行环境有权力杀死处于暂停状态的 Activity 对象,即释放 Activity 对象占有的内存空间。

3) 停止状态(stop)

当 Activity 对象的窗口失去焦点,在设备屏幕上被完全遮挡,比如其他的处于活动状态的 Activity 对象的窗口或其他应用程序的窗口全部遮挡了当前 Activity 对象的窗口,那么当前 Activity 对象处于停止状态。处于停止状态的 Activity 对象不能和用户进行交互式操作,但仍然驻留在内存中,可以恢复到运行状态。需要注意的是,当运行环境发现内存空间处于极度消耗状态时,运行环境有权力杀死处于停止状态的 Activity 对象,即释放 Activity 对象占有的内存空间。

4) 死亡状态(die)

如果一个 Activity 对象释放了自己占用的内存,那么就进入了死亡状态。一个处于暂停或停止状态的 Activity 对象可能被系统杀死,进入死亡状态,一个 Activity 对象也可以主动调用 finish()方法使自己进入死亡状态。进入死亡状态的 Activity 对象无法回到活动状态,应用程序必须重新加载死亡的 Activity 对象,才能使 Activity 对象回到活动状态。

3. Activity 对象在生命周期中涉及的方法

一个 Activity 对象在它的生命周期中会涉及下列 6 个方法的调用:

- public void onCreate(Bundle savedInstanceState);
- public void onStart();
- public void onResume();
- public void onPause();
- public void onStop();
- public void onDestroy()。

在 Activity 对象的生命周期内,onStart(),onStop,onPause(),onResume()方法都可能被多次调用,但 onCreate(Bundle savedInstanceState)和 onDestroy()只被调用一次。

1) public void onCreate(Bundle savedInstanceState)方法

Activity 对象由运行环境负责创建并加载到内存。当一个 Activity 对象被创建并加载到内存后,该对象立刻调用 public void onCreate(Bundle savedInstanceState)方法。在编写 Activity 的子类时,子类需要重写父类 Activity 的 onCreate 方法,并且要把访问权限升级到 public 权限。重写 onCreate 方法的目的是完成 Activity 对象的初始化工作,比如向 Activity 对象添加视图等。在重写 onCreate 方法时,需要用 super 关键字调用被隐藏的父类的 onCreate 方法,以便将自己的一些重要信息传递给运行环境,如下所示:

```
public class FirstActivity extends Activity {
    public void onCreate(Bundle savedInstanceState) {
        super.onCreate(savedInstanceState);   //调用隐藏的 onCreate
        //用户的其他代码,例如:
        setContentView(R.layout.main);
    }
}
```

在 Activity 对象的生命周期内,onCreate 方法只被调用一次。

2) public void onStart()方法

处于活动状态的 Activity 对象调用过 onCreate 方法之后,就会立刻调用 onStart()方法。用户可以根据程序的需要,确定是否在 onStart()方法中编写代码。比如,用户曾在 onCreate 方法中创建了一个图片自动播放器,希望 Activity 对象处于活动状态时播放图片,那么就可以在 onStart()方法中启动图片自动播放器。

需要特别注意的是,用户重写 onStart()方法时,必须首先调用父类的 onStart()方法(即首先调用被隐藏的 onStart()方法),否则程序将出现运行异常。如下所示:

```
public void onStart() {
    super.onStart(); //必须首先调用被隐藏的 onStart()方法
    //用户的其他代码
}
```

3) public void onResume()方法

处于活动状态的 Activity 对象调用过 onStart()方法之后,就会立刻调用 onResume()方法。用户可以根据程序的需要,确定是否在 onResume()方法中编写代码。

需要特别注意的是,重写 onResume()方法时必须首先调用父类的 onResume()方法(即首先调用被隐藏的 onResume()方法),否则程序将出现运行异常。如下所示:

```
public void onResume() {
    super.onResume(); //必须首先调用被隐藏的 onResume()方法
    //用户的其他代码
}
```

4) public void onPause()方法

Activity 对象进入暂停状态时就会调用 onPause()方法,当 Activity 对象从暂停状态变成活动状态时,会再次调用 onResume()方法。用户可以根据程序的需要,确定是否在 onPause()方法中编写代码,比如,暂停状态的 Activity 对象可能会被系统杀死,那么应当在 onPause()方法中及时保存重要的数据。

需要特别注意的是,重写 onPause()方法时必须首先调用父类的 onPause()方法(即首先调用被隐藏的 onPause()方法),否则程序将出现运行异常。如下所示:

```
public void onPause() {
    super.onPause(); //必须首先调用被隐藏的 onPause()方法
    //用户的其他代码
}
```

5) public void onStop()方法

Activity 对象进入停止状态时就会调用 onStop()方法,当 Activity 对象从停止状态变成活动状态时,会再次调用 onStart()方法。用户可以根据程序的需要,确定是否在 onPause()方法中编写代码。比如,用户曾在 onStart()方法中启动了图片自动播放器,当 Activity 对象变成停止状态时,就不应该继续播放图片(Activity 对象的窗口被遮挡了,即使播放,用户也看不见),那么就应该在 onStop()方法中停止播放图片。

需要特别注意的是,重写 onStop()方法时必须首先调用父类的 onStop()方法(即首先

调用被隐藏的 onStop()方法,否则程序将出现运行异常。如下所示:

```
public void onStop() {
    super.onStop();  //必须首先调用被隐藏的 onStop()方法
    //用户的其他代码
}
```

6) public void onDestroy()方法

Activity 对象进入死亡状态时就会调用 onDestroy()方法,用户一般不需要重写该方法。

需要特别注意的是,如果用户重写 onDestroy()方法,重写方法必须首先调用父类的 onDestroy()方法(即首先调用被隐藏的 onDestroy()方法),否则程序将出现运行异常。如下所示:

```
public void onDestroy() {
    super.onDestroy()  //必须首先调用被隐藏的 onDestroy()方法
    //用户的其他代码
}
```

2.5 视图资源

为 Activity 对象构建视图有两个方式:一种方式是在和 Activity 对象相关的 Activity 的子类中编写构建 View 视图的 Java 代码。第二种方式是使用视图资源中的 XML 文件来构建视图。在今后的学习和程序设计中基本都使用第二种方式,但明白第一种方式有利于更好地理解第二种方式。因此,我们首先介绍第一种方式,然后再介绍第二种方式。

1. 在 Activity 的子类中构建 View 视图

Activity 对象的外观除了标题外,重要的就是视图部分。Activity 对象的窗口分为两部分,顶部用来显示 Activity 对象的标题,标题下方的空间用于放置 View 视图。

Android SDK 提供的 View 类(在 andoid.view 包中)作用类似于 Java SE 中 java.awt.Component 类的作用(后续的许多章节将陆续学习怎样使用 View 的子类)。View 类的一个重要子类是 ViewGroup(在 andoid.view 包中),作用类似于 Java SE 中 java.awt.Container 类的作用。允许在 ViewGroup 视图中添加 View 视图,由于 ViewGroup 类是 View 类的子类,因此,ViewGroup 视图也可添加 ViewGroup 视图(如图 2.8 所示)。

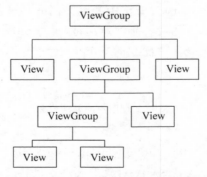

图 2.8 ViewGroup 与 View 视图可以形成的关系

Activity 对象可以使用 setContentView(View view)方法向 Activity 对象使用的窗口放置视图。需要注意的是,Activity 对象的窗口中只能放置一个 View 视图,因此需使用视图的嵌套来放置多个 View 视图。

例如,LinearLayout 类型的视图是 ViewGroup 视图(LinearLayout 是 ViewGroup 的子

类)。LinearLayout 视图把添加它的多个 View 视图按添加的先后顺序排列在一行或一列中。是否排列在一行或一列中取决于 LinearLayout 视图的方向(Orientation),当视图的方向是 HORIZONTAL(水平)时,LinearLayout 视图把添加它的多个 View 视图排列在一行中,当视图的方向是 VERTICAL(垂直)时,LinearLayout 视图把添加它的多个 View 视图排列在一列中(有关 LinearLayout 视图的更多细节将在第 4 章中讲解)。TextView(TextView 是 View 的子类)视图可以显示文本,但用户不可以编辑这些文本。

2. 视图资源

1) 视图资源的位置和引用方式

Android 开发环境要求把和视图相关的 XML 文件存放在工程的\res\layout 目录中。编译工程时,系统 R 类会使用名字为 layout 的静态内部类管理 XML 文件,在静态内部类 layout 中分别用一个 int 型变量绑定和视图相关的每一个 XML 文件,该 int 型变量的名字和 XML 文件的名字相同,其值被 R 认为是 XML 视图文件的资源 ID(layoutResID)。R 类可以用如下方式获得这个 ID:

R.layout.XML 文件名

Activity 对象可以使用 setContentView(int layoutResID)方法设置视图,在使用这个方法时,只需将 XML 视图文件的资源 ID 传递给方法的参数,例如,假设 XML 文件是 tom.xml,Activity 对象设置视图的方式是:

setContentView(R.layout.tom);

工程的\res\layout 目录是专门用来存放视图资源的 XML 文件,Android 应用程序中的 Java 代码使用系统提供的 R 类引用视图资源的 XML 文件,如图 2.9 所示。

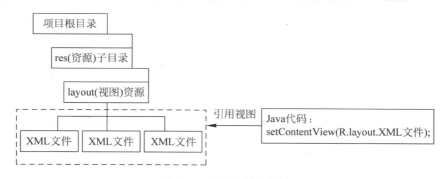

图 2.9 视图资源及使用

Android 应用程序通过视图资源来构建视图有以下两点好处:

- 应用程序中的多个 Activity 对象可以引用相同的 XML 文件来构建 View 视图,Java 代码可方便地更改所引用的 XML 文件来更换 Activity 对象的视图,提高了开发效率。
- 将视图有关的数据和实现分离,即 XML 文件描述视图的有关数据,Java 代码实现视图的外观。当需要修改视图相关的数据时,只要修改 XML 文件,不必修改 Java 代码,当需要修改外观时,不必修改 XML 文件,只需修改 Java 代码。

2) 系统自动提供的 R 类

编译工程后,工程的根目录下产生一个名字为 gen 的目录,gen 目录下的包名对应的子目录中有名字为 R.java 的文件,该文件被自动产生,而且编译后的 .class 文件也是应用程序的一部分。需要注意的是,R.java 是程序设计中专用的特殊 Java 文件,由系统自动创建,用户不可以修改 R.java 的内容。

3) 视图相关的 XML 文件的结构

我们已经知道 ViewGroup 视图中可以添加 View 视图,由于 ViewGroup 类是 View 类的子类,因此,ViewGroup 视图也可添加 ViewGroup 视图(见前面的图 2.6)。如果读者熟悉 XML 语言,就很容易理解视图相关的 XML 文件的结构,这里的标记名称都是某个 View 类的子类的名字,即代表一个 View 视图,标记中的属性都是类的有关成员。如果一个标记是另一个标记的子标记,那么该标记对应的视图就被添加到另一个标记对应的视图中。

视图相关的 XML 文件的根标记的名字必须对应某个 ViewGroup 类的子类(如 LinearLayout 类),其他标记的名字可以是 View 类的子类(如 TextView、Button、EditText、…LinearLayout 类等),如图 2.10 所示。

需要特别注意的是,视图相关的 XML 文件中不要写错视图对应的类的名字,否则项目可以 debug 通过,但运行时将发生异常,系统将终止程序的继续运行。需要再次强调的是,XML 文件和 Java 文件不同,默认的是 UTF-8 编码,因此在保存 XML 文件时必须将编码选择为"UTF-8"、保存类型选择为"所有文件"。

注:视图相关的 XML 文件的名字只能使用小写英文字母(a~z)、数字(0~9)、点(.)和下划线(_),不可以使用大写的英文字母。

3. 示例

下面的例子 2-1 通过在 Activity 的子类中编写具体的 Java 代码向 Activity 对象的窗口中添加视图,运行效果如图 2.11 所示。

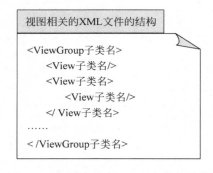

图 2.10 视图相关的 XML 文件的结构

图 2.11 Activity 对象中的视图

例子 2-1

(1) 创建名字是 ch2_1 的工程,主要 Activity 子类的名字为 Example2_1,使用的包名为 ch2.one。用命令行进入 D:\2000,创建工程 D:\2000＞android create project -t 3 -n ch2_1 -p ./ch2_1 -a Example2_1 -k ch2.one。

(2) Java 源文件中的 Example2_1 类要向 Activity 对象的窗口添加一个 ViewGroup 视图,再向这个 ViewGroup 类型添加若干个 View 视图。修改工程的 \src\ch2\one 目录下的 Example2_1.java 文件,修改后的内容如下:

Example2_1. java

```java
package ch2.one;
import android.app.Activity;
import android.os.Bundle;
import android.widget.*;
public class Example2_1 extends Activity {
    LinearLayout layoutH, layoutV;
    TextView label;
    EditText text;
    Button   button;
    public void onCreate(Bundle savedInstanceState) {
        super.onCreate(savedInstanceState);
        layoutV = new LinearLayout(this);    //创建当前 Activity 对象的 LinearLayout 视图
        setContentView(layoutV);
        layoutV.setOrientation(LinearLayout.VERTICAL);
        label = new TextView(this);          //创建当前 Activity 对象的 TextView 视图
        text = new EditText(this);
        button = new Button(this);
        label.setText("input content:");
        text.setText("today");
        button.setText("Enter");
        layoutV.addView(label);
        layoutV.addView(text);
        layoutV.addView(button);
    }
}
```

（3）启动 ADV，进入工程的根目录，用快捷方式编译工程、安装应用程序到 ADV（有关知识点参见 1.5 节）。对于本例子，用命令行进入 D:\2000\ch2_1，执行如下命令：

`D:\2000>ch2_1> ant debug install`

例子 2-2

例子 2-2 通过在 Activity 的子类中引用和视图相关的 XML 文件向 Activity 对象的窗口中添加视图，运行效果如前面的图 2.7 所示。

（1）创建名字是 ch2_2 的工程，主要 Activity 子类的名字是 Example2_2，使用的包名是 ch2.two。用命令行进入 D:\2000，创建工程 D:\2000＞android create project -t 3 -n ch2_2 -p ./ch2_2 -a Example2_2 -k ch2.two。

（2）将和视图相关的 XML 文件 ch2_2.xml 保存到工程的\res\layout 目录中，在保存 XML 文件时将编码选择为 UTF-8，文件类型选择为"所有文件"（这一点务必切记）。

ch2_2. xml

```xml
<?xml version="1.0" encoding="utf-8"?>
<LinearLayout xmlns:android="http://schemas.android.com/apk/res/android"
    android:orientation="vertical"
    android:layout_width="match_parent"
    android:layout_height="wrap_content"   >
    <TextView
```

```
            android:layout_width = "match_parent"
            android:layout_height = "wrap_content"
            android:text = "input content:"    />
        <EditText
            android:layout_width = "match_parent"
            android:layout_height = "wrap_content"
            android:text = "today"    />
        <Button
            android:layout_width = "match_parent"
            android:layout_height = "wrap_content"
            android:text = "Enter"    />
</LinearLayout>
```

（3）修改工程的\src\ch2\two 目录下的 Example2_2.java 文件，在 Example2_2 中使用系统提供的 R 类引用和视图相关的 ch2_2.xml，修改后的内容如下：

Example2_2.java

```
package ch2.two;
import android.app.Activity;
import android.os.Bundle;
public class Example2_2 extends Activity {
    public void onCreate(Bundle savedInstanceState) {
        super.onCreate(savedInstanceState);
        setContentView(R.layout.ch2_2); //使用 R 类引用和视图相关的 ch2_2.xml
    }
}
```

（4）启动 ADV，进入工程的根目录，用快捷方式编译工程、安装应用程序到 ADV（有关知识点参见 1.5 节）。对于本例子，用命令行进入 D:\2000\ch2_2，执行如下命令：

```
D:\2000 > ch2_2 > ant debug install
```

2.6 值 资 源

在 Android 程序设计中，可以将其他的资源（XML 文件）以及 Java 代码需要的一些值，比如视图中需要的名字、颜色值，应用程序的标题等放在规定格式的 XML 文件中，以便其他的资源以及 Java 代码使用这些值。

1. 值资源的位置和引用方式

Android 开发环境要求把值资源相关的 XML 文件存放在工程的\res\values 目录中，其他的资源就可以引用该 XML 文件中的值。例如，视图相关的 XML 文件需要红、绿、蓝三种颜色，那么可以事先编写下列格式的值资源 XML 文件：

```
<?xml version = "1.0" encoding = "utf-8"?>
<resources>
    <color name = "red_color"># FF0000</color>
    <color name = "green_color"># 00FF00</color>
    <color name = "blue_color"># 0000FF</color>
</resources>
```

上述文件必须保存到工程的\res\values 目录中。在保存时,建议文件的名字为 color. xml。color. xml 文件中的根标记的名称必须为 resources,子标记名称是 Android 开发系统事先规定好的某些标记的名称,比如 color,string,integer 等,子标记必须有名字为 name 的属性,name 的属性值由用户指定。

其他资源 XML 文件使用如下格式引用值资源 XML 文件中的子标记的文本内容(值):

@标记名/name 的属性值

例如:

@color/red_color //引用 name 的属性值是 red_color 的 color 标记中文本内容:#FF0000

Java 代码中,某些视图也可以使用自己的方法引用值资源 XML 文件中的子标记的文本内容。系统的 R 类将值资源中的值绑定到一个整数,使用 R 类可获得该整数 R. 标记. name 的属性值。例如,R. color. red_color。所以只要把这个整数传递给方法的参数,该方法就可以使用值资源中的值了。查询类库时,这样的方法的参数都是 int resid,例如一个 TextView 视图 text 调用 setText(int resid)方法可以使用值资源 strings. xml 中的值 t. ext. setText(R. string. app_name)。

工程的\res\values 目录是专门用来存放值资源 XML 文件,Android 应用程序中的 Java 代码或其他资源文件引用值资源 XML 文件中给出的值,如图 2.12 所示。

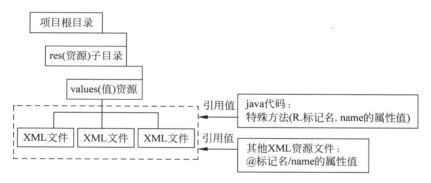

图 2.12 值资源及使用

2. 和值相关的 XML 文件的结构

和值相关的 XML 文件的根标记的名字必须是 resources,子标记名称是 Android 开发系统事先规定好的某些标记的名称,比如 color,string 等子标记,子标记必须有名字为 name 的属性,name 的属性值由用户指定。

在进行项目设计时,值相关的一个 XML 文件中应当只包含一个 Android 开发系统事先规定好的标记的名称,并且该 XML 文件的名字和其中规定好的标记的名称相同。当然可以把所有的和值相关的 XML 文件合成一个 XML 文件(该 XML 文件的名字可任意给定),但这不是程序设计所提倡的。

值相关的 XML 文件中可以使用的常用标记有 color,integer,string,array,style 等。后续章节会陆续使用这些标记(array 的用法参见 3.8 节,style 的用法参见 3.14 节)。

注:值资源相关的 XML 文件的名字只能使用小写英文字母(a~z),数字(0~9),下划

线(_),不可以使用大写的英文字母。

3. 示例

下面的例子 2-3 的 Java 代码和视图资源使用了值资源。运行效果如图 2.13 所示。

图 2.13　使用值资源

例子 2-3

(1) 创建名字为 ch2_3 的工程,主要 Activity 子类的名字为 Example2_3,使用的包名为 ch2.three。用命令行方式进入 D:\2000,创建名字为 ch2_3 的工程 D:\2000＞android create project -t 3 -n ch2_3 -p ./ch2_3 -a Example2_3 -k ch2.three。

(2) 向值资源中增加 XML 文件。将下列 color.xml 保存到工程的\res\values 目录中(在保存时将编码选择为 UTF-8,文件类型选择为"所有文件"),并修改\res\values 目录中已有的 strings.xml(修改后,用另存方式替换原来的文件,并将编码选择为 UTF-8,文件类型选择为"所有文件")。color.xml 以及修改后的 strings.xml 文件的内容如下:

color.xml

```xml
<?xml version = "1.0" encoding = "utf-8"?>
<resources>
    <color name = "red_color">#FF0000</color>
    <color name = "green_color">#00FF00</color>
    <color name = "blue_color">#0000FF</color>
</resources>
```

strings.xml

```xml
<?xml version = "1.0" encoding = "utf-8"?>
<resources>
    <string name = "app_name">你好,我喜欢!</string>
    <string name = "yellow_dog">可爱的小黄狗</string>
    <string name = "bird">飞翔吧,小鸟</string>
</resources>
```

(3) 视图资源。下列视图资源中的 XML 文件 ch2_3.xml 使用了值资源中 color 标记的文本内容(颜色的值),以及 string 标记中的文本内容。将下列 ch2_3.xml 保存到工程的\res\layout 目录中。

ch2_3.xml

```xml
<?xml version = "1.0" encoding = "utf-8"?>
<LinearLayout xmlns:android = "http://schemas.android.com/apk/res/android"
    android:orientation = "horizontal"
    android:layout_width = "match_parent"
    android:layout_height = "match_parent"
    android:background = "@color/red_color"  >
    <TextView
        android:layout_width = "wrap_content"
        android:layout_height = "wrap_content"
        android:text = "@string/yellow_dog"
        android:background = "@color/blue_color"   />
    <Button
        android:layout_width = "wrap_content"
```

```
        android:layout_height = "wrap_content"
        android:text = "@string/bird"
        android:background = "@color/green_color"   />
</LinearLayout>
```

（4）修改 ch2_3 工程的\src\ch2\three 目录下的 Example2_3.java 文件,在 Example2_3.java 中使用系统提供的 R 类引用和视图相关的 ch2_3.xml,修改后的内容如下：

Example2_3.java

```
package ch2.three;
import android.app.Activity;
import android.os.Bundle;
public class Example2_3 extends Activity {
    public void onCreate(Bundle savedInstanceState) {
        super.onCreate(savedInstanceState);
        setContentView(R.layout.ch2_3); //引用视图资源 ch2_3.xml
    }
}
```

（5）启动 ADV,进入工程的根目录,用快捷方式编译工程、安装应用程序到 ADV(有关知识点参见 1.5 节)。对于本例子,用命令行进入 D:\2000\ch2_3,执行如下命令：

```
D:\2000 > ch2_3 > ant debug install
```

2.7 图 像 资 源

在 Android 程序设计中,可以将 Java 代码部分或其他资源需要的一些图像,比如,应用程序图标上的图像、视图上的图像等存放到工程指定的目录中,以便程序代码或其他的资源 XML 文件使用这些图像。

1. 图像资源的位置和引用方式

可以把名字相同,仅大小(像素)不同的 4 幅图像(格式是 png,jpg,gif 或 bmp),比如名字都为 gamePic 的 4 幅图像(4 幅图像的大小不同,格式也可以不同),分别存放在工程的下列目录中：

\res\drawable - hdpi
\res\drawable - ldpi
\res\drawable - mdpi
\res\drawable - xhdpi

应用程序会根据运行环境的屏幕的分辨率选择引用某个目录下的图像。

如果不希望应用程序根据图像的大小不同选择其中之一,那么也可以在工程的 res 目录下新建子目录 drawable,然后把图像存放到该目录下。

其他 XML 资源文件使用如下格式引用图像资源中的图像(如图 2.14 所示)：

@drawable/图像名字

例如：

@drawable/gamePic

Java 代码中,某些视图也可以使用自己的方法引用图资源文件中的图像(如图 2.14 所示)。系统的 R 类将图像资源中的图像绑定到一个整数,使用 R 类可获得该整数:R.drawable.图像名字。例如,R.drawable.gamePic。所以只要把这个整数传递给方法的参数,该方法就可以使用图像资源中的图像了。查询类库时,这样的方法的参数都是 int resid,例如 setBackgroundResource(int resid)。视图 View 调用 setBackgroundResource(int resid)方法可以把背景设置成一幅图像 gamePic:

view.setBackgroundResource(R.drawable.gamePic);

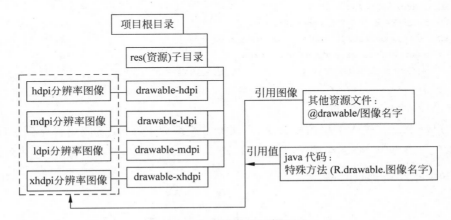

图 2.14 图像资源及使用

注:图像文件的名字只能使用小写英文字母(a～z),数字(0～9),下划线(_),不可以使用大写的英文字母。

2. 示例

以下例子 2-4 使用图像资源修改了应用程序图标上的图像,效果如图 2.15 所示。

图 2.15 修改图标

例子 2-4

(1) 创建名字为 ch2_4 的工程,使用的包名为 ch2.four,主要的 Activity 对象相关的类名为 Example2_4。用命令行方式进入 D:\2000,创建名字为 ch2_4 的工程 D:\2000>android create project -t 3 -n ch2_4 -p ./ch2_4 -a Example2_4 -k ch2.four。

(2) 向图像资源增加图像文件。将名字为 yaya 的图像存放到图像资源中,可以将名字为 yaya 的大小不同的 4 幅图像分别保存到工程的 \res\drawable-hdpi(建议大小是 72*72)、\res\drawable-ldpi(建议大小是 36*36)、\res\drawable-mdpi(建议大小是 48*48)、\res\drawable-xhdpi(建议大小是 96*96)目录中。也可以在\res 中新建一个名字为 drawable 的子目录,将名字为 yaya 的图像保存到\res\drawable 目录中(建议图像大小是 50*56)。

(3) 修改 AndroidManifest.xml 配置文件。AndroidManifest.xml 配置文件位于工程的根目录中,AndroidManifest.xml 中的 <application...>…</application> 标记中的 android:icon 属性值指定应用程序图标上使用的图像,android:icon 属性值可以是图像资源

中的图像。AndroidManifest.xml 中默认让 icon 取值如下：

```
android:icon = "@drawable/ic_launcher"
```

将 icon 取值修改如下：

```
android:icon = "@drawable/yaya "
```

（4）启动 ADV，进入工程的根目录，用快捷方式编译工程、安装应用程序到 ADV（有关知识点参见 1.5 节）。对于本例子，用命令行进入 D:\2000\ch2_4，执行如下命令：

```
D:\2000 > ch2_4 > ant debug install
```

2.8 获取资源

当编译器（Debug）发现程序使用了资源文件（\res 目录或子目录下的文件），就会将这些文件打包在应用程序中（apk 文件中）。编译器（Debug）会为\res 目录或子目录下的文件在系统的 R.java 文件生成资源 ID（参见 2.5 节），程序可以使用资源的 ID 访问该资源。

1. Context 与 Resource 类

android.content 包中的 Context 类以及 android.view 中的 View 类提供了获取资源的方法 public Resources getResources()，该方法返回一个 Resources 类的实例（Resource 类在 android.content.res 包中），使用该实例可以访问已有的资源。

应用程序中许多常用的类都是 Context 或 View 类的子孙类，如 Application，Activity，Service 等类都是 Context 的子类；TextView，Button，ImageView 等都是 View 的子类。因此，当应用程序需要使用资源时，可以使用 Resources 类的实例访问所需要的资源。

Resources 类提供了许多获得资源的方法，这里就不一一列举了，在今后的章节中，根据具体的学习内容再做介绍。比如，Resources 类的实例使用 Drawable getDrawable(int id)方法来使用图像资源 id 得到一个 Drawable 对象，例如：

```
Drawable drawable = getResources().getDrawable(R.drawable.flower);
```

Resources 类的实例使用 String getString(int id)方法来使用值资源 id 得到一个 String 对象，例如：

```
Sting title = getResources().getString(R.string.title);
```

2. 示例

以下例子 2-5 使用图像资源设置 Activity 中 Button 视图的背景是鲜花，使用值资源重新设置了 Activity 的标题是"你好 Android"。效果如图 2.16 所示。

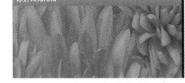

图 2.16　获取资源

例子 2-5

（1）创建名字为 ch2_5 的工程，使用的包名为 ch2.five，主要的 Activity 对象相关的类名为 Example2_5。用命令行方式进入 D:\2000，创建名字为 ch2_5 的工程 D:\2000＞android create project -t 3 -n ch2_5 -p ./ch2_4 -a Example2_5 -k ch2.five。

（2）修改值资源中的 strings.xml 文件，即工程的\res\values 下 strings.xml 文件，需要

再次强调的是,XML 文件和 Java 文件不同,默认的是 UTF-8 编码,因此在保存(或另存)XML 文件时必须将编码选择为 UTF-8、保存类型选择为"所有文件"。修改后的 strings.xml 文件的内容如下:

strings.xml

```xml
<?xml version = "1.0" encoding = "utf-8"?>
<resources>
    <string name = "app_name">Example2_5</string>
    <string name = "title">你好 Android</string>
</resources>
```

(3) 向图像资源增加图像文件。将名字为 flower 的图像(格式可以是 jpg,png,gif 或 bmp)存放到图像资源中,比如工程的 res\drawable-hdpi 目录中,或在工程的 res 目录下新建子目录 drawable,然后把图像存放到该目录下。

(4) 修改 ch2_5 工程的\src\ch2\five 目录下的 Example2_5.java 文件,在 Example2_5.java 中使用 Resources 对象来使用图像资源和值资源,修改后的内容如下:

Example2_5.java

```java
package ch2.five;
import android.app.Activity;
import android.os.Bundle;
import android.content.res.Resources;
import android.graphics.drawable.Drawable;
import android.widget.*;
public class Example2_5 extends Activity {
    public void onCreate(Bundle savedInstanceState) {
        super.onCreate(savedInstanceState);
        Button b = new Button(this);
        setContentView(b);
        Resources source = getResources();
        String title = source.getString(R.string.title);
        Drawable drawable = source.getDrawable(R.drawable.flower);
        b.setBackground(drawable);
        setTitle(title);
    }
}
```

(5) 启动 ADV,进入工程的根目录,用快捷方式编译工程、安装应用程序到 ADV(有关知识点参见 1.5 节)。对于本例子,用命令行进入 D:\2000\ch2_5,执行如下命令:

```
D:\2000>ch2_5>ant debug install
```

习 题 2

1. 一个应用程序中只可以有一个 Activity 对象吗?

2. 假设创建工程时使用的包名是 sun.moon,那么程序中的 Java 源文件使用的包名是什么,java 源文件应当保存在工程的哪个目录中?

3. Activity 对象的窗口分为几部分？都用来放置什么？

4. 与视图相关的 XML 文件，即视图资源，需要保存在工程的哪个目录中，保存时应当选择怎样的编码？XML 文件的名字中是否可以有大写字母？

5. 模仿例子 2-2，创建名字为 xiti2_5 的工程，主要的 Activity 子类的名字为 Xiti2_5，使用的包名为 xiti.five。该工程使用视图资源 XML 文件构建应用程序的 Activity 对象的视图，要求视图资源 XML 文件中提供 TextView 视图和 Button 视图。

6. 值（values）资源 XML 文件应当保存在工程的哪个目录中？值（values）资源 XML 文件的根标记必须是什么标记？

7. 创建一个工程，并修改工程\res\values 目录中已有的 strings.xml（修改后，用另存方式替换原来的文件，并将编码选择为 UTF-8，文件类型选择为"所有文件"）。修改后的 strings.xml 文件的内容如下：

strings.xml

```
<?xml version = "1.0" encoding = "utf-8"?>
<resources>
    <string name = "app_name">我编写的 Android 应用程序</string>
</resources>
```

然后用 ant debug install 命令编译工程、安装应用程序到 AVD。

8. 程序想使用的图像文件可以保存在工程的哪个目录中？

9. 怎样得到一个 Resources 的实例？

10. 参考例子 2-5，在 Activity 对象的视图区域显示 Android 的 Logo（小机器人）。

第 3 章　　常用 View 视图

主要内容：
- View 视图的常用属性与度量值视图；
- TextView 视图；
- EditText 视图；
- Button 视图；
- ToggleButton 视图；
- CheckBox 视图；
- RadioButton 视图；
- Spinner 视图；
- ListView 视图；
- 动态创建 Spinner 视图和 ListView 视图；
- GridView 视图；
- ScrollView 视图。

在第 2 章的 2.5 节介绍了为 Activity 对象构建视图的两种方式：一种方式是在和 Activity 对象相关的类中编写构建 View 视图的 Java 代码，第二种方式是编写和视图相关的 XML 文件。第二种方式是常用方式，本章将使用第二种方式介绍常用的 View 视图。

在本章中需要把和视图相关的 XML 文件放在工程的 \res\layout 目录中。程序的 Java 代码使用系统提供的 R 类引用这个 XML 文件来构建视图，例如，假设 XML 文件是 tom.xml，引用方式如下：

setContentView(R.layout.tom);

本章介绍的常用的 View 视图都是 View 类的子孙类，均在 android.wedget 包中（View 类在 andriod.view 包中）。另外，为了暂时学习的方便，本章使用 LinearLayout 视图作为 ViewGroup 视图，把即将要学习的 View 视图添加到 LinearLayout 视图中。

另外，需要再次强调的是，XML 文件和 Java 文件不同，默认的是 UTF-8 编码，因此在保存 XML 文件时必须将编码选择为"UTF-8"，保存类型选择为"所有文件"。视图相关的 XML 文件的名字只能使用小写英文字母（a～z），数字（0～9），点（.）和下划线（_），不可以使用大写的英文字母。

注：在进行 View 视图设计时，我们重点介绍视图的功能用途和常用的重要属性，也不罗列有关类的方法，建议读者经常查看 docs 帮助文档（参见 1.6 节）。

3.1　View 视图的常用属性与度量值

1. 视图的子孙关系

Android 采用 View 类和 ViewGroup 类来形成视图之间的嵌套关系（Java 语言采用的是 Component 类和 Container 类），ViewGroup 类是 View 类的子类，ViewGroup 视图中既可以有 View 视图，也可以有 ViewGroup 视图，这样就可以形成视图的嵌套。如果一个 View 视图被直接添加（嵌套）在一个 ViewGroup 视图中，就称后者是前者的父（parant）视图，前者是后者的子（child）视图。如果一个 View 视图被间接地添加（嵌套）在一个 ViewGroup 视图中，就称后者是前者的祖先视图，前者是后者的子孙视图。

另外，为了方便起见，我们使用了 Android 帮助文档中表示类与子类之间父子关系的符号∟，即一个类使用符号∟指向自己的子类（不是 UML 图的表示法）。例如表示 TextView 视图是 View 类的一个子类，表示如下：

```
android.view.View
    ∟ android.widget.TextView
```

2. 一个标记对应一个视图

视图相关的 XML 文件中使用标记对应 View 视图，即使用不同的标记对应 View 类的不同子类，用标记中的属性对应类的成员。和视图相关的 XML 文件的根标记必须对应一个 View 视图或 ViewGroup 视图。例如，下列 XML 文件中＜LinearLayout＞标记对应 ViewGroup 类的 LinearLayout 子类、＜Button＞标记对应 View 类的 Button 子类。

```
<?xml version = "1.0" encoding = "utf-8"?>
< LinearLayout xmlns:android = "http://schemas.android.com/apk/res/android"
    android:orientation = "vertical">
    < Button
        android:layout_width = "match_parent"
        android:layout_height = "wrap_content"
        android:text = "确定"
    />
</LinearLayout >
```

如果一个标记是另一个标记的子标记，那么该标记对应的视图就被嵌套（添加）到另一个标记对应的视图中，即该标记对应的视图是另一个标记对应的视图的子视图。

标记中可以有哪些属性以及属性值的范围已经由系统做好了有关规定（XML 文件使用名称空间和 schemas 进行了约定）。如果读者不是很熟悉 XML 语言，可以简单地把 XML 文件中的 xmlns:android = "http://schemas.android.com/apk/res/android"理解为系统做好的约定。因此，我们在 XML 文件中设置属性的值时，需使用名称空间的前缀 android 表示我们的属性是属于约定部分的属性，只有这样才能有效地设置视图的有关属性。另外，不要在 XML 文件中出现系统未约定的标记，否则项目可以通过编译，但会发生运行异常，运行环境会终止程序的运行。

3. 度量值

属性中的某些值经常使用 px,dp,sp,in,mm 等度量单位，解释如下。

px：全称是 pixel，表示屏幕的一个像素。

dp：全称是 density-independent pixels，是设置视图大小时建议使用的单位。

sp：全称是 scaled pixels，是设置字体大小时建议使用的单位。

mm：全称是 millimeter，表示毫米。

in：全称是 inch，表示英寸。

4. View 视图的几个常用属性

由于需要经常使用属性控制视图的外观，因此这里介绍 View 视图的常用属性，由于其他的视图类都是 View 类的子类，因此，后续介绍的视图都可以使用本节介绍的属性设置视图的外观。

1）android:layout_width

作用：设置视图在父视图中的宽度。

取值："match_parent"、"wrap_content"或具体数值，例如，"200dp"或"200px"。

当取值是"match_parent"时，例如，android:layout_width="match_parent"表示当前视图的宽度将填满父视图剩余的宽度，当取值是"wrap_content"时，例如，android:layout_width="wrap_content"，如果当前视图有子视图，表示当前视图的宽度刚好能满足它的子视图的要求。如果当前视图没有子视图，表示当前视图的宽度刚好能显示其中的文本。

注：在 API Level 8 版本之前，使用"fill_parent"，之后的版本将"fill_parent"列为过时参数值(deprecated)，目前仍能使用，API Level 8 版本之后使用"math_parent"代替"fill_parent"。

2）android:layout_height

作用：设置视图在父视图中的高度。

取值："match_parent"、"wrap_content"或具体数值，例如："200dp"或"200px"。

3）android:background

作用：设置视图的背景颜色或背景图像。

取值：颜色值(RGB 颜色模式)或一幅图像。颜色值是用 6 位数字组成的十六进制数，并需要将十六进制数用♯做前缀。6 位数字从左向右，头 2 位表示 R 值的大小（红）、中间 2 位表示 G 值的大小（绿），最后 2 位表示 B 值大小（蓝）。例如，黄颜色的 R,G,B 的值（十进制）分别是 255,255,0，那么用 6 位数字组成的十六进制数表示黄颜色就是"♯FFFF00"，例如，为了让视图的背景色是天蓝色，可以取值如下：

android:background="♯87CEEB"

图像可以是图像资源中的一幅图像（参见 2.7 节），例如，假设 dog.jpg 是资源图像中的一幅图像，那么 android:background 可以取值 android:background="@drawable/dog"。

4）android:layout_gravity

作用：设置视图在父视图中的对齐方式。

取值："top"、"bottom"、"left"、"right"、"center"，以及这些值的组合值，例如"left|center"、"left|bottom"等。例如：

android:layout_gravity="center"

5）android:padding

作用：设置视图的边距。android:padding 同时设置视图 4 个边的边距，而 android:

paddingBottom、android:paddingLeft、android:paddingRight、android:paddingTop 等分别设置底边、左边、右边和顶边的边距。

取值：度量值。例如：

```
android:padding = "8dp"
```

6）android:id

作用：确定视图的 ID 标识。在某些时候，Java 代码部分需要根据视图的 ID 寻找 XML 文件中给出的视图，以便做进一步的编码。

取值：一个用字符串表示的整数，字符串由用户来指定，所代表的整数由系统的 R 类负责指定。给视图指定 ID 的格式如下：

```
android:id = "@ + id/ID值"
```

例如，下列 XML 文件中的 Button 视图的 ID 是 save_button。

```
< Button
    android:id = "@ + id/save_button"    <!-- 指定视图的 ID -->
    android:layout_width = "wrap_content"
    android:layout_height = "wrap_content"
    android:text = "保存"
/>
```

在 Java 代码中，比如在 Activity 子类重写的 onCreate 方法中，可以使用 findViewById (int resid)方法找到 XML 文件中给出的 Button 视图（使用系统提供的 R 类将视图 ID 传递给方法的参数 resid）：

```
Button saveButton = (Button)findViewById(R.id.save_button);
```

7）android:alpha

一个视图可以设置 android:alpha 属性的值（取值 0～1 的浮点数），例如：

```
android:alpha = "0.6"
```

当 android:alpha 属性值是 0，视图完全透明，是 1，视图完全不透明。

3.2 TextView 视图

TextView 视图的继承关系如下：

```
android.view.View
    └ android.widget.TextView
```

TextView 视图的特点是显示文本，但不允许用户编辑它上面的文本，其作用相当于标签的作用。

1. 视图的常用属性

在 XML 文件中使用<TextView>标记对应 TextView 视图，用<TextView>标记的属性值设置视图的有关数据。

表 3.1 是常用属性以及属性值的意义。

表 3.1 TextView 视图的常用属性

属性及样例	取 值	作 用
android:autoLink android:autoLink="web"	"web"、"phone"、 "email"、"map"、 "all"、"none"	自动识别链接。当 TextView 视图上的文字包含 Web 网址、Email 地址、phone 号码或 map 地图时，就将这部分特殊的文本显示为可单击的链接。用户单击操作链接，系统将启动相应的模块，比如单击 phone 号码，将启动拨打电话的模块
android:textColor android:textColor="#FFAA55"	6 位十六进制数表示的颜色	设置 TextView 视图上文字的颜色，默认颜色是黑色。颜色采用 RGB 颜色模式
android:textColorLink android:textColorLink="#00FF00"	6 位十六进制数表示的颜色	设置 TextView 视图上文字中超链接文字的颜色（默认颜色是蓝色）
android:textSize android:textSize="16sp"	单位是 sp 的正整数	设置 TextView 视图上文字的大小
android:textStyle android:textStyle="bold|italic"	"normal"、"bold"、 "italic"，也可以用|同时取几个值	设置 TextView 视图上文字的字型
android:fontFamily android:fontFamily="Arial"	标准字体的名称	设置 TextView 视图上文字的字体
android:text android:text="Hello"	字符串（字符串中可以有转义字符）	设置 TextView 视图上的文字内容
android:gravity android:gravity="center"	"top"、"bottom"、"left"、 "right"、"center"…	设置文本在视图中的对齐方式

2. 示例

例子 3-1 使用了 TextView 视图的部分属性，运行效果如图 3.1 所示（图 3.1(b)的效果需要保证手机或 AVD 所在 PC 与 Internet 相连）。

(a) TextView 视图　　　　(b) 单击视图上的超链接后

图 3.1　运行效果

例子 3-1

（1）创建名字为 ch3_1 的工程，主要 Activity 子类的名字为 Example3_1，使用的包名为 ch3.one。用命令行进入 D:\2000，创建工程 D:\2000＞android create project -t 3 -n ch3_1 -p ./ch3_1 -a Example3_1 -k ch3.one。

(2) 增加视图资源。创建工程后将名字为 flower.jpg 的图像存放到工程的图像资源中（有关知识点参见 2.7 节）。

(3) 将下列和视图相关的 XML 文件 ch3_1.xml 保存到工程的 \res\layout 目录中。

ch3_1.xml

```xml
<?xml version = "1.0" encoding = "utf-8"?>
<LinearLayout xmlns:android = "http://schemas.android.com/apk/res/android"
    android:orientation = "vertical"
    android:layout_width = "match_parent"
    android:layout_height = "match_parent"
    android:background = "@drawable/flower"   >
    <TextView
        android:id = "@+id/my_textViewOne"
        android:autoLink = "all"
        android:textStyle = "bold|italic"
        android:fontFamily = "Arial"
        android:layout_width = "wrap_content"
        android:layout_height = "wrap_content"
        android:background = "#87CEEB"
        android:textSize = "20sp"
        android:text = "http://www.sina.com\n电话:020-12987"   />
    <TextView
        android:id = "@+id/my_textViewTwo"
        android:textColor = "#806400"
        android:layout_width = "wrap_content"
        android:layout_height = "wrap_content"
        android:layout_gravity = "center"
        android:text = "您好,很高兴认识你."
        android:textSize = "25sp"
        android:alpha = "0.8"   />
</LinearLayout>
```

(4) 修改工程 \src\ch3\one 目录下的 Example3_1.java 文件,修改后的内容如下：

Example3_1.java

```java
package ch3.one;
import android.app.Activity;
import android.os.Bundle;
import android.widget.*;
import android.graphics.Color;
public class Example3_1 extends Activity {
    public void onCreate(Bundle savedInstanceState) {
        super.onCreate(savedInstanceState);
        setContentView(R.layout.ch3_1);
        TextView tViewOne = (TextView)findViewById(R.id.my_textViewOne);
        tViewOne.setTextColor(Color.RED);   //改为红颜色的字(不包括超链接文字)
        TextView tViewTwo = (TextView)findViewById(R.id.my_textViewTwo);
        CharSequence cSequence = tViewTwo.getText();
        String str = cSequence.toString();
        str = str + " how are you, nice to meet you";
        tViewTwo.setText(str);
```

    }
}

（5）启动 AVD，进入工程的根目录，用快捷方式编译工程、安装应用程序到 AVD（有关知识点参见 1.5 节）。对于本例子，用命令行进入 D:\2000\ch3_1，执行如下命令：

```
D:\2000 > ch3_1 > ant debug install
```

3.3 EditText 视图

EditText 视图的继承关系如下：

```
android.view.View
    ↳ android.widget.TextView
        ↳ android.widget.EditText
```

EditText 视图的特点是允许用户编辑文本。当 EditText 视图处于有焦点状态后，手机下方会出现虚拟键盘，用户可使用这个虚拟键盘在 EditText 视图中编辑文本。在设计 EditText 视图时，如果不进行特别的约定，EditText 视图默认允许用户输入多行文本。在设计 EditText 视图时可设置该视图在父视图中显示的文本的最大行数，以及初始时显示的文本的最少行数等。

1. 视图的常用属性

在 XML 文件中使用＜EditText＞标记对应 EditText 视图，用＜EditText＞标记的属性值设置视图的有关数据。表 3.2 是常用属性以及属性值的意义。

表 3.2　EditText 视图的常用属性

属性及样例	取　　值	作　　用
android:digits android:digits="abc"	一个字符串	设置 EditText 视图中只可以输入属性值中的字符，例如输入的文本中只可以有 a，b，c 字符
android:inputType android:inputType="time"	"number"，"numberDecimal"，"date"，"time"，"datetime"	设置 EditText 视图中可以输入的数据类型。当设置成 date 时可以输入数字和"/"，当设置成 time 时可以输入数字和"："
android:singleLine android:singleLine="true"	"true"，"false"	设置 EditText 视图中是否是单行模式，默认是"false"
android:password android:password="true"	"true"，"false"	设置 TextView 视图作为密码输入视图
android:phoneNumber android:phoneNumber="true"	"true"，"false"	设置 TextView 视图只可以输入电话号码
android:lines android:lines="7"	正整数	设置 TextView 视图上显示的文本的行数
android:maxLines android:maxLines="20"	正整数	设置 TextView 视图至多显示的文本的行数
android:minLines android:minLines="20"	正整数	设置 TextView 视图上至少显示的文本的行数

需要注意的是,当 android:inputType 取值是"number","numberDecimal","date","time"或"datetime"时,TextView 视图将自动变成单行输入模式。

2. 处理 TextChanged 事件

当 EditText 视图中的文本内容发生改变时,将触发 TextChanged(文本变化)事件,处理 EditText 视图上的 TextChanged 事件的步骤如下。

(1)确定 EditText 视图。首先给出需要处理 TextChanged 事件的 EditText 视图,例如:

EditText tEdit = (EditText)findViewById(R.id.my_edit);

(2)确定监视器。需要使用实现 TextWatcher 接口(在 android.text 包中)的类的对象作为监视器,EditText 视图通过如下方法注册监视器:

public void addTextChangedListener(TextWatcher watcher);

(3)监视器重写 TextWatcher 接口的 afterTextChanged、beforeTextChanged 和 onTextChanged。当 EditText 视图中的文本内容发生改变时,监视器就会调用重写的这 3 个方法。

3. 示例

例子 3-2 使用了 EditText 视图的部分属性。运行效果如图 3.2 所示。

例子 3-2

(1)创建名字为 ch3_2 的工程,主要 Activity 子类的名字为 Example3_2,使用的包名为 ch3.two。用命令行进入 D:\2000,创建工程 D:\2000>android create project -t 3 -n ch3_2 -p ./ch3_2 -a Example3_2 -k ch3.two。

图 3.2　EditText 视图

(2)将下列和视图相关的 XML 文件保存到工程的\res\layout 目录中。

ch3_2.xml

```
<?xml version = "1.0" encoding = "utf-8"?>
<LinearLayout xmlns:android = "http://schemas.android.com/apk/res/android"
    android:orientation = "vertical"
    android:layout_width = "match_parent"
    android:layout_height = "match_parent"
    android:background = "#87CEEB">
    <TextView
        android:background = "#555555"
        android:layout_width = "wrap_content"
        android:layout_height = "wrap_content"
        android:text = "输入出生日期:" />
    <EditText
        android:layout_width = "match_parent"
        android:layout_height = "wrap_content"
        android:inputType = "date" />
    <TextView
        android:background = "#555555"
```

```xml
        android:layout_width = "wrap_content"
        android:layout_height = "wrap_content"
        android:text = "输入电话:" />
    <EditText
        android:layout_width = "match_parent"
        android:layout_height = "wrap_content"
        android:digits = "1234567890 - " />
    <TextView
        android:background = "#555555"
        android:layout_width = "wrap_content"
        android:layout_height = "wrap_content"
        android:text = "输入密码:" />
    <EditText
        android:layout_width = "match_parent"
        android:layout_height = "wrap_content"
        android:password = "true" />
    <TextView
        android:background = "#009900"
        android:layout_width = "wrap_content"
        android:layout_height = "wrap_content"
        android:text = "输入简历:" />
    <EditText
        android:layout_width = "match_parent"
        android:layout_height = "wrap_content"
        android:lines = "3" />
</LinearLayout>
```

(3) 修改工程\src\ch3\two 目录下的 Example3_2.java 文件,修改后的内容如下:

Example3_2.java

```java
package ch3.two;
import android.app.Activity;
import android.os.Bundle;
public class Example3_2 extends Activity {
    public void onCreate(Bundle savedInstanceState) {
        super.onCreate(savedInstanceState);
        setContentView(R.layout.ch3_2);
    }
}
```

(4) 启动 AVD,进入工程的根目录,用快捷方式编译工程、安装应用程序到 AVD(有关知识点参见 1.5 节)。对于本例子,用命令行进入 D:\2000\ch3_2,执行如下命令:

```
D:\2000 > ch3_2 > ant debug install
```

下面例子 3-3 处理了 EditText 视图上的 TextChanged 事件,当 EditText 视图上的文本发生变化后,程序计算出 EditText 视图中的数字的代数和以及平均值,并将这些结果放置在 TextView 视图中。运行效果如图 3.3 所示。

图 3.3 处理 EditText 视图上的文本变化事件

例子 3-3

(1) 创建名字为 ch3_3 的工程，主要 Activity 子类的名字为 Example3_3，使用的包名为 ch3.three。用命令行进入 D:\2000，创建工程 D:\2000＞android create project -t 3 -n ch3_3 -p ./ch3_3 -a Example3_3 -k ch3.three。

(2) 将下列和视图相关的 XML 文件 ch3_3.xml 保存到工程的 \res\layout 目录中。

ch3_3.xml

```xml
<?xml version = "1.0" encoding = "utf-8"?>
<LinearLayout xmlns:android = "http://schemas.android.com/apk/res/android"
    android:orientation = "vertical"
    android:layout_width = "match_parent"
    android:layout_height = "match_parent"
    android:background = "#87CEEB">
    <TextView
        android:background = "#B3C1BF"
        android:textColor = "#0000FF"
        android:layout_width = "match_parent"
        android:layout_height = "wrap_content"
        android:text = "输入数字,用空格分隔: " />
    <EditText
        android:layout_width = "match_parent"
        android:layout_height = "wrap_content"
        android:inputType = "numberDecimal|textMultiLine"
        android:id = "@+id/my_edit" />
    <TextView
        android:background = "#B3C1BF"
        android:textColor = "#0000FF"
        android:layout_width = "match_parent"
        android:layout_height = "wrap_content"
        android:id = "@+id/my_text" />
</LinearLayout>
```

(3) 修改工程 \src\ch3\three 目录下的 Example3_3.java 文件，修改后的内容如下：

Example3_3.java

```java
package ch3.three;
import android.app.Activity;
import android.os.Bundle;
import android.widget.*;
import android.text.*;
import java.util.*;
public class Example3_3 extends Activity {
    TextView tText;
    EditText tEdit;
    public void onCreate(Bundle savedInstanceState) {
        super.onCreate(savedInstanceState);
        setContentView(R.layout.ch3_3);
        tText = (TextView)findViewById(R.id.my_text);
        tEdit = (EditText)findViewById(R.id.my_edit);
        TextWatcher  watcher = new EditListner();
```

```
        tEdit.addTextChangedListener(watcher);      //监视器是 EditListner 类的对象
    }
    public class EditListner implements TextWatcher {    //实现接口的内部类
        public void afterTextChanged(Editable s) {
            int count = 0;
            Scanner scanner = new Scanner(tEdit.getText().toString());
            scanner.useDelimiter("[^0123456789.]+");  //scanner 设置分隔标记
            double sum = 0;
            while(scanner.hasNext()){
                try{   double price = scanner.nextDouble();
                       count++;
                       sum = sum + price;
                }
                catch(InputMismatchException exp){
                       String t = scanner.next();
                }
                String totalSum = String.format("%.2f",sum);
                String aver = String.format("%.2f",sum/count);
                tText.setText("total sum = " + totalSum + "\n aver = " + aver);
            }
        }
        public void beforeTextChanged(CharSequence s, int start, int count, int after)
        {}
        public void onTextChanged (CharSequence s, int start, int before, int count)
        {}
    }
}
```

（4）启动 AVD，进入工程的根目录，用快捷方式编译工程、安装应用程序到 AVD（有关知识点参见 1.5 节）。对于本例子，用命令行进入 D:\2000\ch3_3，执行如下命令：

D:\2000 > ch3_3 > ant debug install

3.4 Button 视图

Button 视图的继承关系如下：

android.view.View
  └ android.widget.TextView
      └ android.widget.Button

Button 视图的特点是允许用户 Click（单击）它。Button 视图上可以触发 Click 事件，程序可以通过处理 Click 事件对用户单击 Button 视图做出相应的响应。

1. 视图的常用属性

在 XML 文件中使用＜Button＞标记对应 Button 视图，即用 Button 标记的属性值设置视图的有关数据，Button 视图的常用属性都是从 TextView 继承而来的。

2. 在程序代码中注册监视器、处理 Click 事件

在代码中处理 Button 视图上的 Click 事件的步骤如下。

(1) 确定 Button 视图。首先给出需要处理 Click 事件的 Button 视图，例如：

```
Button button = (Button)findViewById(R.id.my_button_add);
```

(2) 确定监视器。需要使用实现 View.OnClickListener 接口的类的对象作为 Button 视图的监视器，Button 视图通过如下方法注册监视器：

```
public void setOnClickListener(View.OnClickListener listener);
```

(3) 监视器重写 View.OnClickListener 接口中的 onClick(View v)方法。

例如：

```
button.setOnClickListener(new View.OnClickListener() {
            public void onClick(View v) {
                //Perform action on click
            }
});
```

当 Button 视图上触发 Click 事件后，监视器调用 onClick(View view)方法，该方法中的参数 view 存放着当前触发 Click 事件的 Button 视图的引用。

3. 在视图资源文件中约定 Click 事件

可以在 Button 的视图资源文件中设置 android:onClick 属性的值，约定 Button 视图触发 Click 事件后，Activity 对象要执行的方法，步骤如下。

1) 在 Button 的视图资源文件中设置 android:onClick 属性的值

android:onClick 属性的值是用户在 Activity 的子类中定义的一个方法的名字，即 Activity 对象需要调用的方法，该方法的参数必须是 View 类型（方法名由用户定义），例如：

```
<Button
    android:text = "button"
    android:onClick = "handle"   <!-- 约定需要调用执行的 handle 方法 -->
/>
```

2) 定义出约定的方法

需要在 Activity 的子类中，比如 MyActivity 子类，定义视图文件中 android:onClick 属性曾约定的方法，该方法的参数必须是 View 类型，例如：

```
public class MyActivity extends Activity {
    public void onCreate(Bundle savedInstanceState) {
        super.onCreate(savedInstanceState);
        setContentView(R.layout.main);
    }
    public void handle(View view) {     //根据视图的约定所定义的 handle 方法
        ....
    }
}
```

通过视图文件约定 Click 事件后，系统认为 Button 视图上的 Click 事件的监视器是当前的 Activity 对象。当 Button 视图上触发 Click 事件后，Activity 对象调用视图文件中约定的方法，比如 handle(View view)方法，该方法中的参数 view 就是当前触发 Click 事件的

Button 视图。如果代码部分，没有按视图文件中 android:onClick 属性的约定定义出正确的方法，程序可以编译通过，但运行时，用户单击按钮将发生运行异常，系统将终止程序的运行。

4. 示例

下面的例子 3-4 是一个简单的计算器，在代码中处理了 addButton 视图和 subButton 上的 Click 事件，在视图文件中约定了 mutiButton 视图和 divButton 上的 Click 事件。运行效果如图 3.4 所示。

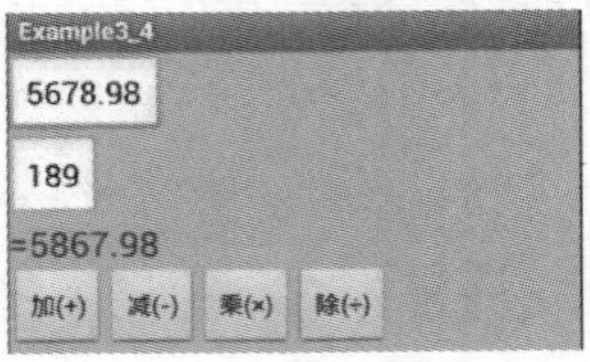

例子 3-4

（1）创建名字为 ch3_4 的工程，主要 Activity 子类的名字为 Example3_4，使用的包名为 ch3.four。用命令行进入 D:\2000，创建工程 D:\2000＞android create project -t 3 -n ch3_4 -p ./ch3_4 -a Example3_4 -k ch3.four。

图 3.4　简单的计算器

（2）增加视图资源。将下列和视图相关的 XML 文件保存到工程的 \res\layout 目录中。

ch3_4.xml

```xml
<?xml version = "1.0" encoding = "utf-8"?>
<LinearLayout xmlns:android = "http://schemas.android.com/apk/res/android"
    android:orientation = "vertical"
    android:layout_width = "match_parent"
    android:layout_height = "match_parent"
    android:background = "#87CEEB"  >
    <EditText
        android:layout_width = "wrap_content"
        android:layout_height = "wrap_content"
        android:id = "@+id/my_edit_1"
        android:text = "100"
        android:inputType = "numberDecimal"   />
    <EditText
        android:layout_width = "wrap_content"
        android:layout_height = "wrap_content"
        android:id = "@+id/my_edit_2"
        android:text = "200"
        android:inputType = "numberDecimal"   />
    <TextView
        android:layout_width = "wrap_content"
        android:layout_height = "wrap_content"
        android:id = "@+id/my_text"
        android:textSize = "20sp"
        android:textColor = "#0000FF"
        android:text = " = "    />
    <LinearLayout
        android:orientation = "horizontal"
        android:layout_width = "match_parent"
        android:layout_height = "match_parent">
```

```xml
<Button
    android:layout_width = "wrap_content"
    android:layout_height = "wrap_content"
    android:id = "@+id/my_button_add"
    android:text = "加(+)"        />
<Button
    android:layout_width = "wrap_content"
    android:layout_height = "wrap_content"
    android:id = "@+id/my_button_sub"
    android:text = "减(-)"        />
<Button
    android:layout_width = "wrap_content"
    android:layout_height = "wrap_content"
    android:text = "乘(×)"
    android:id = "@+id/my_button_muti"
    android:onClick = "selfDestruct"        />
<Button
    android:layout_width = "wrap_content"
    android:layout_height = "wrap_content"
    android:text = "除(÷)"
    android:id = "@+id/my_button_div"
    android:onClick = "selfDestruct"        />
</LinearLayout>
</LinearLayout>
```

(3) 修改工程\src\ch3\four 目录下的 Example3_4.java 文件，修改后的内容如下：

Example3_4.java

```java
package ch3.four;
import android.app.Activity;
import android.os.Bundle;
import android.widget.*;
import android.view.*;
public class Example3_4 extends Activity implements View.OnClickListener {
    TextView tText;
    EditText tEdit_1,tEdit_2;
    Button buttonAdd,buttonSub,buttonMuti,buttonDiv;
    public void onCreate(Bundle savedInstanceState) {
        super.onCreate(savedInstanceState);
        setContentView(R.layout.ch3_4);
        tText = (TextView)findViewById(R.id.my_text);
        tEdit_1 = (EditText)findViewById(R.id.my_edit_1);
        tEdit_2 = (EditText)findViewById(R.id.my_edit_2);
        buttonAdd = (Button)findViewById(R.id.my_button_add);
        buttonSub = (Button)findViewById(R.id.my_button_sub);
        buttonMuti = (Button)findViewById(R.id.my_button_muti);
        buttonDiv = (Button)findViewById(R.id.my_button_div);
        buttonAdd.setOnClickListener(this);
        buttonSub.setOnClickListener(this);
    }
    public void onClick(View view) {
```

```java
            String s1 = tEdit_1.getText().toString();
            String s2 = tEdit_2.getText().toString();
            if(view == buttonAdd) {
                computer(s1,s2,'+');
            }
            if(view == buttonSub) {
                computer(s1,s2,'-');
            }
        }
        public void selfDestruct(View view) {
            String s1 = tEdit_1.getText().toString();
            String s2 = tEdit_2.getText().toString();
            if(view == buttonMuti) {
                computer(s1,s2,'*');
            }
            if(view == buttonDiv) {
                computer(s1,s2,'/');
            }
        }
        void computer(String s1,String s2,char op) {
            double n1 = 1,n2 = 1,result = 1;
            n1 = Double.parseDouble(s1);
            n2 = Double.parseDouble(s2);
            switch(op) {
                case '+':   result = n1 + n2;
                            break;
                case '-':   result = n1 - n2;
                            break;
                case '*':   result = n1 * n2;
                            break;
                case '/':   result = n1/n2;
                            break;
            }
            tText.setText(" = " + result);
        }
    }
```

（4）启动 AVD，进入工程的根目录，用快捷方式编译工程、安装应用程序到 AVD（有关知识点参见 1.5 节）。对于本例子，用命令行进入 D:\2000\ch3_4，执行如下命令：

D:\2000 > ch3_4 > ant debug install

3.5　ToggleButton 视图

ToggleButton 视图的继承关系如下：

android.widget.Button
　　┗ android.widget.CompoundButton
　　　　┗ android.widget.ToggleButton

和普通的 Button 视图相比较，ToggleButton 外观可以提供两种状态，一种是选中（checked）状态，另一种是未选中状态（unchecked），用户单击 ToggleButton 视图可以切换它的状态。

1. 视图的常用属性

android:disabledAlpha：设置按钮在未选中状态（unchecked）时的 alpha 值。属性取值 0～1 之间的小数，取值 0，按钮完全透明，取值 1，按钮完全不透明。

android:textOff：设置未选中状态（unchecked）时的文本。取值是字符串。

android:textOn：设置选中状态（checked）时的文本。取值是字符串。

2. 在程序代码中注册监视器、处理 Click 事件

在代码中处理 ToggleButton 视图上的 Click 事件的步骤如下。

（1）确定 ToggleButton 视图。首先给出需要处理 Click 事件的 ToggleButton 视图，例如：

```
ToggleButton toggle = (ToggleButton) findViewById(R.id.togglebutton);
```

（2）确定监视器。需要使用实现 CompoundButton.OnCheckedChangeListener 接口的类的对象作为 ToggleButton 视图的监视器，ToggleButton 视图通过如下方法注册监视器：

```
public void setOnCheckedChangeListener(CompoundButton.OnCheckedChangeListener listener);
```

（3）监视器重写 CompoundButton.OnCheckedChangeListener 接口中的 public void onCheckedChanged(CompoundButton buttonView, boolean isChecked) 方法。

例如：

```
toggle.setOnCheckedChangeListener(new CompoundButton.OnCheckedChangeListener() {
    public void onCheckedChanged(CompoundButton buttonView, boolean isChecked) {
        if (isChecked) {
            //选中状态时需要执行的代码
        } else {
            //未选中状态时需要执行的代码
        }
    }
});
```

当 ToggleButton 视图上触发 Click 事件后，监视器调用 onCheckedChanged 方法，该方法中的参数 buttonView 存放着当前触发 Click 事件的 ToggleButton 视图的引用，参数 isChecked 的值是 ToggleButton 视图的当前状态。

3. 在视图资源文件中约定 Click 事件

可以在 ToggleButton 视图资源文件中设置 android:onClick 属性的值，约定 ToggleButton 视图触发 Click 事件后，Activity 对象要执行的方法，这一点和 Button 视图相同，有关步骤参见前面的 3.4 节。

4. 示例

下面的例子 3-5 使用 ToggleButton 视图来决定是否显示一幅图像。在视图文件中约定了 ToggleButton 视图上的 Click 事件。运行效果如图 3.5(a)和图 3.5(b)所示。

例子 3-5

（1）创建名字为 ch3_5 的工程，主要 Activity 子类的名字为 Example3_5，使用的包名

(a) 从checked切换到unchecked　　　(b) 从unchecked切换到checked

图 3.5　运行效果

为 ch3.five。用命令行进入 D:\2000，创建工程 D:\2000＞android create project -t 3 -n ch3_5 -p ./ch3_5 -a Example3_5 -k ch3.five。

（2）增加图像资源。将名字是 car.jpg 的图像保存到图像资源中（有关知识点参见 2.7 节）。

（3）增加视图资源。将下列和视图相关的 XML 文件保存到工程的\res\layout 目录中。

ch3_5.xml

```xml
<?xml version = "1.0" encoding = "utf-8"?>
<LinearLayout xmlns:android = "http://schemas.android.com/apk/res/android"
    android:orientation = "vertical"
    android:layout_width = "match_parent"
    android:layout_height = "match_parent"
    android:background = "#87CEEB">
    <ToggleButton
      android:layout_width = "wrap_content"
      android:layout_height = "wrap_content"
      android:textOn = "显示图像"
      android:textOff = "不显示图像"
      android:onClick = "showImage" />
    <TextView
      android:id = "@+id/showPic"
      android:layout_width = "wrap_content"
      android:layout_height = "wrap_content"
        android:background = "@drawable/car" />
</LinearLayout>
```

（4）修改工程\src\ch3\five 目录下的 Example3_5.java 文件，修改后的内容如下：

Example3_5.java

```java
package ch3.five;
import android.app.Activity;
import android.os.Bundle;
import android.widget.*;
import android.view.*;
public class Example3_5 extends Activity {
    TextView showPic;
```

```java
    public void onCreate(Bundle savedInstanceState) {
        super.onCreate(savedInstanceState);
        setContentView(R.layout.ch3_5);
        showPic = (TextView)findViewById(R.id.showPic);
    }
    public void showImage(View view) {
        CompoundButton b = (CompoundButton)view;
        boolean isChecked = b.isChecked();
        if (isChecked) {
            showPic.setBackground(null);
        }
        else {
            showPic.setBackgroundResource(R.drawable.car);
        }
    }
}
```

(5) 启动 AVD，进入工程的根目录，用快捷方式编译工程、安装应用程序到 AVD(有关知识点参见 1.5 节)。对于本例子，用命令行进入 D:\2000\ch3_5，执行如下命令：

```
D:\2000 > ch3_5 > ant debug install
```

3.6　CheckBox 视图

CheckBox 视图的继承关系如下：

```
android.widget.Button
    ↳ android.widget.CompoundButton
        ↳ android.widget.CheckBox
```

CheckBox 视图是我们熟悉的复选框，其特点是为用户提供两种状态，一种是选中状态，另一种是未选中状态，用户单击复选框可以切换复选框的状态。CheckBox 视图上可以触发 Click 事件，程序可以通过处理 Click 事件对用户在 CheckBox 视图上给出的选择做出响应。

1. 视图的常用属性

在 XML 文件中使用＜CheckBox＞标记对应 CheckBox 视图，即用 CheckBox 标记的属性值设置视图的有关数据，CheckBox 视图的常用属性都是从 CompoundButton 继承而来的。

2. 在程序代码中处理 Click 事件

在代码中处理 CheckBox 视图上的 Click 事件的步骤如下。

(1) 确定 CheckBox 视图。首先给出需要处理 Click 事件的 CheckBox 视图，例如：

```
CheckBox box = (CheckBox)findViewById(R.id.myBox);
```

(2) 确定监视器。需要使用实现 View.OnClickListener 接口的类的对象作为 CheckBox 视图的监视器，CheckBox 视图通过如下方法注册监视器：

```
public void setOnClickListener(View.OnClickListener listener)
```

(3) 监视器重写 View.OnClickListener 接口中的 onClick(View v)方法。

例如：

```
box.setOnClickListener(new View.OnClickListener() {
    public void onClick(View v) {
        //Perform action on click
    }
});
```

当 CheckBox 视图上触发 Click 事件后（选择状态发生变化），监视器调用 onClick(View view)方法,该方法中的参数 view 存放着当前触发 Click 事件的 CheckBox 视图的引用。

3. 在视图资源文件中约定 Click 事件

可以在 CheckBox 视图资源文件中设置 android:onClick 属性的值，约定 CheckBox 视图触发 Click 事件后，Activity 对象要执行的方法，这一点和 Button 视图相同，有关步骤参见前面的 3.4 节。

4. 示例

下面的例子 3-6 是让用户使用 CheckBox 视图选出自己喜欢的动物，在视图文件中约定了 4 个 CheckBox 视图上的 Click 事件。运行效果如图 3.6 所示。

图 3.6 选择喜欢的动物

例子 3-6

(1) 创建名字为 ch3_6 的工程，主要 Activity 子类的名字为 Example3_6，使用的包名为 ch3.six。用命令行进入 D:\2000，创建工程 D:\2000＞android create project -t 3 -n ch3_6 -p ./ch3_6 -a Example3_6 -k ch3.six。

(2) 增加视图资源。将下列和视图相关的 XML 文件保存到工程的 \res\layout 目录中。

ch3_6.xml

```xml
<?xml version = "1.0" encoding = "utf-8"?>
<LinearLayout xmlns:android = "http://schemas.android.com/apk/res/android"
    android:orientation = "vertical"
    android:layout_width = "match_parent"
    android:layout_height = "match_parent"
    android:background = "#87CEEB">
    <LinearLayout  android:orientation = "horizontal"
        android:layout_width = "match_parent"
        android:layout_height = "wrap_content">
        <CheckBox
          android:layout_width = "wrap_content"
          android:layout_height = "wrap_content"
          android:id = "@+id/box_tiger"
          android:text = "老虎"
          android:textSize = "12sp"
          android:textColor = "#FF00FF"
          android:onClick = "find"  />
        <CheckBox
```

```xml
            android:layout_width = "wrap_content"
            android:layout_height = "wrap_content"
            android:id = "@ + id/box_lion"
            android:text = "狮子"
            android:textSize = "12sp"
            android:textColor = "#FF0000"
            android:onClick = "find"   />
    <CheckBox
            android:layout_width = "wrap_content"
            android:layout_height = "wrap_content"
            android:id = "@ + id/box_cat"
            android:text = "小猫"
            android:textSize = "12sp"
            android:textColor = "#0000FF"
            android:onClick = "find"   />
    <CheckBox
            android:layout_width = "wrap_content"
            android:layout_height = "wrap_content"
            android:id = "@ + id/box_dog"
            android:text = "小狗"
            android:textSize = "12sp"
            android:textColor = "#000000"
            android:onClick = "find"   />
    <Button
            android:layout_width = "wrap_content"
            android:layout_height = "wrap_content"
            android:textSize = "12sp"
            android:text = "重新选择"
            android:onClick = "undo"   />
</LinearLayout>
<TextView
        android:layout_width = "match_parent"
        android:layout_height = "wrap_content"
        android:background = "#00FF00"
        android:textColor = "#000000"
        android:id = "@ + id/mess_text"   />
</LinearLayout>
```

（3）修改工程\src\ch3\six 目录下的 Example3_6.java 文件，修改后的内容如下：

Example3_6.java

```java
package ch3.six;
import android.app.Activity;
import android.os.Bundle;
import android.widget.*;
import android.view.*;
public class Example3_6 extends Activity {
    TextView tText;
    CheckBox box_tiger,box_lion,box_cat,box_dog;
    public void onCreate(Bundle savedInstanceState) {
        super.onCreate(savedInstanceState);
```

```java
        setContentView(R.layout.ch3_6);
        tText = (TextView)findViewById(R.id.mess_text);
        box_tiger = (CheckBox)findViewById(R.id.box_tiger);
        box_lion = (CheckBox)findViewById(R.id.box_lion);
        box_cat = (CheckBox)findViewById(R.id.box_cat);
        box_dog = (CheckBox)findViewById(R.id.box_dog);
    }
    public void find(View view) {
        CheckBox box = (CheckBox)view;
        String name = box.getText().toString();
        if(box.isChecked()){
            String content = tText.getText().toString();
            if(!content.contains(name))
                tText.append(name + "\n");
        }
        else {
            String content = tText.getText().toString();
            if(content.contains(name)) {
                content = content.replaceAll(name + "\n","");
                tText.setText(content);
            }
        }
    }
    public void undo(View view) {
        tText.setText("");
        box_dog.setChecked(false);
        box_cat.setChecked(false);
        box_tiger.setChecked(false);
        box_lion.setChecked(false);
    }
}
```

（4）启动 AVD，进入工程的根目录，用快捷方式编译工程、安装应用程序到 AVD（有关知识点参见 1.5 节）。对于本例子，用命令行进入 D:\2000\ch3_6，执行如下命令：

```
D:\2000 > ch3_6 > ant debug install
```

3.7 RadioButton 视图

RadioButton 视图的继承关系如下：

```
android.widget.Button
        ↳ android.widget.CompoundButton
             ↳ android.widget.RadioButton
```

RadioButton 视图是我们熟悉的单选框，其特点是为用户提供两种状态，一种是选中状态，另一种是未选中状态。需要把多个单选框归到同一组中，即添加到一个 RadioGroup 视图中（RadioGroup 视图是 LinearLayout 视图的子类），在同一时刻，在 RadioGroup 视图中只能有一个单选框处于选中状态。和 CheckBox 不同的是，当单选框是未选中状态时，用户单击单选框可以让单选框处于选中状态,但是,如果单选框已经处于选中状态,用户单击该单选框不能将它切换到未选中状态,用户必须单击同组中(在同一 RadioGroup 视图中)的其他单选框才能切换当前单选框的状态。

RadioButton 视图上可以触发 Click 事件,程序可以通过处理 Click 事件对用户在 RadioButton 视图上给出的选择做出响应。

1. 视图的常用属性

在 XML 文件中使用<RadioButton>标记对应 RadioButton 视图,即用 RadioButton 标记的属性值设置视图的有关数据,RadioButton 视图的常用属性都是从 CompoundButton 视图继承而来的。

2. 在程序代码中处理 Click 事件

在代码中处理 RadioButton 视图上的 Click 事件的步骤如下。

(1) 确定 RadioButton 视图。首先给出需要处理 Click 事件的 RadioButton 视图,例如:

```
RadioButton  radiobutton = (RadioButton)findViewById(R.id.myRadiobutton);
```

(2) 确定监视器。需要使用实现 View.OnClickListener 接口的类的对象作为 RadioButton 视图的监视器。RadioButton 视图通过如下方法注册监视器:

```
public void setOnClickListener(View.OnClickListener listener);
```

(3) 监视器重写 View.OnClickListener 接口中的 onClick(View v)方法。例如:

```
radiobutton.setOnClickListener(new View.OnClickListener() {
        public void onClick(View v) {
            //Perform action on click
        }
});
```

当 RadioButton 视图上触发 Click 事件后(选择状态发生变化),监视器调用 onClick (View view)方法,该方法中的参数 view 存放着当前触发 Click 事件的 RadioButton 视图的引用。

3. 在视图资源文件中约定 Click 事件

可以在 RadioButton 视图资源文件中设置 android:onClick 属性的值,约定 RadioButton 视图触发 Click 事件后,Activity 对象要执行的方法,这一点和 Button 视图相同,有关步骤参见 3.4 节。

4. 示例

下面的例子 3-7 是让用户使用 RadioButton 视图浏览动物的图片,在视图文件中约定了 4 个 RadioButton 视图上的 Click 事件。运行效果如图 3.7 所示。

图 3.7 浏览动物图片

例子 3-7

(1) 创建名字为 ch3_7 的工程,主要 Activity 子类的名字为 Example3_7,使用的包名为 ch3.seven。用命令行进入 D:\2000,创建工程 D:\2000> android create project -t 3 -n ch3_7 -p ./ch3_7 -a Example3_7 -k ch3.seven。

(2) 增加图像资源。将名字为 cat.jpg,dog.jpg,lion.jpg 和 tiger.jpg 的图像保存到图

像资源中(有关知识点参见 2.7 节)。

(3) 增加视图资源。将下列和视图相关的 XML 文件保存到工程的\res\layout 目录中。

ch3_7.xml

```xml
<?xml version = "1.0" encoding = "utf-8"?>
<LinearLayout xmlns:android = "http://schemas.android.com/apk/res/android"
    android:orientation = "vertical"
    android:layout_width = "match_parent"
    android:layout_height = "match_parent"
    android:background = "#87CEEB">
    <RadioGroup
     android:layout_width = "match_parent"
     android:layout_height = "wrap_content"
     android:orientation = "horizontal">
        <RadioButton
            android:layout_width = "wrap_content"
            android:layout_height = "wrap_content"
            android:id = "@+id/button_tiger"
            android:text = "老虎"
            android:onClick = "seePicture"    />
        <RadioButton
            android:layout_width = "wrap_content"
            android:layout_height = "wrap_content"
            android:id = "@+id/button_lion"
            android:text = "狮子"
            android:onClick = "seePicture"    />
        <RadioButton
            android:layout_width = "wrap_content"
            android:layout_height = "wrap_content"
            android:id = "@+id/button_cat"
            android:text = "小猫"
            android:onClick = "seePicture" />
        <RadioButton
            android:layout_width = "wrap_content"
            android:layout_height = "wrap_content"
            android:id = "@+id/button_dog"
            android:text = "小狗"
            android:onClick = "seePicture" />
    </RadioGroup>
    <TextView
        android:layout_width = "100dp"
        android:layout_height = "100dp"
        android:layout_gravity = "center"
        android:id = "@+id/show_pic" />
</LinearLayout>
```

(4) 修改工程\src\ch3\seven 目录下的 Example3_7.java 文件,修改后的内容如下:

Example3_7. java

```
package ch3.seven;
import android.app.Activity;
import android.os.Bundle;
import android.widget.*;
import android.view.*;
public class Example3_7 extends Activity {
    TextView showPic;
    RadioButton button_tiger,button_lion,button_cat,button_dog;
    public void onCreate(Bundle savedInstanceState) {
        super.onCreate(savedInstanceState);
        setContentView(R.layout.ch3_7);
        showPic = (TextView)findViewById(R.id.show_pic);
        button_tiger = (RadioButton)findViewById(R.id.button_tiger);
        button_lion = (RadioButton)findViewById(R.id.button_lion);
        button_cat = (RadioButton)findViewById(R.id.button_cat);
        button_dog = (RadioButton)findViewById(R.id.button_dog);
    }
    public void seePicture(View view) {
        RadioButton button = (RadioButton)view;
        if(button == button_tiger)
            showPic.setBackgroundResource(R.drawable.tiger);
        if(button == button_cat)
            showPic.setBackgroundResource(R.drawable.cat);
        if(button == button_lion)
            showPic.setBackgroundResource(R.drawable.lion);
        if(button == button_dog)
            showPic.setBackgroundResource(R.drawable.dog);
    }
}
```

（5）启动 AVD，进入工程的根目录，用快捷方式编译工程、安装应用程序到 AVD（有关知识点参见 1.5 节）。对于本例子，用命令行进入 D:\2000\ch3_7，执行如下命令：

D:\2000>ch3_7>ant debug install

3.8　Spinner 视图

Spinner 视图的继承关系如下：

android.view.ViewGroup
　　↳ android.widget.AdapterView < T extends android.widget.Adapter >
　　　　↳ android.widget.AbsSpinner
　　　　　　↳ android.widget.Spinner

Spinner 视图是我们熟悉的下拉列表，其特点是为用户提供多项选择。

1. 视图的常用属性

在 XML 文件中使用＜Spinner＞标记对应 Spinner 视图，即用 Spinner 标记的属性值设置视图的有关数据,表 3.3 是 Spinner 视图的一些常用属性。

表 3.3 Spinner 视图的常用属性

属性及样例	取值	作用
android:spinnerMode android:spinnerMode="dropdown"	"dropdown", "dialog"	设置下拉列表的展开方式:dropdown 样式或 dialog 样式
android:popupBackground android:popupBackground="#f03a65"	颜色值	当下拉列表的展开方式是 dropdown 样式时,设置下拉列表展开时的背景颜色
android:dropDownWidth android:dropDownWidth="200dp"	度量值	当下拉列表的展开方式是 dropdown 样式时,设置下拉列表展开时的宽度
android:dropDownHorizontalOffset android:dropDownHorizontalOffset="2sp"	度量值	当下拉列表的展开方式是 dropdown 样式时,设置下拉列表展开时的水平偏移量
android:dropDownVerticalOffset android:dropDownVerticalOffset="2sp"	度量值	当下拉列表的展开方式是 dropdown 样式时,设置下拉列表展开时的垂直偏移量
android:entries android:entries="@array/font_lis	值资源	设置下拉列表中的选项
android:prompt android:prompt="@string/app_name"	值资源	当下拉列表的展开方式是 dialog 样式时,设置下拉列表展开时的提示标题

Spinner 视图中最重要的属性就是 android:entries。在设置该属性值之前,必须首先在值资源中给出下拉列表的选项,即需要在值资源中编写一个值资源类型的 XML 文件(参见 2.6 节),该文件必须包含名字为 array 的标记,array 的 item 子标记中的文本内容用于指定下拉列表的选项,例如:

```xml
<?xml version = "1.0" encoding = "utf-8"?>
<resources>
    <array name = "font_list">
        <item>宋体</item>
        <item>楷体</item>
        <item>黑体</item>
    </array>
</resources>
```

该 XML 文件的名字建议为 array.xml。然后,指定 android:entries 属性的值是值资源某个 array 标记,例如:

```
android:entries = "@array/font_lis"
```

2. 在程序代码中处理选择事件

Spinner 视图可以触发选择事件。在代码中处理选择事件的步骤如下。

(1) 确定 Spinner 视图。首先给出需要处理选择事件的 Spinner 视图,例如:

```
Spinner spinner = (Spinner)findViewById(R.id.my_list);
```

(2) 确定监视器。需要使用实现 AdapterView.OnItemSelectedListener 接口的类的对象作为 Spinner 视图的监视器。Spinner 视图通过如下方法注册监视器:

```
public void setOnItemSelectedListener(AdapterView.OnItemSelectedListener listener);
```

（3）监视器重写 AdapterView.OnItemSelectedListener 接口中的 onItemSelected 方法和 onNothingSelected 方法。例如：

```
public void onItemSelected(AdapterView parent,View view,int pos,long id) {
        //参数 parent 存放着当前 Spinner 视图的引用,参数 pos 的值是被选中的选项的索引值
        Object item = parent.getItemAtPosition(pos);//返回选中的选项
}
public void onNothingSelected(AdapterView<?> parent) {
        //Another interface callback
}
```

当 Spinner 视图某个选项处于选中状态，监视器调用 onItemSelected(AdapterView parent, View view, int pos, long id)方法，该方法中的参数 parent 存放着当前 Spinner 视图的引用，参数 pos 的值是被选中的选项的索引值（索引值从 0 开始）。

3. 通过 Click 事件避免选择事件带来的问题

Spinner 视图的首项（索引值是 0 的项）默认处于选中状态，因此如果程序代码部分处理了选择事件，那么程序就会立刻执行 onItemSelected 方法。如果希望用户选择某项后再进行有关操作，代码部分可以不处理选择事件，而是处理一个按钮的 onClick 事件。当处理 onClick 事件时，再根据 Spinner 视图中的选项情况进行相关的操作。

4. 示例

下面的例子 3-8 使用 Spinner 视图，并通过处理选择事件让用户浏览汽车的信息。运行效果如图 3.8 所示。

例子 3-8

（1）创建名字为 ch3_8 的工程，主要 Activity 子类的名字为 Example3_8，使用的包名为 ch3.eight。用命令行进入 D:\2000，创建工程 D:\2000\ch3_8>android create project -t 3 -n ch3_8 -p ./ch3_8 -a Example3_8 -k ch3.eight。

图 3.8 浏览汽车信息

（2）增加和修改值资源。将下列 array.xml 保存到值资源中，即保存到工程的\res\values 目录中（有关知识点参见 2.6 节），修改值资源中 strings.xml 文件，修改后的内容见如下的 strings.xml。

array.xml

```
<?xml version = "1.0" encoding = "utf-8"?>
<resources>
    <array name = "car_list">
        <item>audi</item>
        <item>jeep</item>
        <item>ford</item>
    </array>
</resources>
```

strings.xml

```
<?xml version = "1.0" encoding = "utf-8"?>
<resources>
    <string name = "app_name">Example3_8</string>
```

```xml
    <string name = "audi">Audi 汽车\n 价格：32 万元。\n 产地：德国</string>
    <string name = "jeep">Jeep 越野车\n 价格：28 万元。\n 产地：美国</string>
    <string name = "ford">Ford 汽车\n 价格：16 万元。\n 产地：美国</string>
</resources>
```

（3）增加视图资源。将下列和视图相关的 XML 文件保存到工程的\res\layout 目录中。

ch3_8.xml

```xml
<?xml version = "1.0" encoding = "utf-8"?>
<LinearLayout xmlns:android = "http://schemas.android.com/apk/res/android"
    android:orientation = "vertical"
    android:layout_width = "match_parent"
    android:layout_height = "match_parent"
    android:background = "#87CEEB">
    <Spinner
        android:layout_width = "wrap_content"
        android:layout_height = "wrap_content"
        android:spinnerMode = "dropdown"
        android:popupBackground = "#f03a65"
        android:dropDownWidth = "200sp"
        android:dropDownVerticalOffset = "2sp"
        android:dropDownHorizontalOffset = "2sp"
        android:id = "@+id/my_list"
        android:entries = "@array/car_list" />
    <TextView
        android:id = "@+id/my_text"
        android:layout_width = "220dp"
        android:layout_height = "220dp"
        android:layout_gravity = "center"
        android:textColor = "#0000FF"
        android:background = "#876789" />
</LinearLayout>
```

（4）修改工程\src\ch3\eight 目录下的 Example3_8.java 文件，修改后的内容如下：

Example3_8.java

```java
package ch3.eight;
import android.app.Activity;
import android.os.Bundle;
import android.widget.*;
import android.view.*;
public class Example3_8 extends Activity implements AdapterView.OnItemSelectedListener {
    TextView textView;
    Spinner spinner;
    public void onCreate(Bundle savedInstanceState) {
        super.onCreate(savedInstanceState);
        setContentView(R.layout.ch3_8);
        spinner = (Spinner)findViewById(R.id.my_list);
        spinner.setOnItemSelectedListener(this);
        textView = (TextView)findViewById(R.id.my_text);
```

```
        }
        public void onItemSelected(AdapterView parent,View view,int pos,long id) {
            String selectedItem = parent.getItemAtPosition(pos).toString();
            if(selectedItem.equals("audi"))
                textView.setText(R.string.audi);
            if(selectedItem.equals("jeep"))
                textView.setText(R.string.jeep);
            if(selectedItem.equals("ford"))
                textView.setText(R.string.ford);
        }
        public void onNothingSelected(AdapterView<?> parent) {
            textView.append("nothing");
        }
    }
```

(5) 启动 AVD,进入工程的根目录,用快捷方式编译工程、安装应用程序到 AVD(有关知识点参见 1.5 节)。对于本例子,用命令行进入 D:\2000\ch3_8,执行如下命令:

```
D:\2000 > ch3_8 > ant debug install
```

下面的例子 3-9 是一个四则运算器,用户使用 EditText 视图(文本框)输入数字,使用 Spinner 视图(下拉列表)选择运算符号(但程序不处理选择事件),用户单击 Button 视图(＝按钮)得到计算结果(在 Button 的视图资源中约定了 onClick 事件)。运行效果如图 3.9 所示。

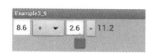

图 3.9 四则运算

例子 3-9

(1) 创建名字为 ch3_9 的工程,主要 Activity 子类的名字为 Example3_9,使用的包名为 ch3.nine。用命令行进入 D:\2000,创建工程 android create project -t 3 -n ch3_9 -p ./ch3_9 -a Example3_9 -k ch3.nine。

(2) 增加和修改值资源。将下列 array.xml 保存到值资源中,即保存到工程的\res\values 目录中(有关知识点参见 2.6 节)。

array.xml

```
<?xml version = "1.0" encoding = "utf-8"?>
<resources>
    <array name = "yunsuan_list">
        <item> + </item>
        <item> - </item>
        <item> * </item>
        <item>/</item>
    </array>
</resources>
```

(3) 增加视图资源。将下列和视图相关的 XML 文件保存到工程的\res\layout 目录中。

ch3_9.xml

```
<?xml version = "1.0" encoding = "utf-8"?>
```

```xml
<LinearLayout xmlns:android="http://schemas.android.com/apk/res/android"
    android:layout_width="match_parent"
    android:layout_height="match_parent"
    android:background="#87CEEB">
    <EditText
        android:layout_width="wrap_content"
        android:layout_height="wrap_content"
        android:id="@+id/my_edit_1"
        android:text="100"
        android:inputType="numberDecimal" />
    <Spinner
        android:layout_width="wrap_content"
        android:layout_height="wrap_content"
        android:spinnerMode="dropdown"
        android:dropDownWidth="70dp"
        android:id="@+id/yunsuan_list"
        android:entries="@array/yunsuan_list" />
    <EditText
        android:layout_width="wrap_content"
        android:layout_height="wrap_content"
        android:id="@+id/my_edit_2"
        android:text="200"
        android:inputType="numberDecimal" />
    <Button
        android:layout_width="wrap_content"
        android:layout_height="wrap_content"
        android:id="@+id/my_button"
        android:text=" = "
        android:onClick="compute" />
    <TextView
        android:layout_width="wrap_content"
        android:layout_height="wrap_content"
        android:id="@+id/my_text"
        android:textSize="20sp"
        android:textColor="#0000FF" />
</LinearLayout>
```

（4）修改工程\src\ch3\nine 目录下的 Example3_9.java 文件，修改后的内容如下：

Example3_9.java

```java
package ch3.nine;
import android.app.Activity;
import android.os.Bundle;
import android.widget.*;
import android.view.*;
public class Example3_9 extends Activity {
    TextView tText;
    EditText tEdit_1,tEdit_2;
    Spinner spinner;
    public void onCreate(Bundle savedInstanceState) {
        super.onCreate(savedInstanceState);
```

```java
        setContentView(R.layout.ch3_9);
        tText = (TextView)findViewById(R.id.my_text);
        tEdit_1 = (EditText)findViewById(R.id.my_edit_1);
        tEdit_2 = (EditText)findViewById(R.id.my_edit_2);
        spinner = (Spinner)findViewById(R.id.yunsuan_list);
    }
    public void compute(View view) {
        String s1 = tEdit_1.getText().toString();
        String s2 = tEdit_2.getText().toString();
        String fuhao = spinner.getSelectedItem().toString();
        if(fuhao.equals("+")) {
            jisuan(s1,s2,'+');
        }
        if(fuhao.equals("-")) {
            jisuan(s1,s2,'-');
        }
        if(fuhao.equals("*")) {
            jisuan(s1,s2,'*');
        }
        if(fuhao.equals("/")) {
            jisuan(s1,s2,'/');
        }
    }
    void jisuan(String s1,String s2,char op) {
        double n1 = 1, n2 = 1, result = 1;
        n1 = Double.parseDouble(s1);
        n2 = Double.parseDouble(s2);
        switch(op) {
            case '+':   result = n1 + n2;
                        break;
            case '-':   result = n1 - n2;
                        break;
            case '*':   result = n1 * n2;
                        break;
            case '/':   result = n1/n2;
                        break;
        }
        tText.setText("" + result);
    }
}
```

（5）启动 AVD。进入工程的根目录，用快捷方式编译工程、安装应用程序到 AVD（有关知识点参见 1.5 节）。对于本例子，用命令行进入 D:\2000\ch3_9，执行如下命令：

D:\2000 > ch3_9 > ant debug install

3.9 ListView 视图

ListView 视图的继承关系如下：

android.view.ViewGroup
　　└ android.widget.AdapterView < T extends android.widget.Adapter >

```
            └ android.widget.AbsListView
                    └ android.widget.ListView
```

ListView 视图是我们熟悉的滚动列表,其特点是为用户提供多项选择。

1. 视图的常用属性

在 XML 文件中使用<ListView>标记对应 ListView 视图,即用 ListView 标记的属性值设置视图的有关数据,表 3.4 是 ListView 视图的一些常用属性。

表 3.4 ListView 视图的常用属性

属性及样例	取 值	作 用
android:divider android:divider="#00FFAA"	颜色值或图像资源	设置选项之间的分隔线
android:dividerHeight android:dividerHeight="6dp"	度量值	设置选项之间的分隔线的高度
android:entries android:entries="@array/font_lis"	值资源	设置滚动列表中的选项
android:listSelector android:listSelector="@drawable/ic_launcher"	颜色值或图像资源	设置被选中项的背景颜色或背景图像

ListView 视图中最重要的属性就是 android:entries。在设置该属性值之前,必须首先在值资源中给出滚动列表的选项,即需要在值资源中编写一个值资源类型的 XML 文件(参见 2.6 节),该文件必须包含名字为 array 的标记,array 的 item 子标记中的文本内容用于指定下拉列表的选项,例如:

```
<?xml version = "1.0" encoding = "utf-8"?>
<resources>
    <array name = "university_list">
        <item>清华大学</item>
        <item>北京大学</item>
        <item>天津大学</item>
    </array>
</resources>
```

该 XML 文件的名字建议为 array.xml。然后,指定 android:entries 属性的值是值资源某个 array 标记,例如:

```
android:entries = "@array/ university_list"
```

2. 在程序代码中处理选择事件

ListView 视图可以触发单击选项事件。在代码中处理单击选项事件的步骤如下。
(1) 确定 ListView 视图。首先给出需要处理选择事件的 ListView 视图,例如:

```
ListView  spinner = (ListView)findViewById(R.id.my_list);
```

(2) 确定监视器。需要使用实现 AdapterView.OnItemClickListener 接口的类的对象作为 ListView 视图的监视器。ListView 视图通过如下方法注册监视器:

```
public void setOnItemClickListener(AdapterView.OnItemClickListener listener);
```

(3) 监视器重写 AdapterView.OnItemClickListener 接口中的 onItemClick 方法。例如：

```
public void onItemClick (AdapterView parent,View view,int pos,long id) {
    //参数 parent 存放着当前 ListView 视图的引用,参数 pos 的值是被单击的选项的索引值
    Object item = parent.getItemAtPosition(pos);//返回选中的选项
}
```

当 ListView 视图某个选项被单击后，监视器调用 onItemClick(AdapterView parent, View view,int position,long id)方法，该方法中的参数 parent 存放着当前 ListView 视图的引用，参数 pos 的值是被单击的选项的索引值。

3. 示例

下面的例子 3-10 使用 ListView 视图，并通过处理单击选项事件让用户浏览大学的信息。运行效果如图 3.10 所示。

例子 3-10

(1) 创建名字为 ch3_10 的工程，主要 Activity 子类的名字为 Example3_10，使用的包名为 ch3.ten。用命令行进入 D:\2000,创建工程 D:\2000＞android create project -t 3 -n ch3_10 -p ./ch3_10 -a Example3_10 -k ch3.ten。

(2) 增加和修改值资源。将下列 array.xml 保存到值资源中，即保存到工程的\res\values 目录中(有关知识点参见 2.6 节)，修改值资源中的 strings.xml 文件，修改后的内容见如下的 strings.xml。

图 3.10 浏览大学信息

array.xml

```xml
<?xml version = "1.0" encoding = "utf-8"?>
<resources>
    <array name = "university_list">
        <item>清华大学</item>
        <item>北京大学</item>
        <item>天津大学</item>
    </array>
</resources>
```

strings.xml

```xml
<?xml version = "1.0" encoding = "utf-8"?>
<resources>
    <string name = "app_name">Example3_10</string>
    <string name = "Beijing">北京大学\n重点大学,985 院校。\n 所在地：北京市</string>
    <string name = "TsingHua">清华大学\n重点大学,985 院校。\n 所在地：北京市</string>
    <string name = "Tianjing">天津大学\n重点大学,985 院校。\n 所在地：天津市</string>
</resources>
```

(3) 增加图像资源。将名字为 choice.jpg 的图像保存到图像资源中(有关知识点参见 2.7 节)。

（4）增加视图资源。将下列和视图相关的 XML 文件保存到工程的\res\layout 目录中。

ch3_10.xml

```xml
<?xml version="1.0" encoding="utf-8"?>
<LinearLayout xmlns:android="http://schemas.android.com/apk/res/android"
    android:orientation="vertical"
    android:layout_width="match_parent"
    android:layout_height="match_parent"
    android:background="#87CEEB">
    <ListView
        android:layout_width="wrap_content"
        android:layout_height="wrap_content"
        android:divider="#0000FF"
        android:dividerHeight="6dp"
        android:background="#22bbcc"
        android:id="@+id/my_list"
        android:listSelector="@drawable/choice"
        android:entries="@array/university_list" />
    <TextView
        android:id="@+id/my_text"
        android:layout_width="match_parent"
        android:layout_height="120dp"
        android:textColor="#0000FF"
        android:background="#a7f7a9"
        android:padding="20dp" />
</LinearLayout>
```

（5）修改工程\src\ch3\ten 目录下的 Example3_10.java 文件，修改后的内容如下：

Example3_10.java

```java
package ch3.ten;
import android.app.Activity;
import android.os.Bundle;
import android.widget.*;
import android.view.*;
public class Example3_10 extends Activity implements AdapterView.OnItemClickListener {
    TextView textView;
    ListView  listView;
    public void onCreate(Bundle savedInstanceState) {
        super.onCreate(savedInstanceState);
        setContentView(R.layout.ch3_10);
        listView = (ListView)findViewById(R.id.my_list);
        listView.setOnItemClickListener(this);
        textView = (TextView)findViewById(R.id.my_text);
    }
    public void onItemClick(AdapterView parent, View view, int pos, long id) {
        if(pos == 0)
            textView.setText(R.string.TsingHua);
        if(pos == 1)
            textView.setText(R.string.Beijing);
```

```
            if(pos == 2)
                textView.setText(R.string.Tianjing);
        }
    }
```

(6) 启动 AVD，进入工程的根目录，用快捷方式编译工程、安装应用程序到 AVD(有关知识点参见 1.5 节)。对于本例子，用命令行进入 D:\2000\ch3_10，执行如下命令：

```
D:\2000 > ch3_10 > ant debug install
```

3.10　动态创建 Spinner 视图和 ListView 视图

1. 动态创建 Spinner 视图

我们已经知道，Spinner 视图用下拉列表方式为用户提供多项选择，所谓动态创建 Spinner 视图，是指在程序的 Java 代码部分指定 Spinner 视图中的选项。比如，程序可能需要将产生的线性表结构中的各个数据分别作为 Spinner 视图中的各个选项，那么就需要动态地指定 Spinner 视图中的选项。

动态创建 Spinner 视图的关键是需要一个实现 Adapter 接口的类的实例，习惯将这样的类称作一个适配器，适配器的作用是将某些数据转换为 Spinner 视图中的选项(起着适配的作用)。动态创建 Spinner 视图的步骤如下。

1) 适配器

android.widget 包中的 BaseAdapter 类已经实现了 Adapter 的两个子接口：ListAdapter 和 SpinnerAdapter，因此用户程序只要编写 BaseAdapter 的子类，并按要求重写需要的方法即可，比如如何将数据转换成 Spinner 视图中的选项的相关方法等。

- public int getCount()。

重写 public int getCount()方法，以便适配器确定 Spinner 视图中的选项的数目，如果用户需要 Spinner 视图中有 3 条选项，那么可按如下简单重写 public int getCount()方法：

```
public int getCount() {
    return 3;
}
```

- public Object getItem (int position)。

重写 public Object getItem (int position)，以便适配器确定 Spinner 视图中位置索引是 position 上的选项，如果 Spinner 视图中有 3 条选项(getCount()返回的值是 3)，如果需要 Spinner 视图中有"清华大学"、"北京大学"和"天津大学"3 个选项，那么可如下简单重写 public Object getItem(int position)方法：

```
public String getItem (int position) {
    if(position == 0)
        return "清华大学";
    else if(position == 1)
        return "北京大学";
    else if(position == 2)
        return "天津大学";
```

}

- public long getItemId (int position)。

重写 public long getItemId (int position)，以便适配器为 Spinner 视图中位置索引是 position 上的选项指定一个新的索引 id。一般返回 position 即可。

```
public long getItemId (int position) {
    return position;
}
```

- public View getView (int position, View convertView, ViewGroup parent)。

必须为选项配置一个视图，该视图的目的是显示选项的有关信息，可以是选项的全部信息或部分信息。为选项配置的视图上的信息也可根据需要由用户任意定义，不会影响选项本身，因为该视图仅仅是为选项提供了一个外观而已。比如，必须用某个视图显示选项"清华大学"的有关信息，那么这个视图上可以显示"清华大学"，也可以显示"TsingHua University"，甚至可以显示任意一个字符串。其中的参数 position 是下拉列表的索引值。以下是一个重写的示例：

```
public View getView (int position, View convertView, ViewGroup parent) {
    TextView text = new TextView(context);
    text.setText(getItem (position));   //用 TextView 视图显示下拉列表中的第 position 个
                                        选项
    return text;
}
```

2）Spinner 视图使用适配器

在 Java 代码部分获得一个 Spinner 视图，例如：

```
Spinner list = findByViewId(R.id.my_list);
```

然后 Spinner 视图调用 void setAdapter(SpinnerAdapter adapter)方法将一个适配器的实例，比如 adapter，传递给 Spinner 视图：

```
void setAdapter(adapter);
```

那么适配器 adapter 就会设置好 Spinner 视图中的选项。

2. 动态创建 ListView 视图

由于 BaseAdapter 类已经实现了 Adapter 的两个子接口 ListAdapter 和 SpinnerAdapter，因此动态创建 ListView 视图与动态创建 Spinner 视图类似，不再赘述。

3. 示例

下面的例子 3-11 动态创建 Spinner 视图，并通过处理选择事件，浏览汽车信息。运行效果如图 3.11 所示。

例子 3-11

（1）创建名字为 ch3_11 的工程，主要 Activity 子类的名字为 Example3_11，使用的包名为 ch3.eleven。用命令行进入 D:\2000，创建工程 D:\2000＞android create

图 3.11　动态创建 Spinner 视图

project -t 3 -n ch3_11 -p ./ch3_11 -a Example3_11 -k ch3.eleven。

(2) 增加视图资源。将下列和视图相关的 XML 文件保存到工程的\res\layout 目录中。

ch3_11.xml

```xml
<?xml version = "1.0" encoding = "utf-8"?>
<LinearLayout xmlns:android = "http://schemas.android.com/apk/res/android"
    android:orientation = "vertical"
    android:layout_width = "match_parent"
    android:layout_height = "match_parent"
    android:background = "#87CEEB">
    <Spinner
        android:layout_width = "wrap_content"
        android:layout_height = "wrap_content"
        android:spinnerMode = "dropdown"
        android:popupBackground = "#f03a65"
        android:dropDownWidth = "200sp"
        android:dropDownVerticalOffset = "2sp"
        android:dropDownHorizontalOffset = "2sp"
        android:id = "@+id/my_list" />
    <TextView
        android:id = "@+id/my_text"
        android:layout_width = "match_parent"
        android:layout_height = "220dp"
        android:layout_gravity = "center"
        android:textColor = "#0000FF"
        android:textSize = "22sp"
        android:background = "#876789" />
</LinearLayout>
```

(3) 将下列适配器的 Java 源文件 MyAdapter.java 以及 Car.java 源文件存放到工程\src\ch3\eleven 目录下，并修改工程\src\ch3\eleven 目录下的 Example3_11.java 文件，修改后的内容见如下的 Example3_11.java。

Car.java

```java
package ch3.eleven; //包名必须与 Example3_11 的相同,即创建工程时给出的包名
public class Car {
    String name, madeTime;
    double price;
    public Car(String name, String madeTime, double price) {
        this.name = name;
        this.madeTime = madeTime;
        this.price = price;
    }
}
```

MyAdapter.java

```java
package ch3.eleven; //包名必须与 Example3_11 的相同
import android.content.Context;
```

```java
import android.app.*;
import android.os.*;
import android.widget.*;
import android.view.*;
import java.util.*;
public class MyAdapter extends  BaseAdapter {
    ArrayList<Car> list;
    Context context;
    public void setContext(Context context) {
       this.context = context;
    }
    public void setArrayList(ArrayList<Car> list) {
       this.list = list;
    }
    public int getCount() {
       return list.size();
    }
    public Car getItem (int position) {
       Car item = list.get(position);
       return item;
    }
    public long getItemId (int position) {
       return position;
    }
    public View getView (int position, View convertView, ViewGroup parent) {
       TextView text = new TextView(context);
       text.setTextSize(1,30);
       Car car = getItem(position);
       text.setText(car.name);
       return text;
    }
}
```

Example3_11.java

```java
package ch3.eleven;
import android.content.Context;
import android.app.Activity;
import android.os.Bundle;
import android.widget.*;
import android.view.*;
import java.util.*;
public class Example3_11 extends Activity  implements AdapterView.OnItemSelectedListener {
   TextView textView;
   Spinner spinner;
   MyAdapter adapter;     //适配器
   ArrayList<Car> list;
   public void onCreate(Bundle savedInstanceState) {
      super.onCreate(savedInstanceState);
      setContentView(R.layout.ch3_11);
      spinner = (Spinner)findViewById(R.id.my_list);
```

```
        list = new ArrayList<Car>();
        list.add(new Car("audi","2012-09-12",31.8));
        list.add(new Car("jeep","2010-01-24",58.7));
        list.add(new Car("fort","2011-07-31",17.5));
        list.add(new Car("bmo","2013-11-09",125.5));
        adapter = new MyAdapter();
        adapter.setContext(this);
        adapter.setArrayList(list);
        spinner.setAdapter(adapter);
        spinner.setOnItemSelectedListener(this);
        textView = (TextView)findViewById(R.id.my_text);
    }
    public void onItemSelected(AdapterView parent,View view,int pos,long id) {
        Car car = (Car)parent.getItemAtPosition(pos);
        textView.setText("");
        textView.append("car name:" + car.name + "\n");
        textView.append("made time:" + car.madeTime + "\n");
        textView.append("price:" + car.price + "\n");
    }
    public void onNothingSelected(AdapterView<?> parent) {
        textView.append("nothing");
    }
}
```

(4) 启动 AVD,进入工程的根目录,用快捷方式编译工程、安装应用程序到 AVD(有关知识点参见 1.5 节)。对于本例子,用命令行进入 D:\2000\ch3_11,执行如下命令:

```
D:\2000 > ch3_11 > ant debug install
```

下面的例子 3-12 动态创建 ListView 视图,并给每个选项附加上一个机器人的小图标。运行效果如图 3.12 所示。

例子 3-12

(1) 创建名字为 ch3_12 的工程,主要 Activity 子类的名字为 Example3_12,使用的包名为 ch3.twelve。用命令行进入 D:\2000,创建工程 D:\2000 > android create project -t 3 -n ch3_12 -p ./ch3_12 -a Example3_12 -k ch3.twelve。

图 3.12 动态创建 ListView 视图

(2) 增加视图资源。将下列和视图相关的 XML 文件保存到工程的 \res\layout 目录中。

ch3_12.xml

```
<?xml version = "1.0" encoding = "utf-8"?>
<LinearLayout xmlns:android = "http://schemas.android.com/apk/res/android"
        android:orientation = "vertical"
        android:layout_width = "match_parent"
        android:layout_height = "match_parent"
        android:background = "#87CEEB">
    <ListView
        android:layout_width = "wrap_content"
```

```
            android:layout_height = "wrap_content"
            android:divider = "#0000FF"
            android:dividerHeight = "6dp"
            android:background = "#777777"
            android:id = "@+id/my_list" />
    <TextView
            android:id = "@+id/my_text"
            android:layout_width = "match_parent"
            android:layout_height = "120dp"
            android:textColor = "#0000FF"
            android:background = "#a7f7a9"
            android:textSize = "25sp"
            android:padding = "10dp"    />
</LinearLayout>
```

(3) 将下列适配器的 Java 源文件 MyAdapter.java 存放到工程\src\ch3\twelve 目录下,并修改工程\src\ch3\twelve 目录下的 Example3_12.java 文件,修改后的内容见如下的 Example3_12.java。

MyAdapter.java

```
package ch3.twelve;   //包名必须与 Example3_12 的相同
import android.content.Context;
import android.app.*;
import android.os.*;
import android.widget.*;
import android.view.*;
import android.graphics.drawable.Drawable;
import java.util.*;
public class MyAdapter extends  BaseAdapter {
    ArrayList<String> list;
    Context context;
    public void setContext(Context context) {
        this.context = context;
    }
    public void setArrayList(ArrayList<String> list) {
        this.list = list;
    }
    public int getCount() {
        return list.size();
    }
    public String getItem (int position) {
        String item = list.get(position);
        return item;
    }
    public long getItemId (int position) {
        return 0;
    }
    public View getView (int position, View convertView, ViewGroup parent) {
        LinearLayout layout;
        layout = new LinearLayout(context);
```

```java
        TextView image = new TextView(context);
        TextView text = new TextView(context);
        text.setTextSize(1,25);
        text.setText(" -- " + getItem (position) + " -- ");
        image.setBackgroundResource(R.drawable.ic_launcher);
        layout.addView(image);
        layout.addView(text);
        return layout;
    }
}
```

Example3_12.java

```java
package ch3.twelve;
import android.app.*;
import android.os.Bundle;
import android.widget.*;
import android.view.*;
import java.util.*;
public class Example3_12 extends Activity  implements  AdapterView.OnItemClickListener {
    TextView textView;
    ListView listView;
    MyAdapter adapter;     //适配器
    ArrayList<String> listItem;
    public void onCreate(Bundle savedInstanceState) {
        super.onCreate(savedInstanceState);
        setContentView(R.layout.ch3_12);
        listView = (ListView)findViewById(R.id.my_list);
        listView.setOnItemClickListener(this);
        listItem = new ArrayList<String>();
        listItem.add("Andoid Design");
        listItem.add("Andoid Develop");
        listItem.add("Andoid Training");
        listItem.add("Andoid API Guides");
        adapter = new MyAdapter();
        adapter.setContext(this);
        adapter.setArrayList(listItem);
        listView.setAdapter(adapter);
        textView = (TextView)findViewById(R.id.my_text);
    }
    public void onItemClick(AdapterView parent,View view,int pos,long id) {
        String selectedItem = parent.getItemAtPosition(pos).toString();
        textView.setText(selectedItem);
    }
}
```

（4）启动 AVD，进入工程的根目录，用快捷方式编译工程、安装应用程序到 AVD（有关知识点参见 1.5 节）。对于本例子，用命令行进入 D:\2000\ch3_12，执行如下命令：

```
D:\2000 > ch3_12 > ant debug install
```

3.11 GridView 视图

GridView 视图的继承关系如下：

```
android.view.ViewGroup
   └ android.widget.AdapterView< T extends android.widget.Adapter >
       └ android.widget.AbsListView
           └ android.widget.GridView
```

GridView 视图用二维表格的形式为用户提供选项，每个格子中放置一个选项。

1. 视图的常用属性
- android:columnWidth：设置列的宽度，取值为度量值。
- android:numColumns：设置列的数目，取值为正整数或-1（等价为"auto_fit"）。如果取-1 或"auto_fit"，例如 android:numColumns＝"auto_fit"，表示根据空间的大小设置恰当的列数。
- android:stretchMode：设置伸展模式，以便 GridView 视图填满横向的剩余空间。取值为"none"、"spacingWidth"、"columnWidth"和"spacingWidthUniform"。
- android:verticalSpacing：设置行之间的垂直间距，取值为度量值。
- android:horizontalSpacing：设置列之间的水平间距，取值为度量值。

2. 动态创建 GridView 视图

只能用动态方式创建 GridView 视图中的选项。步骤和方法与动态创建 ListView 视图类似，只需注意的是，GridView 是二维表格形式，视图中选项的索引按行依次进行，比如 GridView 设置了 3 列，一共有 6 个选项，那么第一行的 3 个选项的索引就是 0、1、2，第二行上的三个选项的索引就是 3、4、5。

3. 示例

下面的例子 3-13 使用 GridView 视图浏览学生的基本信息，即在 GridView 视图的方格中填入学生的姓名，用户单击姓名可以查看该学生的诸如出生日期、生源地等信息。程序运行效果图 3.13 所示。

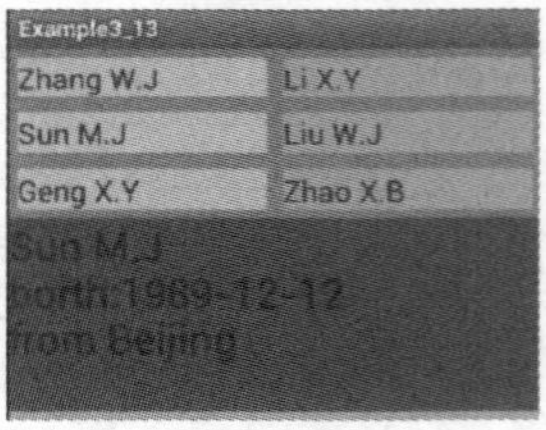

图 3.13 动态创建 GridView 视图

例子 3-13

（1）创建名字为 ch3_13 的工程，主要 Activity 子类的名字为 Example3_13，使用的包名为 ch3.thirteen。用命令行进入 D:\2000，创建工程 D:\2000＞android create project -t 3 -n ch3_13 -p ./ch3_13 -a Example3_13 -k ch3.thirteen。

（2）增加视图资源。将下列和视图相关的 XML 文件保存到工程的\res\layout 目录中。

ch3_13.xml

```
< LinearLayout xmlns:android = "http://schemas.android.com/apk/res/android"
    android:orientation = "vertical"
    android:layout_width = "match_parent"
```

```xml
            android:layout_height = "match_parent"
            android:background = "#87CEEB">
    <GridView
            android:id = "@+id/gridview"
            android:layout_width = "match_parent"
            android:layout_height = "wrap_content"
            android:columnWidth = "100dp"
            android:numColumns = "2"
            android:verticalSpacing = "10dp"
            android:horizontalSpacing = "10dp"
            android:stretchMode = "columnWidth"
            android:background = "#99999a"
            android:gravity = "center" />
    <TextView
            android:id = "@+id/my_text"
            android:layout_height = "wrap_content"
            android:layout_width = "match_parent"
            android:layout_gravity = "center"
            android:textColor = "#0000FF"
            android:textSize = "25sp"
            android:background = "#876789" />
</LinearLayout>
```

（3）将下列适配器的 Java 源文件 MyAdapter.java 以及 Student.java 源文件存放到工程\src\ch3\thirteen 目录下，并修改工程\src\ch3\thirteen 目录下的 Example3_13.java 文件，修改后的内容见如下的 Example3_13.java。

MyAdapter.java

```java
package ch3.thirteen;                    //包名必须与 Example3_13 的相同
import android.content.Context;
import android.app.*;
import android.os.*;
import android.widget.*;
import android.view.*;
import java.util.*;
import android.graphics.Color;
public class MyAdapter extends    BaseAdapter {
    ArrayList<Student> list;
    Context context;
    public void setContext(Context context) {
        this.context = context;
    }
    public void setArrayList(ArrayList<Student> list) {
        this.list = list;
    }
    public int getCount() {
        return list.size();
    }
    public Student getItem (int position) {
        Student stu = list.get(position);
```

```java
            return stu;
        }
        public long getItemId (int position) {
            return 0;
        }
        public View getView (int position, View convertView, ViewGroup parent) {
            Student stu = list.get(position);
            TextView showMess = new TextView(context);
            showMess.setTextSize(1,18);
            showMess.setTextColor(Color.BLUE);
            if(position % 2 == 0)
                showMess.setBackgroundColor(Color.rgb(200,255,0));
            else
                showMess.setBackgroundColor(Color.rgb(0,255,250));
            showMess.setText(stu.name);
            return showMess;
        }
}
```

Student.java

```java
package ch3.thirteen;              //包名必须与 Example3_13 的相同
public class Student {
    String name,borth,location;
    Student(String name,String borth,String location) {
        this.name = name;
        this.borth = borth;
        this.location = location;
    }
}
```

Example3_13.java

```java
package ch3.thirteen;
import android.app.Activity;
import android.os.Bundle;
import android.widget.*;
import android.view.*;
import java.util.*;
public class Example3_13 extends Activity implements   AdapterView.OnItemClickListener {
    TextView textView;
    GridView gridView;
    MyAdapter adapter;     //适配器
    ArrayList<Student> list;
    public void onCreate(Bundle savedInstanceState) {
        super.onCreate(savedInstanceState);
        setContentView(R.layout.ch3_13);
        gridView = (GridView)findViewById(R.id.gridview);
        gridView.setOnItemClickListener(this);
        list = new ArrayList<Student>();
        list.add(new Student("Zhang W.J","borth:1989 - 12 - 12","from Beijing"));
```

```
            list.add(new Student("Li X.Y","borth:1990-10-12","from Shanghai"));
            list.add(new Student("Sun M.J","borth:1989-12-12","from Beijing"));
            list.add(new Student("Liu W.J","borth:1992-09-25","from Tianjing"));
            list.add(new Student("Geng X.Y","borth:1991-01-18","from Xian"));
            list.add(new Student("Zhao X.B","borth:1991-08-31","from Shenyang"));
            adapter = new MyAdapter();
            adapter.setContext(this);
            adapter.setArrayList(list);
            gridView.setAdapter(adapter);
            textView = (TextView)findViewById(R.id.my_text);
        }
        public void onItemClick(AdapterView parent,View view,int pos,long id) {
            Student stu = (Student)parent.getItemAtPosition(pos);
            textView.setText("");
            textView.append(stu.name + "\n");
            textView.append(stu.borth + "\n");
            textView.append(stu.location + "\n");
        }
}
```

（4）启动 AVD，进入工程的根目录，用快捷方式编译工程、安装应用程序到 AVD（有关知识点参见 1.5 节）。对于本例子，用命令行进入 D:\2000\ch3_13，执行如下命令：

D:\2000 > ch3_13 > ant debug install

3.12　ScrollView 视图

ScrollView 类的继承关系如下：

```
android.view.ViewGroup
    ↳ android.widget.FrameLayout
        ↳ android.widget.ScrollView
```

当一个视图的纵向尺寸较大时，可能会导致用户无法看到视图的全貌（毕竟手机的屏幕大小是固定的），这时就可以把这样的视图事先放到 ScrollView 视图中，即将当前视图作为 ScrollView 的子视图。

1. 视图的作用

ScrollView 视图提供垂直滚动条，用户可用滚动方式和其中的子视图进行交互。需要特别注意的是，ListView 以及 GridView 等视图本身自带滚动条，因此不要把 ListView 以及 GridView 等视图直接作为 ScrollView 视图的子视图。

只能向 ScrollView 视图添加一个子视图，该子视图将位于 ScrollView 视图的顶部（top），但子视图也可以使用 android:layout_gravity 属性指定自己在父视图中的位置。经常向 ScrollView 视图添加一个 LinearLayout 视图，LinearLayout 视图的特点是将其中的子视图按顺序放置在一行或一列中（在 4.1 节将详细学习 LinearLayout 视图）。

2. 示例

如果 TextView 视图中有很多行的文本，就会导致 TextView 视图在垂直方向变得过大,下面的例子3-14将TextView视图放到了ScrollView视图中。运行效果如图3.14所示。

例子 3-14

(1) 创建名字为ch3_14的工程,主要Activity子类的名字为Example3_14,使用的包名为ch3.fourteen。用命令行进入D:\2000,创建工程D:\2000>android create project -t 3 -n ch3_14 -p ./ch3_14 -a Example3_14 -k ch3.fourteen。

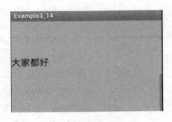

图 3.14 ScrollView 视图

(2) 将下列和视图相关的XML文件保存到工程的\res\layout目录中。

ch3_14.xml

```xml
<?xml version = "1.0" encoding = "utf-8"?>
<LinearLayout xmlns:android = "http://schemas.android.com/apk/res/android"
    android:orientation = "vertical"
    android:layout_width = "match_parent"
    android:layout_height = "match_parent">
    <!--
        ScrollView - 滚动条控件
        scrollbarStyle - 滚动条的样式
    -->
    <ScrollView android:id = "@ + id/scrollView"
        android:layout_width = "match_parent"
        android:layout_height = "200px"
        android:scrollbarStyle = "outsideOverlay"
        android:background = "#87CEEB">
        <TextView android:layout_width = "match_parent"
            android:layout_height = "100dp"
            android:textSize = "20sp"
            android:textColor = "#000000"
            android:id = "@ + id/textView"
            android:text = "Hello\n\n\n\n 你好\n\n\n\n\n\n\n\n\n\n 大家都好\n\n\n\n"/>
    </ScrollView>
</LinearLayout>
```

(3) 修改工程\src\ch3\fourteen目录下的Example3_14.java文件,修改后的内容如下:

Example3_14.java

```java
package ch3.fourteen;
import android.app.Activity;
import android.os.Bundle;
public class Example3_14 extends Activity {
    public void onCreate(Bundle savedInstanceState) {
        super.onCreate(savedInstanceState);
        setContentView(R.layout.ch3_14);
    }
}
```

(4) 启动AVD,进入工程的根目录,用快捷方式编译工程、安装应用程序到AVD(有关知识点参见1.5节)。对于本例子,用命令行进入D:\2000\ch3_14,执行如下命令:

D:\2000 > ch3_14 > ant debug install

3.13　HorizontalScrollView 视图

HorizontalScrollView 类的继承关系如下：

android.view.ViewGroup
　　└ android.widget.FrameLayout
　　　　└ android.widget.HorizontalScrollView

当一个视图的横向尺寸较大时，可能会导致用户无法看到视图的全貌（毕竟手机的屏幕大小是固定的），这时就可以把这样的视图事先放到 HorizontalScrollView 视图中，即将当前视图作为 HorizontalScrollView 视图的子视图。

1. 视图的作用

HorizontalScrollView 视图提供水平滚动条，用户可用滚动方式和其中的子视图进行交互。经常向 HorizontalScrollView 视图添加一个 LinearLayout 视图，LinearLayout 视图的特点是将其中的子视图按顺序放置在一行或一列中（在 4.1 节将详细学习 LinearLayout 视图）。

2. 示例

下面的例子 3-15 中将多个 ImageButton 视图（在 5.3 节将介绍这个专门用于显示图像的视图）放到 LinearLayout 视图中，然后再将 LinearLayout 视图放到 HorizontalScrollView 视图。运行效果如图 3.15 所示。

例子 3-15

（1）创建名字为 ch3_15 的工程，主要 Activity 子类的名字为 Example3_15，使用的包名为 ch3.fifteen。用命令行进入 D:\2000，创建工程 D:\2000 > android create project -t 3 -n ch3_15 -p ./ch3_15 -a Example3_15 -k ch3.fifteen。

图 3.15　HorizontalScrollView 视图

（2）增加视图资源。将名字为 flower1.jpg ～ flower6.jpg 的 6 幅图像文件存放到工程的图像资源中（有关知识点参见 2.7 节）。

（3）将下列和视图相关的 XML 文件保存到工程的 \res\layout 目录中。

ch3_15.xml

```
<?xml version = "1.0" encoding = "utf-8"?>
<HorizontalScrollView xmlns:android = "http://schemas.android.com/apk/res/android"
  android:layout_width = "match_parent"
  android:layout_height = "match_parent">
  <LinearLayout
    android:orientation = "horizontal"
    android:layout_width = "match_parent"
```

```
            android:layout_height = "match_parent">
         < ImageButton android:layout_width = "wrap_content"
            android:layout_height = "match_parent"
            android:src = "@drawable/flower1" />
         < ImageButton android:layout_width = "wrap_content"
            android:layout_height = "match_parent"
            android:src = "@drawable/flower2" />
         < ImageButton android:layout_width = "wrap_content"
            android:layout_height = "120dp"
            android:src = "@drawable/flower3" />
          < ImageButton android:layout_width = "wrap_content"
            android:layout_height = "match_parent"
            android:src = "@drawable/flower4" />
         < ImageButton android:layout_width = "wrap_content"
            android:layout_height = "match_parent"
            android:src = "@drawable/flower5" />
         < ImageButton android:layout_width = "wrap_content"
            android:layout_height = "match_parent"
            android:src = "@drawable/flower6" />
    </LinearLayout >
</HorizontalScrollView >
```

（4）修改工程\src\ch3\fifteen 目录下的 Example3_15.java 文件，修改后的内容如下：

Example3_15.java

```
package ch3.fifteen;
import android.app.Activity;
import android.os.Bundle;
public class Example3_15 extends Activity {
    public void onCreate(Bundle savedInstanceState) {
        super.onCreate(savedInstanceState);
        setContentView(R.layout.ch3_15);
    }
}
```

（5）启动 AVD，进入工程的根目录，用快捷方式编译工程、安装应用程序到 AVD（有关知识点参见 1.5 节）。对于本例子，用命令行进入 D:\2000\ch3_15，执行如下命令：

```
D:\2000 > ch3_15 > ant debug install
```

3.14 使用样式资源简化视图文件

1. 样式（style）

如果一个视图中需要许多子视图，这些子视图的许多属性值都是一样的，那么为每个子视图都设置这些属性的值，就会显得视图文件很冗余。

我们可能经常遇到这样的问题，一个视图中需要许多 TextView 子视图，这些子视图需要的字体颜色、大小以及许多属性值都是相同的，那么就可以考虑建立一个样式相关的 XML 文件（属于值资源文件，有关知识点参见 2.6 节）。样式相关的 XML 文件中使用 style

标记(元素)表示一种样式,style 标记使用 item 子标记,表示 style 代表的样式由哪些属性所组成。以下是一个和样式有关的 XML 文件,该文件需要事先保存到项目的值资源中。

style.xml

```xml
<?xml version = "1.0" encoding = "utf-8"?>
<resources>
    <style name = "myStyle">      <!-- name 的值是 myStyle,引用该样式时需要这个值 -->
        <item name = "android:textSize">26dp</item>
        <item name = "android:layout_width">100dp</item>
        <item name = "android:layout_height">100dp</item>
        <item name = "android:textColor">#FF00FF</item>
        <item name = "android:background">#87CEEB</item>
        <item name = "android:padding">10dp</item>
    </style>
</resources>
```

2. 视图文件使用样式

视图相关的 XML 文件通过引用样式可以减少不必要的冗余,不仅减少编辑量,而且也便于样式的维护和统一更改。

视图相关的 XML 文件引用样式 style 的语法格式:

style = "@style/style 标记的 name 属性的值"

例如:

style = "@style/myStyle"

以下的 XML 视图文件使用了 style.xml 中的样式。

main.xml

```xml
<?xml version = "1.0" encoding = "utf-8"?>
<LinearLayout xmlns:android = "http://schemas.android.com/apk/res/android"
    android:orientation = "vertical"
    android:layout_width = "fill_parent"
    android:layout_height = "fill_parent">
    <TextView
        style = "@style/myStyle"
        android:text = "How are you" />
    <TextView
        style = "@style/myStyle"
        android:text = "你好" />
    <TextView
        style = "@style/myStyle"
        android:text = "大家好" />
</LinearLayout>
```

习 题 3

1. 在视图相关的 XML 文件中,一个 View 视图使用 android:layout_width 和 android:layout_height 属性的作用是什么?

2. 如果想让 TextView 视图中的文本居中显示,应当设置 android:gravity 属性的值还是 android:layout_gravity 属性的值是"center"。

3. EditView 视图中使用 android:inputType 属性的作用是什么？该属性可以取哪些值？

4. EditView 视图上可以触发怎样的事件？请编写一个应用程序,处理 EditView 视图上触发的事件,当用户在 EditView 视图中编辑文本时,程序使用一个 TextView 视图显示用户编辑的文本的长度(可参考例子 3-3)。

5. 怎样在视图资源文件中约定 Button 视图上的 Click 事件？请编写一个应用程序,在视图资源文件中约定 Button 视图触发 Click 事件后,Activity 对象要执行的方法,当用户单击 Button 视图时,程序使用一个 TextView 视图显示"Hello Button"(可参考例子 3-4)。

6. ToggleButton 视图的特点是什么？请编写一个应用程序,当用户单击 ToggleButton 视图后,如果此时 ToggleButton 视图是选中(checked)状态,程序使用一个 TextView 视图显示"from unchecked to checked",如果此时 ToggleButton 视图是未选中状态(unchecked),程序使用一个 TextView 视图显示"from checked to unchecked"(可参考例子 3-5)。

7. ListView 视图的特点是什么？请编写一个应用程序,当用户单击 ListView 视图中的某个选项后,程序使用一个 TextView 视图显示选项的名称和索引位置(可参考例子 3-10)。

8. ScrollView 和 HorizontalScrollView 视图的特点是什么？在什么情况下适合使用 ScrollView 视图,在什么情况下适合使用 HorizontalScrollView 视图？

第 4 章 常用的 ViewGroup 视图

主要内容：
- LinearLayout 视图；
- RelativeLayout 视图；
- TableLayout 视图；
- TabHost 视图；
- GridLayout 视图；
- FrameLayout 视图；
- AbsoluteLayout 视图。

在第 3 章介绍了常用的 View 视图，如 TextView 视图、EditText 视图和 Button 视图等，本章介绍常用的 ViewGroup 视图。ViewGroup 视图是 View 类的子类，ViewGroup 视图中既可以有 View 视图，也可以有 ViewGroup 视图，因此人们习惯地将 ViewGroup 视图称为容器型视图，View 视图称为组件式视图，容器型视图中可以放置组件型视图。不同的 ViewGroup 视图采用不同的布局策略来放置 View 视图，应用程序可根据需求，用适当的 ViewGroup 视图提供和用户进行交互的程序界面。本章介绍的常用 ViewGroup 视图都是 ViewGroup 类的子孙类，均在 android.wedget 包中（ViewGroup 类在 andriod.view 包中）。

在本章中需要把和视图相关的 XML 文件放在工程的\res\layout 目录中。程序的 Java 代码使用系统提供的 R 类引用这个 XML 文件来构建视图，例如，假设 XML 文件是 tom.xml，引用方式如下：

setContentView(R.layout.tom);

另外，需要再次强调的是，XML 文件和 Java 文件不同，默认的是 UTF-8 编码，因此在保存 XML 文件时必须将编码选择为"UTF-8"，保存类型选择为"所有文件"。视图相关的 XML 文件的名字只能使用小写英文字母（a～z），数字（0～9），点（.）和下划线（_），不可以使用大写的英文字母。

注： 在进行 ViewGroup 视图设计时，我们重点介绍视图的功能用途和常用的重要属性，也不罗列有关类的方法，建议经常查看 docs 帮助文档（参见 1.6 节）

4.1 LinearLayout 视图

LinearLayout 视图的继承关系如下：

android.view.View

```
└ android.view.ViewGroup
    └ android.widget.LinearLayout
```

LinearLayout 视图的特点是将其中的子视图按顺序放置在一行或一列中。

1. 行布局或列布局

LinearLayout 视图是否按行或列来放置它的子视图（是否是行或列布局）由 LinearLayout 视图的 android:orientation 属性值来决定，当 android:orientation 属性值是"vertical"时，表示是列布局，将按顺序将子视图放置在一列中，当属性值是"horizontal"时，表示是行布局，将按顺序将子视图放置在一行中。

LinearLayout 视图是行布局时，行中的子视图按列的顺序从左向右排列（即每列上只有一个子视图），每个子视图根据自己的 android:layout_width 属性值去占用行中的剩余空间。LinearLayout 视图是列布局时，列中的子视图按行的顺序从上向下排列（即每行上只有一个子视图），每个子视图根据自己的 android:layout_height 属性值去占用列中的剩余空间。

2. 子视图的权重（layout_weight）属性

LinearLayout 视图在放置组件时，可能由于某种原因导致某些子视图不能显示在 LinearLayout 视图中。比如，如果 LinearLayout 视图是列布局，子视图按行的顺序从上向下排列（即每行上只有一个子视图），由于每个子视图根据自己的 android:layout_height 属性值去占用列中的剩余空间，可能导致某些子视图无空间可用（无行可用）。当出现这种情况时，子视图就需要设置 layout_weight 的值。

layout_weight 的值表示当前子视图在占用剩余空间时具有的"权重"（取值是度量值）。以下就按列放置子视图的情况说明子视图的 layout_weight 的值的作用（按行放置情况类似）。

如果不设置子视图 a 的 layout_weight 的值，该值的默认值是"0"，表示当前子视图 a 按照自己的 layout_height 属性的值去占用列中的剩余的空间。

如果设置子视图 a 的 layout_weight 属性的值是正的度量值，那么同时需要将子视图 a 的 layout_height 属性的值设置为"0dp"（不要为 layout_height 指定其他的值，否则容易引起混乱）。进行如此设置后，当前子视图 a 不按自己的 layout_height 的属性值去占用列的剩余的空间，而是按照 layout_weight 的值去占用列中的剩余空间。比如，假设当前子视图 a 的 layout_weight 的值是"1"，并且子视图 a 后面还有 2 个子视图，分别是 b 和 c，其 layout_weight 的值分别是默认值"0"和"2"。那么当轮到 a 开始占用列的剩余空间时，列的剩余空间首先去掉 b 按照 layout_height 的值占用的空间后，再将此时的剩余空间分成 1+2 份，即按照 a 和 c 的权重划分成 3 份。当前子视图 a 占用 3 份中的 1 份，c 占用 3 份中的 2 份。需要注意的是，如果列的剩余空间去掉 b 按照 layout_height 的值占用的空间后就再没有剩余空间了，那么子视图 a 和 c 就无法显示出来了。

如果 LinearLayout 视图中设置所有子视图的 layout_weight 的属性值都是"1"，那么这些子视图就都能被显示出来。

3. 使用 ScrollView 或 HorizontalScrollView

ScrollView 视图提供垂直滚动条（参见 3.12 节），用户可用滚动方式和其中的子视图进行交互。如果列布局的 LinearLayout 视图中有较多视图时，一个更好的办法是把它放到

ScrollView 视图中。HorizontalScrollView 视图提供水平滚动条（参见 3.13 节），用户可用水平方式和其中的子视图进行交互。如果行布局的 LinearLayout 视图中有较多视图时，一个更好的办法是把它放到 HorizontalScrollView 视图中。

4. 示例

例子 4-1 使用了两个 LinearLayout 视图，每个 LinearLayout 视图中提供一个处理 onClick 事件的按钮视图（在视图的 XML 文件约定处理 onClick 事件的方法，知识点参见 3.4 节），单击该按钮可以切换到另一个 LinearLayout 视图，运行效果如图 4.1 所示。

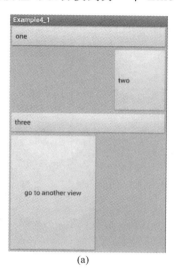 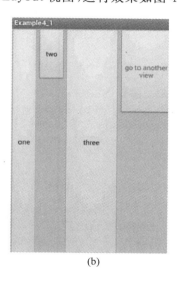

图 4.1 LinearLayout 的列布局和行布局

例子 4-1

（1）创建名字为 ch4_1 的工程，主要 Activity 子类的名字为 Example4_1，使用的包名为 ch4.one。用命令行进入 D:\2000，创建工程 D:\2000＞android create project -t 3 -n ch4_1 -p ./ch4_1 -a Example4_1 -k ch4.one。

（2）将下列和视图相关的 XML 文件保存到工程的\res\layout 目录中。

ch4_1_1.xml

```
<?xml version = "1.0" encoding = "utf-8"?>
<LinearLayout xmlns:android = "http://schemas.android.com/apk/res/android"
    android:layout_width = "match_parent"
    android:layout_height = "match_parent"
    android:background = "#87CEEB"
    android:orientation = "vertical">
    <Button
        android:layout_width = "match_parent"
        android:layout_height = "wrap_content"
        android:gravity = "left|center"
        android:text = "one" />
    <Button
        android:layout_width = "110dp"
```

```xml
            android:layout_height = "0dp"
            android:layout_gravity = "right"
            android:gravity = "left|center"
            android:layout_weight = "2"
            android:text = "two" />
        <Button
            android:layout_width = "match_parent"
            android:layout_height = "wrap_content"
            android:gravity = "left|center"
            android:text = "three" />
        <Button
            android:layout_width = "180dp"
            android:layout_height = "wrap_content"
            android:gravity = "center"
            android:layout_weight = "3"
            android:text = "go to another view"
            android:textColor = "#0000FF"
            android:onClick = "changeTwo" />
</LinearLayout>
```

ch4_1_2.xml

```xml
<?xml version = "1.0" encoding = "utf-8"?>
<LinearLayout xmlns:android = "http://schemas.android.com/apk/res/android"
    android:layout_width = "match_parent"
    android:layout_height = "match_parent"
    android:background = "#87CEEB"
    android:orientation = "horizontal">
    <Button
        android:layout_width = "0dp"
        android:layout_height = "match_parent"
        android:layout_weight = "1"
        android:text = "one" />
    <Button
        android:layout_width = "0dp"
        android:layout_height = "110dp"
        android:layout_gravity = "left"
        android:layout_weight = "1"
        android:text = "two" />
    <Button
        android:layout_width = "0dp"
        android:layout_height = "match_parent"
        android:layout_weight = "2"
        android:text = "three" />
    <Button
        android:layout_width = "0dp"
        android:layout_height = "180dp"
        android:layout_gravity = "left"
        android:layout_weight = "2"
        android:text = "go to another view"
        android:textColor = "#FF0000"
```

```
            android:onClick = "changeOne"    />
</LinearLayout>
```

（3）修改工程\src\ch4\one 目录下的 Example4_1.java 文件，修改后的内容如下：

Example4_1.java

```
package ch4.one;
import android.app.Activity;
import android.os.Bundle;
import android.widget.*;
import android.view.*;
public class Example4_1 extends Activity   {
    public void onCreate(Bundle savedInstanceState) {
        super.onCreate(savedInstanceState);
        setContentView(R.layout.ch4_1_1);
    }
    public void changeTwo(View view) {
        setContentView(R.layout.ch4_1_2);      //切换到视图 ch4_1_2
    }
    public void changeOne(View view) {
        setContentView(R.layout.ch4_1_1);      //切换到视图 ch4_1_1
    }
}
```

（4）启动 AVD，进入工程的根目录，用快捷方式编译工程、安装应用程序到 AVD（有关知识点参见 1.5 节）。对于本例子，用命令行进入 D:\2000\ch4_1，执行如下命令：

D:\2000 > ch4_1 > ant debug install

4.2 RelativeLayout 视图

RelativeLayout 视图的继承关系如下：

android.view.View
　　↳ android.view.ViewGroup
　　　　↳ android.widget.RelativeLayout

RelativeLayout 视图的特点是将其中的子视图按相对位置来摆放。一个子视图可以用它周围的某个子视图作参照物（anchor），给出其相对这个参照物的相对位置，比如设置当前子视图在其参照物的左面、右面、上面或下面等。一个子视图也可以给出其在父视图区域中的相对位置，比如设置当前子视图在父视图区域的左区域、右区域、上区域、下区域或中心区域等。

对于 RelativeLayout 视图中的子视图，如果不设置它的相对位置，该子视图默认的相对位置是左边与 RelativeLayout 视图的左边对齐，上边与 RelativeLayout 视图的上边对齐。

1. 相对父视图的区域给出相对位置

RelativeLayout 视图中的子视图可以通过设置相对属性值，设置自己在父视图区域中的相对位置。

- android:layout_alignParentBottom：取值是"true"或"false"，当取值是"true"时，当前子视图的底边与父视图的底边对齐。
- android:layout_alignParentLeft：取值是"true"或"false"，当取值是"true"时，当前子视图的左边与父视图的左边对齐。
- android:layout_alignParentRight：取值是"true"或"false"，当取值是"true"时，当前子视图的右边与父视图的右边对齐。
- android:layout_alignParentTop：取值是"true"或"false"，当取值是"true"时，当前子视图的上边与父视图的上边对齐。
- android:layout_centerHorizontal：取值是"true"或"false"，当取值是"true"时，当前子视图的中心与父视图的水平中心相同。
- android:layout_centerVertical：取值是"true"或"false"，当取值是"true"时，当前子视图的中心与父视图的垂直中心相同。
- android:layout_centerInParent：取值是"true"或"false"，当取值是"true"时，当前子视图的中心与父视图的水平中心和垂直中心相同，即和父视图的中心相同。

2. 相对其他子视图给出相对位置

RelativeLayout 视图中的子视图可以通过设置相对属性值，设置自己在父视图区域中的相对其他子视图的位置。

- android:layout_above：取值是某个子视图的 id，当前子视图的下边刚好在视图 id 的上方（下边与子视图的 id 的上边对齐），例如，取值"@＋id/bird"，当前子视图在 id 是"bird"的子视图的上方。
- android:layout_below：取值是某个子视图的 id，当前子视图的上边在视图 id 的下方（上边与子视图的 id 的下边对齐），例如，取值"@＋id/dog"，当前子视图在 id 是"dog"的子视图的下方。
- android:layout_alignLeft：取值是某个子视图的 id，当前子视图的左边与视图 id 的左边对齐，比如与 id 是"cat"的子视图的左边对齐。
- android:layout_alignRight：取值是某个子视图的 id，当前子视图的右边与视图 id 的右边对齐。
- android:layout_alignTop：取值是某个子视图的 id，当前子视图的上边与视图 id 的上边对齐。
- android:layout_alignBottom：取值是某个子视图的 id，当前子视图的下边与视图 id 的下边对齐。
- android:layout_toLeftOf：取值是某个子视图的 id，当前子视图在视图 id 的左边（右边与子视图的 id 的左边对齐）。
- android:layout_toRightOf：取值是某个子视图的 id，当前子视图在视图 id 的右边。

3. 子视图的高度与宽度

由于子视图需要根据相对属性值来确定自己的位置，因此在某些情况下，子视图 android:layout_width 或 android:layout_height 属性值设置的子视图的宽或高将失去作用，子视图的大小会自动进行调整，以满足相对位置的要求。视图 b 以视图 cat 为参照物，

让自己在视图 cat 的下方,左边与父视图的左边对齐,下边与父视图的底边对齐:

  android:layout_width = "30dp"
  android:layout_height = "50dp"
  android:layout_below = "@id/cat"
  android:layout_alignParentLeft = "true"
  android:layout_alignParentBottom = "true"

那么子视图 b 的高的设置就失效了,高会被调整,以满足让 b 在 cat 的下方,同时和父视图的底边对齐的位置要求。如果子视图 b 再让自己的右边和父视图的右边对齐:

  android:layout_alignParentRight = "true"

那么子视图 b 的宽的设置就失效了,宽会被调整,以满足让 b 的左边与父视图的左边对齐,右边和父视图的右边对齐的要求。

4. 子视图的重叠

  因为使用的是相对位置,可能会发生多个子视图的位置出现重叠。例如,子视图 b 在子视图 a 的下方 android:layout_below = "@id/a",子视图 c 在子视图 b 的下方 android:layout_below = "@id/b",子视图 d 在子视图 a 的下方 android:layout_below = "@id/a"。

  那么子视图 d 就会和子视图 b,甚至子视图 c 发生重叠(如果子视图 d 有较高的高度),如图 4.2 所示。当视图发生重叠时,后面的视图会遮挡先前的视图的部分区域或全部区域,因此可以考虑设置视图的透明度,以便消除遮挡带来的不利影响。一个视图可以设置 android:alpha 属性的值(取值 0～1 的浮点数),例如:

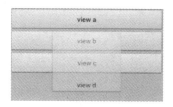

图 4.2 子视图出现重叠

  android:alpha = "0.6"

当 android:alpha 属性值是 0,视图完全透明,是 1,完全不透明(图 4.2 中的子视图 d 设置的 android:alpha 属性值是 0.6)。

5. 避免出现参照错误

  需要特别注意的是,当一个子视图还没有被添加到父视图之前,其他视图不能用它作参照物,否则会导致程序终止执行。例如,下列子视图 button1 错误地参照了还没有添加到父视图的 button2:

  < Button
    android:id = "@ + id/button1"
    android:layout_width = "50dp"
    android:layout_height = "wrap_content"
    android:layout_alignParentLeft = "true"
    android:layout_toLeftOf = "@ + id/button2" <!-- 这是一个错误参照 -->
    android:text = "door" />
  < Button
    android:id = "@ + id/button2"
    android:layout_width = "50dp"
    android:layout_height = "wrap_content"
    android:layout_toRightOf = "@ + id/button1"

```
android:text = "good" />
```

6. 示例

以下例子 4-2 使用 RelativeLayout 放置了几个子视图，效果如图 4.3 所示。

例子 4-2

（1）创建名字为 ch4_2 的工程，主要 Activity 子类的名字为 Example4_2，使用的包名为 ch4.two。用命令行进入 D:\2000，创建工程 D:\2000>android create project -t 3 -n ch4_2 -p ./ch4_2 -a Example4_2 -k ch4.two。

（2）将下列和视图相关的 XML 文件 ch4_2.xml 保存到工程的\res\layout 目录中。

ch4_2.xml

```xml
<?xml version = "1.0" encoding = "utf-8"?>
<RelativeLayout xmlns:android = "http://schemas.android.com/apk/res/android"
    android:layout_width = "match_parent"
    android:layout_height = "match_parent"
    android:background = "#87CEEB">
  <Button
      android:id = "@+id/button1"
      android:layout_width = "80dp"
      android:layout_height = "wrap_content"
      android:layout_centerHorizontal = "true"
      android:text = "button1" />
  <Button
      android:id = "@+id/button2"
      android:layout_width = "60dp"
      android:layout_height = "wrap_content"
      android:layout_toLeftOf = "@+id/button1"
      android:layout_alignParentLeft = "true"
      android:text = "button2" />
  <Button
      android:id = "@+id/button3"
      android:layout_width = "60dp"
      android:layout_height = "wrap_content"
      android:layout_toRightOf = "@+id/button1"
      android:layout_alignParentRight = "true"
      android:text = "button3" />
  <Button
      android:id = "@+id/button4"
      android:layout_width = "match_parent"
      android:layout_height = "50dp"
      android:layout_below = "@+id/button1"
      android:text = "button4" />
  <Button
      android:id = "@+id/button5"
```

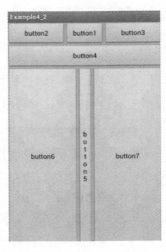

图 4.3　RelativeLayout 布局

```
            android:layout_width = "30dp"
            android:layout_height = "match_parent"
            android:layout_below = "@ + id/button4"
            android:layout_centerHorizontal = "true"
            android:text = "button5"
            />
        <Button
            android:id = "@ + id/button6"
            android:layout_width = "60dp"
            android:layout_height = "match_parent"
            android:layout_below = "@ + id/button4"
            android:layout_toLeftOf = "@ + id/button5"
            android:layout_alignParentLeft = "true"
            android:text = "button6" />
        <Button
            android:id = "@ + id/button6"
            android:layout_width = "60dp"
            android:layout_height = "match_parent"
            android:layout_below = "@ + id/button4"
            android:layout_toRightOf = "@ + id/button5"
            android:layout_alignParentRight = "true"
            android:text = "button7" />
</RelativeLayout>
```

(3) 修改工程\src\ch4\two 目录下的 Example4_2.java 文件,修改后的内容如下:

Example4_2.java

```
package ch4.two;
import android.app.Activity;
import android.os.Bundle;
public class Example4_2 extends Activity {
    public void onCreate(Bundle savedInstanceState) {
        super.onCreate(savedInstanceState);
        setContentView(R.layout.ch4_2);
    }
}
```

(4) 启动 AVD,进入工程的根目录,用快捷方式编译工程、安装应用程序到 AVD(有关知识点参见 1.5 节)。对于本例子,用命令行进入 D:\2000\ch4_2,执行如下命令:

D:\2000 > ch4_2 > ant debug install

4.3　TableLayout 视图

TableLayout 视图的继承关系如下:

android.view.ViewGroup
　　└ android.widget.LinearLayout
　　　　└ android.widget.TableLayout

TableLayout 视图的特点是用表格样式放置子视图。

1. TableLayout 视图的行（TableRow 视图）

TableLayout 由若干行组成，每行中可以放置若干个子视图。TableLayout 视图的行由 TableRow 视图来担当，即 TableRow 视图作为 TableLayout 视图的直接子视图，其他视图将作为 TableRow 视图的子视图。也就是说，TableLayout 视图可以首先放置若干个 TableRow 视图，然后每个 TableRow 视图中再放置若干个视图。用户看不到 TableRow 视图，只能看见 TableRow 视图中放置的视图。

以下是 TableLayout 视图的几个常用属性。

- android:collapseColumns：设置隐藏表格的哪些列，多列用逗号隔开，如 android:collapseColumns="0,1,2"（列索引从 0 开始），如果要隐藏所有的列，用 * 代替数字，例如，android:collapseColumns="*"。
- android:shrinkColumns：设置收缩表格的哪些列。如 android:shrinkColumns="0,1,2"，即表格的第 0、1、2 列的内容是收缩的，以适合屏幕，不会被挤出屏幕。如果要收缩所有的列，用 * 代替数字，例如，android:shrinkColumns="*"。
- android:stretchColumns：设置拉伸表格的哪些列。如 android:stretchColumns="0,1,2"，即拉伸表格的第 0、1、2 列，以适合屏幕，不给屏幕留出空隙。如果要拉伸所有的列，用 * 代替数字，例如，android:stretchColumns="*"。

2. TableRow 视图的宽和高

TableRow 视图将作为 TableLayout 中的行，那么 TableRow 视图的 layout_width 属性的值被规定为一定是"match_parent"，layout_height 属性的值一定是"wrap_content"。

3. 示例

以下例子 4-3 使用 TableLayout 放置了几个子视图，效果如图 4.4 所示。

图 4.4　TableLayout 布局

例子 4-3

（1）创建名字为 ch4_3 的工程，主要 Activity 子类的名字为 Example4_3，使用的包名为 ch4.three。用命令行进入 D:\2000，创建工程 D:\2000>android create project -t 3 -n ch4_3 -p ./ch4_3 -a Example4_3 -k ch4.three。

（2）将下列和视图相关的 XML 文件 ch4_3.xml 保存到工程的\res\layout 目录中。

ch4_3.xml

```
<?xml version = "1.0" encoding = "utf-8"?>
<TableLayout xmlns:android = "http://schemas.android.com/apk/res/android"
    android:layout_width = "match_parent"
    android:layout_height = "match_parent"
    android:background = "#87CEEB"
    android:shrinkColumns = "0"
    android:stretchColumns = "1,2">
    <TableRow>
        <Button android:text = "Hello" />
        <Button android:text = "Hello" />
        <Button android:text = "Hello" />
```

```
        </TableRow>
        <TableRow>
            <Button android:text = "Hello" />
            <Button android:text = "Hello" />
            <Button android:text = "Hello" />
        </TableRow>
        <TableRow>
            <Button android:text = "Hello" />
            <Button android:text = "Hello" />
            <Button android:text = "Hello" />
        </TableRow>
</TableLayout>
```

(3) 修改工程\src\ch4\three 目录下的 Example4_3.java 文件,修改后的内容如下:

Example4_3.java

```
package ch4.three;
import android.app.Activity;
import android.os.Bundle;
public class Example4_3 extends Activity {
    public void onCreate(Bundle savedInstanceState) {
        super.onCreate(savedInstanceState);
        setContentView(R.layout.ch4_3);
    }
}
```

(4) 启动 AVD,进入工程的根目录,用快捷方式编译工程、安装应用程序到 AVD(有关知识点参见 1.5 节)。对于本例子,用命令行进入 D:\2000\ch4_3,执行如下命令:

D:\2000 > ch4_3 > ant debug install

4.4 TabHost 视图

TabHost 视图的继承关系如下:

android.view.ViewGroup
 ┗ android.widget.FrameLayout
   ┗ android.widget.TabHost

TabHost 视图是我们熟悉的一种视图形式,习惯地称为"选项卡"视图。TabHost 视图可以让用户分页显示若干个视图,其特点是用选项卡的形式来组织若干个视图。TabHost 视图中有若干个选项卡,用户单击某个选项卡将显示若干个视图中的某一个视图(该选项卡对应的视图)。

1. TabHost 视图的构成

TabHost 视图中必须包含有两个视图,一个是 TabWidget 视图,一个是 FrameLayout 视图。TabWidget 视图负责放置选项卡(Tab),其特点是可以放置多个选项卡,每个选项卡的职责是负责指定需要显示的视图。FrameLayout 视图的职责是负责放置选项卡指定的视图,FrameLayout 视图的特点是可以向其添加多个子视图,但是 FrameLayout 视图只能显示其中的一个子视图，即用户单击哪个选项卡，FrameLayout 就立刻显示当前选项卡指定的视图。

TabHost 视图可以使用自己的某种布局策略，比如 LinearLayout 视图，放置其中的两个视图：TabWidget 和 FrameLayout 视图，比如 TabHost 视图可以把它的 TabWidget 视图放置在自己区域中的上方，将 FrameLayout 视图放置在自己区域中的下方。

需要特别注意的是，在使用 XML 视图文件给出 TabHost 视图时，只能在 XML 文件中指定 TabHost 视图中的 TabWidget 视图和 FrameLayout 视图，无法向 TabWidget 视图中添加选项卡以及指定选项卡负责的视图，这些工作需要在 Java 代码中完成（见稍后的内容）。因为是 Java 代码负责完成添加选项卡，因此系统要求 TabWidget 视图必须给出 id 属性的值，而且该值必须是"@android:id/tabs"：android:id＝"@android:id/tabs"，FrameLayout 视图必须给出 id 属性的值，而且该值必须是"@android:id/tabcontent"：android:id＝"@android:id/tabcontent"。

以下是一个简单的 TabHost 视图的 XML 文件。

```
<?xml version="1.0" encoding="utf-8"?>
<TabHost xmlns:android="http://schemas.android.com/apk/res/android"
    <LinearLayout
        <TabWidget android:id="@android:id/tabs"
        />       <!-- 必须要有的 TabWidget 视图，负责放置选项卡 -->
        <FrameLayout android:id="@android:id/tabcontent"
        />       <!-- 必须要有的 FrameLayout 视图，负责放置选项卡指定的视图 -->
    </LinearLayout>
</TabHost>
```

2. Java 代码中添加选项卡

在 Java 代码，即 Activity 的子类中，负责完成向 TabHost 视图的 TabWidget 视图里添加选项卡以及制定选项卡所负责的视图。假设需要两个选项卡，二者负责的视图分别是 one.xml 和 two.xml（事先存放到视图资源中）。Java 代码中包含的主要步骤如下：

1) 得到 TabHost 视图

在 Java 代码中使用 TabHost 视图的 id（这里假设 id 值是 tabhost）得到该 TabHost 视图，并让该 TabHost 视图调用 setup 方法完成添加选项卡的初始化工作：

```
TabHost host = (TabHost)findViewById(R.id.tabhost);
    host.setup();
```

当 TabHost 视图调用 setup 方法时，会根据 id 寻找 TabHost 视图中的 TabWidget 视图和 FrameLayout 视图，这就是为什么要求在 TabHost 视图的 XML 文件中，TabWidget 视图必须给出 id 属性的值，而且该值必须是"tabs"：android:id＝"@android:id/tabs"，FrameLayout 视图必须给出 id 属性的值，而且该值必须是"tabcontent"：android:id＝"@android:id/tabcontent"。

2) 绑定选项卡负责的视图 XML 文件

由于 TabHose 对应的视图 XML 文件中并不包含选项卡所负责的视图，因此必须在 Java 代码中将选项卡所负责的视图 XML 文件与 TabHost 视图中的 FrameLayout 视图实施绑定（比如，我们前面假设的 one.xml 和 two.xml），只有这样，FrameLayout 才能显示这些视图。

绑定工作由 LayoutInflater 类(该类在 android.view 包中)的实例负责,代码如下:

```
LayoutInflater myinflater = LayoutInflater.from(this);
myinflater.inflate(R.layout.one,host.getTabContentView());    //绑定 one.xml 到 FrameLayout
myinflater.inflate(R.layout.two,host.getTabContentView());    //绑定 two.xml 到 FrameLayout
```

3) 创建选项卡

选项卡由 TabHost 的内部静态类 TabHost.TabSpec 负责,TabHost 视图调用 public TabHost.TabSpec newTabSpec(String tag)方法可以返回一个选项卡对象,其中参数是选项卡的内部名字,并不显示在选项卡上。

选项卡可以调用方法设置选项卡上的标识(Indicator)以及自己负责的视图(Content)。

TabHost.TabSpec setIndicator(CharSequence label):用参数 label 指定的字符序列作为选项卡上的标识。

TabHost.TabSpec setIndicator(CharSequence label, Drawable icon):用参数 label 指定的字符序列和参数 icon 指定的图像共同作为选项卡上的标识。

TabHost.TabSpec setIndicator(View view):用参数 view 指定的视图作为选项卡上的标识。

TabHost.TabSpec setContent(int viewId):用参数 viewId 代表的视图作为选项卡负责的视图,这里参数 viewId 必须是视图的 id。因此一个视图想成为选项卡所负责的视图,那么该视图对应的 XML 文件中必须为视图指定 id(这里假设 one.xml 中指定的 id 值是 myCat, two.xml 中指定的 id 值是 myDog)。

创建 2 个选项卡的具体代码如下:

```
TabHost.TabSpec specOne = host.newTabSpec("catTab");    //第 1 个选项卡
specOne.setIndicator("I like myCat");
specOne.setContent(R.id.myCat);
TabHost.TabSpec specTwo = host.newTabSpec("dogTab");    //第 2 个选项卡
specTwo.setIndicator("I like myDog");
specTwo.setContent(R.id.myDog);
```

4) 添加选项卡

TabHost 视图调用 void addTab(TabHost.TabSpec tabSpec)方法向自己的 TabWidget 视图中添加选项卡,例如:

```
host.addTab(specOne);
host.addTab(specTwo);
```

TabHost 视图默认使用 FrameLayout 显示第一个选项卡所负责的视图,当用户单击某个选项卡后,FrameLayout 将显示当前被单击的选项卡负责的视图。也可让 TabHost 视图调用 void setCurrentTab(int index)或 void setCurrentTabByTag(String tag)方法设置 FrameLayout 显示选项卡负责的视图。

注意到 newTabSpec(String tag)、setIndicator(CharSequence label)等方法会返回当前选项卡对象,因此,TabHost 对象 host 经常简捷地创建、添加选项卡,代码如下:

```
host.addTab(host.newTabSpec("myCat").setIndicator("I like myCat").setContent(R.id.myCat));
```

注：Android 3.0（API level 11）版本之后，系统为 Activity 对象提供了动作栏（ActionBar），用户可以使用 ActionBar 来实现 TabHost 视图实现的交互行为（参见 6.5 节）。

3. 示例

以下例子 4-4 使用 TabHost 视图显示了一个 LinearLayout 视图和一个 TableLayout 视图，效果如图 4.5 所示。

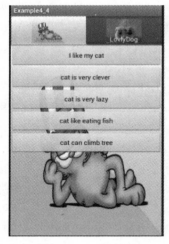
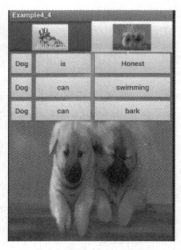

(a) 单击JiafeiCat选项卡效果　　(b) 单击LovelyDog选项卡效果

图 4.5　运行效果

例子 4-4

（1）创建名字为 ch4_4 的工程，主要 Activity 子类的名字为 Example4_4，使用的包名为 ch4.four。用命令行进入 D:\2000，创建工程 D:\2000＞android create project -t 3 -n ch4_4 -p ./ch4_4 -a Example4_4 -k ch4.four。

（2）增加图像资源。选项卡准备用字符串和图像组成自己的标识，因此项目需要提供图像资源。将名字为 cat.jpg、dog.jpg 的图像保存到图像资源中（有关知识点参见 2.7 节）。

（3）将下列和视图相关的 XML 文件 ch4_4.xml（activity 要使用的 TabHost 视图），one.xml 和 two.xml（TabHost 视图中的选项卡所负责的视图）保存到工程的\res\layout 目录中。

ch4_4.xml

```
<?xml version = "1.0" encoding = "utf-8"?>
< TabHost xmlns:android = "http://schemas.android.com/apk/res/android"
        android:id = "@ + id/tabhost"
        android:layout_width = "match_parent"
        android:layout_height = "match_parent">
    < LinearLayout
        android:orientation = "vertical"
        android:layout_width = "match_parent"
        android:layout_height = "match_parent">
        < TabWidget
            android:id = "@android:id/tabs"
```

```xml
            android:layout_width = "match_parent"
            android:layout_height = "wrap_content"
            android:background = "#FF6611"/>      <!-- id 必须是@android:id/tabs -->
        <FrameLayout
            android:id = "@android:id/tabcontent"
            android:layout_width = "match_parent"
            android:layout_height = "match_parent"
            android:background = "#66FFEB">
        </FrameLayout>                            <!-- id 必须是@android:id/tabcontent -->
    </LinearLayout>
</TabHost>
```

one.xml

```xml
<?xml version = "1.0" encoding = "utf-8"?>
<LinearLayout xmlns:android = "http://schemas.android.com/apk/res/android"
    android:layout_width = "match_parent"
    android:layout_height = "match_parent"
    android:background = "@drawable/cat"
    android:orientation = "vertical"
    android:id = "@+id/myCat">    <!-- 必须有 id -->
    <Button android:text = "I like my cat"
            android:layout_width = "match_parent"
            android:layout_height = "wrap_content"/>
    <Button android:text = "cat is very clever "
            android:layout_width = "match_parent"
            android:layout_height = "wrap_content" />
    <Button android:text = "cat is very lazy"
            android:layout_width = "match_parent"
            android:layout_height = "wrap_content"/>
    <Button android:text = "cat like eating fish"
            android:layout_width = "match_parent"
            android:layout_height = "wrap_content" />
    <Button android:text = "cat can climb tree"
            android:layout_width = "match_parent"
            android:layout_height = "wrap_content"/>
</LinearLayout>
```

two.xml

```xml
<?xml version = "1.0" encoding = "utf-8"?>
<TableLayout xmlns:android = "http://schemas.android.com/apk/res/android"
    android:layout_width = "match_parent"
    android:layout_height = "match_parent"
    android:background = "@drawable/dog"
    android:shrinkColumns = "0"
    android:stretchColumns = "1,2"
    android:id = "@+id/myDog">        <!-- 必须有 id -->
    <TableRow>
        <Button android:text = "Dog" />
        <Button android:text = "is" />
        <Button android:text = "Honest" />
```

```xml
        </TableRow>
        <TableRow>
            <Button android:text = "Dog" />
            <Button android:text = "can" />
            <Button android:text = "swimming" />
        </TableRow>
        <TableRow>
            <Button android:text = "Dog" />
            <Button android:text = "can" />
            <Button android:text = "bark" />
        </TableRow>
</TableLayout>
```

(4) 修改工程\src\ch4\four 目录下的 Example4_4.java 文件,修改后的内容如下:

Example4_4.java

```java
package ch4.four;
import android.widget.*;
import android.view.*;
import android.app.*;
import android.os.Bundle;
public class Example4_4 extends Activity {
    public void onCreate(Bundle savedInstanceState) {
        super.onCreate(savedInstanceState);
        setContentView(R.layout.ch4_4);
        TabHost host = (TabHost)findViewById(R.id.tabhost);    //获得 TabHost 视图
        host.setup();                                           //初始化 TabHost 视图
        LayoutInflater myinflater = LayoutInflater.from(this);
        myinflater.inflate(R.layout.one, host.getTabContentView());
                                                                //绑定 one.xml 到 FrameLayout
        myinflater.inflate(R.layout.two, host.getTabContentView());
                                                                //绑定 two.xml 到 FrameLayout
        TabHost.TabSpec specOne = host.newTabSpec("catTab");    //第 1 个选项卡
        specOne.setIndicator("JiaFeiCat",getResources().getDrawable(R.drawable.cat));
        specOne.setContent(R.id.myCat);
        TabHost.TabSpec specTwo = host.newTabSpec("dogTab");    //第 2 个选项卡
        specTwo.setIndicator("LovlyDog",getResources().getDrawable(R.drawable.dog));
        specTwo.setContent(R.id.myDog);
        host.addTab(specOne);
        host.addTab(specTwo);
        host.setCurrentTab(0);         //默认显示第一个选项负责的视图,选项卡索引从 0 开始
                                       //也可用 setCurrentTabByTag("catTab")
    }
}
```

(5) 启动 AVD,进入工程的根目录,用快捷方式编译工程、安装应用程序到 AVD(有关知识点参见 1.5 节)。对于本例子,用命令行进入 D:\2000\ch4_4,执行如下命令:

```
D:\2000 > ch4_4 > ant debug install
```

4.5 GridLayout 视图

GridLayout 视图的继承关系如下：

android.view.View
　└ android.view.ViewGroup
　　　└ android.widget.GridLayout

GridLayout 视图是我们熟悉的网格布局，GridLayout 视图将自己占据的区域分成 m 行 n 列的 $m \times n$ 个网格（cell）的区域，每个网格可以放置一个视图，网格的大小取决放置的子视图的大小。

1. GridLayout 视图的网格数目

GridLayout 视图最基本的属性是 android:rowCount 和 android:columnCount，分别设置 GridLayout 视图中的行数和列数，例如：

```
android:columnCount = "3"
android:rowCount = "3"
```

GridLayout 视图将子视图按行的顺序或列的顺序放置子视图，当 android:orientation 属性值是"horizontal"时，按行的顺序放置子视图，当 android:orientation 属性值是 "vertical"时，按列的顺序放置子视图，android:orientation 属性的默认值是"horizontal"。

GridLayout 视图中子视图的 layout_width 和 layout_height 属性的默认值都是"wrap_content"，除非特别需要，可不必设置子视图的 layout_width 和 layout_height 属性的值。

2. 示例

以下例子 4-5 使用 GridLayout 视图让用户玩"九宫"填数字游戏，即在 3×3 的格子中放置 1～9 数字，使得横、竖以及对角线格子中的数字之和都是 15。当用户玩成功后，格子中的视图变得半透明，使得用户可以完整地看到 GridLayout 视图的图像，效果如图 4.6 所示。

图 4.6　GridLayout 布局

例子 4-5

（1）创建名字为 ch4_5 的工程，主要 Activity 子类的名字为 Example4_5，使用的包名为 ch4.five。用命令行进入 D:\2000，创建工程 D:\2000＞android create project -t 3 -n ch4_5 -p ./ch4_5 -a Example4_5 -k ch4.five。

（2）增加图像资源。程序需要显示一朵花的图像，将名字为 flower.jpg 的图像保存到图像资源中（有关知识点参见 2.7 节）。

（3）将下列和视图相关的 XML 文件 ch4_5.xml 保存到工程的\res\layout 目录中。

ch4_5.xml

```
<?xml version = "1.0" encoding = "utf - 8"?>
< GridLayout xmlns:android = "http://schemas.android.com/apk/res/android"
    android:orientation = "vertical"
    android:background = "@drawable/flower"
```

```xml
            android:layout_width = "wrap_content"
            android:layout_height = "wrap_content"
            android:columnCount = "3"
            android:rowCount = "3">
    < EditText  android:id = "@ + id/edit1"
                android:inputType = "number"
                android:textSize = "60sp"
                android:layout_width = "80dp"
                android:layout_height = "80dp"/>
    < EditText  android:id = "@ + id/edit2"
                android:inputType = "number"
                android:textSize = "60sp"
                android:layout_width = "80dp"
                android:layout_height = "80dp"/>
    < EditText  android:id = "@ + id/edit3"
                android:inputType = "number"
                android:textSize = "60sp"
                android:layout_width = "80dp"
                android:layout_height = "80dp"/>
    < EditText  android:id = "@ + id/edit4"
                android:inputType = "number"
                android:textSize = "60sp"
                android:layout_width = "80dp"
                android:layout_height = "80dp"/>
    < EditText  android:id = "@ + id/edit5"
                android:inputType = "number"
                android:textSize = "60sp"
                android:layout_width = "80dp"
                android:layout_height = "80dp"/>
    < EditText  android:id = "@ + id/edit6"
                android:inputType = "number"
                android:textSize = "60sp"
                android:layout_width = "80dp"
                android:layout_height = "80dp"/>
    < EditText  android:id = "@ + id/edit7"
                android:inputType = "number"
                android:textSize = "60sp"
                android:layout_width = "80dp"
                android:layout_height = "80dp"/>
    < EditText  android:id = "@ + id/edit8"
                android:inputType = "number"
                android:textSize = "60sp"
                android:layout_width = "80dp"
                android:layout_height = "80dp"/>
    < EditText  android:id = "@ + id/edit9"
                android:inputType = "number"
                android:textSize = "60sp"
                android:layout_width = "80dp"
                android:layout_height = "80dp"/>
</GridLayout >
```

(4) 修改工程\src\ch4\five 目录下的 Example4_5.java 文件,修改后的内容如下:
Example4_5.java

```java
package ch4.five;
import android.app.Activity;
import android.os.Bundle;
import android.widget.*;
import android.text.*;
import java.util.*;
public class Example4_5 extends Activity implements TextWatcher{
    EditText edit [];
    int a[];
    public void onCreate(Bundle savedInstanceState) {
        super.onCreate(savedInstanceState);
        setContentView(R.layout.ch4_5);
        edit = new EditText[9];
        a = new int[9];
        edit[0] = (EditText)findViewById(R.id.edit1);
        edit[1] = (EditText)findViewById(R.id.edit2);
        edit[2] = (EditText)findViewById(R.id.edit3);
        edit[3] = (EditText)findViewById(R.id.edit4);
        edit[4] = (EditText)findViewById(R.id.edit5);
        edit[5] = (EditText)findViewById(R.id.edit6);
        edit[6] = (EditText)findViewById(R.id.edit7);
        edit[7] = (EditText)findViewById(R.id.edit8);
        edit[8] = (EditText)findViewById(R.id.edit9);
        for(int i = 0;i < edit.length;i++)
           edit[i].addTextChangedListener(this);
    }
    public void  afterTextChanged (Editable s) {
        for(int i = 0;i < edit.length;i++) {
            try {
                a[i] = Integer.parseInt(edit[i].getText().toString());
            }
            catch(Exception exp) {
                a[i] = 0;
            }
        }
        boolean isRight = true;
        isRight = (a[0] + a[1] + a[2] == 15)&&(a[3] + a[4] + a[5] == 15);
        isRight = isRight&&(a[6] + a[7] + a[8] == 15)&&(a[0] + a[3] + a[6] == 15);
        isRight = isRight&& (a[1] + a[4] + a[7] == 15)&&(a[2] + a[5] + a[8] == 15);
        isRight = isRight&& (a[0] + a[4] + a[8] == 15)&&(a[2] + a[4] + a[6] == 15);
        if(isRight)
           for(int i = 0;i < edit.length;i++)
              edit[i].setAlpha(0.5f);
    }
    public void beforeTextChanged(CharSequence s,int start,int count,int after)
    {}
     public void onTextChanged (CharSequence s,int start,int before,int count)
    {}
}
```

(5) 启动 AVD，进入工程的根目录，用快捷方式编译工程、安装应用程序到 AVD(有关知识点参见 1.5 节)。对于本例子，用命令行进入 D:\2000\ch4_5，执行如下命令：

D:\2000>ch4_5>ant debug install

4.6 FrameLayout 视图

FrameLayout 视图的继承关系如下：

android.view.View
　　└ android.view.ViewGroup
　　　　└ android.widget.FrameLayout

FrameLayout 视图的特点是可以向其添加多个子视图，但 FrameLayout 视图在同一时刻只显示其中的一个子视图，即将一个子视图放在 FrameLayout 视图的最前面显示，默认情况下，FrameLayout 视图的初始状态是将最后添加到它当中的子视图放在 FrameLayout 视图的最前面显示。如果 FrameLayout 视图中的子视图的可见性都是"visible"(android:visibility="visible")，而且没有采用特殊的对齐方式，那么最后一个添加的视图就会遮挡先前的子视图。当然，如果只将某个子视图的 android:visibility 属性的值设置为"visible"，其他的都设置成"invisible"，那么 FrameLayout 视图的初始状态下的可见子视图就是 android:visibility 属性的值设置为"visible"的子视图。

1. 通过事件处理显示 FrameLayout 视图中的子视图

由于 FrameLayout 视图的特点是同一时刻只能显示子视图中的某一个，因此为了能显示用户需要的子视图，就需要在 Java 代码中加入事件处理机制，比如，用户单击某个按钮视图触发 onClick 事件，在处理 onClick 事件的代码中让 FrameLayout 视图显示视图中的某个子视图。FrameLayout 视图中的子视图可以调用 void setVisibility (int visibility)方法设置自己是否可见，当希望 FrameLayout 视图显示它的某个子视图时，需要将其他子视图全部设置成不可见，并将当前子视图设置成可见。setVisibility (int visibility)方法的参数取 View 类中的三个静态常量之一：VISIBLE、INVISIBLE 或 GONE。

2. 示例

以下例子 4-6 使用 FrameLayout 视图，视图中有 4 个 TextView 子视图，这 4 个 TextView 子视图的背景图片分别是春、夏、秋、冬的图像。我们在 FrameLayout 视图的上面又添加了 4 个按钮视图，单击相应的按钮，FrameLayout 视图显示它相应的 TextView 子视图，效果如图 4.7 所示。

例子 4-6

(1) 创建名字为 ch4_6 的工程，主要 Activity 子类的名字为 Example4_6，使用的包名为 ch4.six。用命令行进入 D:\2000，创建工程 D:\2000>android create project -t

图 4.7　FrameLayout 布局

3 -n ch4_6 -p ./ch4_6 -a Example4_6 -k ch4.six。

(2) 增加图像资源。项目需要显示季节相关的图像,将名字为 spring.jpg、summer.jpg、autumn.jpg 和 winter.jpg 的图像保存到图像资源中(有关知识点参见 2.7 节)。

(3) 将下列和视图相关的 XML 文件保存到工程的\res\layout 目录中。

ch4_6.xml

```xml
<?xml version="1.0" encoding="utf-8"?>
<LinearLayout xmlns:android="http://schemas.android.com/apk/res/android"
    android:layout_width="match_parent"
    android:layout_height="match_parent"
    android:orientation="vertical">
    <LinearLayout   android:orientation="horizontal"
        android:layout_width="match_parent"
        android:layout_height="wrap_content">
        <Button android:text="spring"
            android:layout_weight="1"
            android:layout_width="wrap_content"
            android:layout_height="wrap_content"
            android:onClick="changeChildView"/>
        <Button android:text="summer"
            android:layout_weight="1"
            android:layout_width="wrap_content"
            android:layout_height="wrap_content"
            android:onClick="changeChildView"/>
        <Button android:text="autumn"
            android:layout_weight="1"
            android:layout_width="wrap_content"
            android:layout_height="wrap_content"
            android:onClick="changeChildView"/>
        <Button android:text="winter"
            android:layout_weight="1"
            android:layout_width="wrap_content"
            android:layout_height="wrap_content"
            android:onClick="changeChildView"/>
    </LinearLayout>
    <FrameLayout
        android:id="@+id/myFrame"
        android:layout_width="match_parent"
        android:layout_height="match_parent">
        <TextView   android:background="@drawable/spring"
            android:id="@+id/spring"
            android:layout_width="match_parent"
            android:layout_height="match_parent"/>
        <TextView   android:background="@drawable/summer"
            android:id="@+id/summer"
            android:layout_width="match_parent"
            android:layout_height="match_parent"/>
        <TextView   android:background="@drawable/autumn"
            android:id="@+id/autumn"
            android:layout_width="match_parent"
```

```xml
                    android:layout_height = "match_parent"/>
    <TextView   android:background = "@drawable/winter"
                    android:id = "@+id/winter"
                    android:layout_width = "match_parent"
                    android:layout_height = "match_parent"/>
    </FrameLayout>
</LinearLayout>
```

(4) 修改工程\src\ch4\six 目录下的 Example4_6.java 文件,修改后的内容如下:
Example4_6.java

```java
package ch4.six;
import android.widget.*;
import android.view.*;
import android.app.*;
import android.os.Bundle;
public class Example4_6 extends Activity {
    FrameLayout layout;
    TextView spring,summer,autumn,winter;
    public void onCreate(Bundle savedInstanceState) {
        super.onCreate(savedInstanceState);
        setContentView(R.layout.ch4_6);
        layout = (FrameLayout)findViewById(R.id.myFrame);
        spring = (TextView)findViewById(R.id.spring);
        summer = (TextView)findViewById(R.id.summer);
        autumn = (TextView)findViewById(R.id.autumn);
        winter = (TextView)findViewById(R.id.winter);
    }
    public void changeChildView(View v) {
        Button button = (Button)v;
        String str = button.getText().toString();
        if(str.equals("spring")) {
            winter.setVisibility(View.INVISIBLE);
            autumn.setVisibility(View.INVISIBLE);
            summer.setVisibility(View.INVISIBLE);
            spring.setVisibility(View.VISIBLE);
        }
        if(str.equals("summer")) {
            winter.setVisibility(View.INVISIBLE);
            autumn.setVisibility(View.INVISIBLE);
            spring.setVisibility(View.INVISIBLE);
            summer.setVisibility(View.VISIBLE);
        }
        if(str.equals("autumn")) {
            winter.setVisibility(View.INVISIBLE);
            spring.setVisibility(View.INVISIBLE);
            summer.setVisibility(View.INVISIBLE);
            autumn.setVisibility(View.VISIBLE);
        }
```

```
        if(str.equals("winter")) {
            spring.setVisibility(View.INVISIBLE);
            autumn.setVisibility(View.INVISIBLE);
            summer.setVisibility(View.INVISIBLE);
            winter.setVisibility(View.VISIBLE);
        }
    }
}
```

（5）启动 AVD，进入工程的根目录，用快捷方式编译工程、安装应用程序到 AVD（有关知识点参见 1.5 节）。对于本例子，用命令行进入 D:\2000\ch4_6，执行如下命令：

```
D:\2000 > ch4_6 > ant debug install
```

4.7 AbsoluteLayout 视图

AbsoluteLayout 视图的继承关系如下：

```
android.view.View
    ↳ android.view.ViewGroup
        ↳ android.widget.AbsoluteLayout
```

AbsoluteLayout 视图的特点是可以向其添加多个子视图，但这些子视图需要设置自己在父视图中的坐标。AbsoluteLayout 视图的坐标系的原点是左上角，水平向右的方向是坐标系的 x-轴，垂直向下的方向是坐标系的 y-轴。

1. 通过 layout_x 与 layout_y 设置子视图的坐标

AbsoluteLayout 视图的子视图使用 layout_x 和 layout_y 的属性值设置自己视图的左上角在父视图的坐标系中的坐标。layout_x 和 layout_y 的属性值必须是度量值（有关知识点参见 3.1 节），例如：

```
layout_x = "20dp"
layout_y = "28dp"
```

与 LinearLayout，RelativeLayout 等视图相比，AbsoluteLayout 视图缺少灵活性，其原因是子视图的位置是绝对固定的（这也是 AbsoluteLayout 视图名字的主要含义），不会随着父视图的大小的变化发生变化。

2. 示例

魔板游戏是一款经典的智力游戏。魔板由 3×3 格子组成，在前 8 个盒子里随机放置 8 个整数，最后一个盒子是未放置整数的盒子。如果一个盒子里没有放置整数，就称为空盒子。单击任何与空盒子水平或垂直相邻的盒子可以把该盒子中的整数移入到空盒子，使得当前盒子变成为空盒子。通过不断将整数移动到空盒子，使得魔板中整数排列的顺序如图 4.8 所示。以下例子 4-7 在 AbsoluteLayout 视图中用不透明的 Button 视图表示不空的盒子，用透明的 Button 视图表示空盒子，实现了魔板游戏，效果如图 4.9 所示。

图 4.8 排列顺序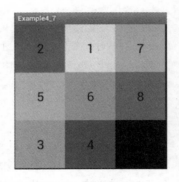

图 4.9 魔板游戏

例子 4-7

(1) 创建名字为 ch4_7 的工程,主要 Activity 子类的名字为 Example4_6,使用的包名为 ch4.seven。用命令行进入 D:\2000,创建工程 D:\2000＞android create project -t 3 -n ch4_7 -p ./ch4_7 -a Example4_7 -k ch4.seven。

(2) 增加值资源。由于许多视图的许多属性值都是一样的,如果为每个视图都设置这些属性的值,就会显得视图文件很冗余,因此需要建立一个和样式有关的值资源,将下列 XML 文件保存到值资源中(有关知识点参见 3.14 节)。

style.xml

```xml
<?xml version = "1.0" encoding = "utf-8"?>
<resources>
    <style name = "myStyle">
        <item name = "android:textSize">26dp</item>
        <item name = "android:layout_width">100dp</item>
        <item name = "android:layout_height">100dp</item>
    </style>
</resources>
```

(3) 将下列和视图相关的 XML 文件保存到工程的\res\layout 目录中。

ch4_7.xml

```xml
<?xml version = "1.0" encoding = "utf-8"?>
<AbsoluteLayout xmlns:android = "http://schemas.android.com/apk/res/android"
    android:layout_width = "match_parent"
    android:layout_height = "match_parent"
    android:id = "@+id/myLayout">
    <Button android:text = "1"
            android:onClick = "moveDigit"
            android:background = "#DD0055"
            android:layout_x = "0dp"
            android:layout_y = "0dp"
            style = "@style/myStyle" />
    <Button android:text = "2"
            android:onClick = "moveDigit"
```

```xml
            android:background = "#FFFF00"
            android:layout_x = "100dp"
            android:layout_y = "0dp"
            style = "@style/myStyle"/>
    <Button android:text = "3"
            android:onClick = "moveDigit"
            android:background = "#77CEEA"
            android:layout_x = "200dp"
            android:layout_y = "0dp"
            style = "@style/myStyle" />
    <Button android:text = "4"
            android:onClick = "moveDigit"
            android:background = "#87FF90"
            android:layout_x = "0dp"
            android:layout_y = "100dp"
            style = "@style/myStyle" />
    <Button android:text = "5"
            android:onClick = "moveDigit"
            android:background = "#00FF00"
            android:layout_x = "100dp"
            android:layout_y = "100dp"
            style = "@style/myStyle" />
    <Button android:text = "6"
            android:onClick = "moveDigit"
            android:background = "#FF00CC"
            android:layout_x = "200dp"
            android:layout_y = "100dp"
            style = "@style/myStyle" />
    <Button android:text = "7"
            android:onClick = "moveDigit"
            android:background = "#F29956"
            android:layout_x = "0dp"
            android:layout_y = "200dp"
            style = "@style/myStyle"/>
    <Button android:text = "8"
            android:onClick = "moveDigit"
            android:background = "#0055FF"
            android:layout_x = "100dp"
            android:layout_y = "200dp"
            style = "@style/myStyle" />
    <Button android:text = "empty"
            android:onClick = "moveDigit"
            android:background = "#AA15FF"
            android:alpha = "0"
            android:layout_x = "200dp"
            android:layout_y = "200dp"
            style = "@style/myStyle" />
</AbsoluteLayout>
```

（4）修改工程\src\ch4\seven 目录下的 Example4_7.java 文件，修改后的内容如下：

Example4_7.java

```java
package ch4.seven;
import android.app.Activity;
import android.os.Bundle;
import android.widget.*;
import android.view.*;
import java.util.*;
public class Example4_7 extends Activity {
    AbsoluteLayout layout;
    Button numberButton,noNumberButton;
    public void onCreate(Bundle savedInstanceState) {
        super.onCreate(savedInstanceState);
        setContentView(R.layout.ch4_7);
        layout = (AbsoluteLayout)findViewById(R.id.myLayout);
        ArrayList<Integer> list = new ArrayList<Integer>();
        for(int i = 1;i < 9;i++) {
            list.add(new Integer(i));
        }
        for(int i = 0;i < 8;i++) {
            Button b = (Button)layout.getChildAt(i);
            int index = (int)(Math.random() * list.size());
            Integer number = list.remove(index);
            b.setText("" + number.intValue());
        }
    }
    public void moveDigit(View v) {
        numberButton = (Button)v;
        String number = numberButton.getText().toString();
        float x0 = numberButton.getX();
        float y0 = numberButton.getY();
        int width = numberButton.getWidth();
        int height = numberButton.getHeight();
        for(int i = 0;i < 9;i++) {
            Button b = (Button)layout.getChildAt(i);
            if(b.isOpaque() == false) {  //如果b是透明的,b就是没有数字的空盒子
                noNumberButton = b;
                break;
            }
        }
        float x1 = noNumberButton.getX();
        float y1 = noNumberButton.getY();
        if(Math.abs(x1 - x0)<= width&&Math.abs(y1 - y0)<= 1) {
            noNumberButton.setText(number);
            noNumberButton.setAlpha(1);     //空盒子变成不空的盒子(不透明)
            numberButton.setAlpha(0);       //当前有数字的盒子变成空盒子
        }
        if(Math.abs(y1 - y0)<= width&&Math.abs(x1 - x0)<= 1) {
            noNumberButton.setText(number);
            noNumberButton.setAlpha(1);     //空盒子变成不空的盒子(不透明)
```

```
            numberButton.setAlpha(0);      //当前有数字的盒子变成空盒子(透明)
        }
    }
}
```

（5）启动 AVD,进入工程的根目录,用快捷方式编译工程、安装应用程序到 AVD(有关知识点参见 1.5 节)。对于本例子,用命令行进入 D:\2000\ch4_7,执行如下命令：

```
D:\2000 > ch4_7 > ant debug install
```

习　题　4

1. 编写一个程序,该程序使用的 XML 视图文件提供了 ScrollView 视图和 LinearLayout 视图,LinearLayout 视图是 ScrollView 视图的子视图。在代码部分再向 LinearLayout 视图添加 20 个 Button 视图。

2. 编写一个程序,该程序使用的 XML 视图文件提供了 RelativeLayout 视图,该 RelativeLayout 视图中有 5 个 Button 视图,请使用相对位置安排这 5 个 Button 视图。

3. 编写一个程序,该程序使用的 XML 视图文件提供了 TableLayout 视图,该 TableLayout 视图中有 6 个 Button 视图,请使用 3 行 2 列安排这 6 个 Button 视图。

4. TabHost 视图中的 TabWidget 视图和 FrameLayout 视图的 id 必须取怎样的值? 为何做这样强制的规定?

5. 编写一个程序,该程序使用的 XML 视图文件提供了 GridLayout 视图,该 GridLayout 视图中有 12 个 Button 视图,请使用 3 行 4 列安排这 12 个 Button 视图。

6. FrameLayout 视图的特点是怎样的?

第 5 章　　常用的专用 View 视图

主要内容：
- DigitalClock 视图、AnalogClock 视图与 CalendarView 视图；
- DatePicker 视图与 TimePicker 视图；
- ImageView 视图与 ImageButton 视图；
- Chronometer 视图；
- Toast 视图；
- ProgressBar 视图；
- VideoView 视图。

本章讲解一些专用 View 视图，比如，显示日历的 CalendarView 视图，显示图像的 ImageView 视图、播放视频/音频的 VideoView 视图，显示网页的 WebView 视图等。

在本章中需要把和视图相关的 XML 文件放在工程的\res\layout 目录中。程序的 Java 代码使用系统提供的 R 类引用这个 XML 文件来构建视图，例如，假设 XML 文件是 tom.xml，引用方式如下：

```
setContentView(R.layout.tom);
```

另外，需要再次强调的是，XML 文件和 Java 文件不同，默认的是 UTF-8 编码，因此在保存 XML 文件时必须将编码选择为"UTF-8"、保存类型选择为"所有文件"。视图相关的 XML 文件的名字只能使用小写英文字母(a~z)，数字(0~9)，下划线(_)，不可以使用大写的英文字母。

注：在进行 View 视图设计时，我们重点介绍视图的功能用途和常用的重要属性，也不罗列有关类的方法，建议经常查看 docs 帮助文档(参见 1.6 节)

5.1　DigitalClock 视图、AnalogClock 视图与 CalendarView 视图

CalendarView 视图的继承关系如下：

```
android.view.ViewGroup
    ↳ android.widget.FrameLayout
        ↳ android.widget.CalendarView
```

CalendarView 视图的特点是将日历全部显示在自己的区域中。CalendarView 视图初始状态是显示当前月的日历，用户可以方便地通过触摸屏幕、向下拖动看下个月的日历、向

上拖动看上个月的日历。

DigitalClock 视图的继承关系如下：

android.view.View
　└ android.widget.TextView
　　└ android.widget.DigitalClock

DigitalClock 视图的特点是用文本方式显示手机当前时间中的时、分、秒，显示的格式是 hh:mm:ss。

AnalogClock 视图的继承关系如下：

android.view.View
　└ android.widget.AnalogClock

AnalogClock 视图的特点是用表盘方式显示手机当前时间中的时和分（没有秒）。

1. CalendarView 常用属性值以及作用

android:firstDayOfWeek：设置日历中每周的第一天是星期几。参数取值范围是 1~7，取值 1 表示每周的第一天是星期日，取值 2 表示每周的第一天是星期一，…取值 7 表示每周的第一天是星期六。

android:focusedMonthDateColor：设置日历中聚焦的日期的颜色。取值是颜色值或图像。

android:unfocusedMonthDateColor：设置日历中未聚焦的日期的颜色。取值是颜色值或图像。

android:selectedWeekBackgroundColor：设置日历中被选中的整个星期（一周）的背景色。取值是颜色值或图像。

android:weekSeparatorLineColor：设置星期之间分隔线的颜色。取值是颜色值或图像。

android:selectedDateVerticalBar：当日历中有选中的号码时，该号码的两边会出现的垂直条，该属性的取值可以改变垂直条的外观。取值是图像。

android:showWeekNumber：设置是否在日历上显示星期的序号。取值"true"或"false"。

android:shownWeekCount：设置是否在日历上显示当前月包含的星期数。取值"true"或"false"。

2. 示例

例子 5-1 使用了 DigitalClock、AnalogClock 和 CalendarView 视图，运行效果如图 5.1 所示。

例子 5-1

（1）创建名字为 ch5_1 的工程，主要 Activity 子类的名字为 Example5_1，使用的包名为 ch5.one。用命令行进入 D:\2000，创建工程 D:\2000>android create project -t 3 -n ch5_1 -p ./ch5_1 -a Example5_1 -k ch5.one。

（2）增加图像资源。准备让选中的当前日期的前后有垂直条标识，因此项目需要提供图像资源。将名字为 bar.jpg 的

图 5.1　日历与时钟视图

图像保存到图像资源中(有关知识点参见 2.7 节)。

(3) 将下列和视图相关的 XML 文件保存到工程的\res\layout 目录中。

ch5_1.xml

```xml
<?xml version = "1.0" encoding = "utf-8"?>
<LinearLayout xmlns:android = "http://schemas.android.com/apk/res/android"
    android:layout_width = "match_parent"
    android:layout_height = "match_parent"
    android:orientation = "vertical">
    <DigitalClock android:layout_width = "match_parent"
        android:layout_height = "wrap_content"
        android:textSize = "28dp"
        android:background = "#888888"
        android:textColor = "#0000FF"
        android:gravity = "center" />
    <AnalogClock  android:background = "#008888"
        android:layout_width = "wrap_content"
        android:layout_height = "wrap_content"
        android:layout_gravity = "center"/>
    <CalendarView  xmlns:android = "http://schemas.android.com/apk/res/android"
        android:layout_width = "match_parent"
        android:layout_height = "match_parent"
        android:background = "#aaaaaa"
        android:firstDayOfWeek = "1"
        android:showWeekNumber = "true"
        android:focusedMonthDateColor = "#FF0000"
        android:selectedWeekBackgroundColor = "#cccccc"
        android:weekSeparatorLineColor = "#0000FF"
        android:selectedDateVerticalBar = "@drawable/bar"
        android:padding = "15dp" />
</LinearLayout>
```

(4) 修改工程\src\ch5\one 目录下的 Example5_1.java 文件，修改后的内容如下：

Example5_1.java

```java
package ch5.one;
import android.app.Activity;
import android.os.Bundle;
public class Example5_1 extends Activity {
    public void onCreate(Bundle savedInstanceState) {
        super.onCreate(savedInstanceState);
        setContentView(R.layout.ch5_1);
    }
}
```

(5) 启动 AVD，进入工程的根目录，用快捷方式编译工程、安装应用程序到 AVD(有关知识点参见 1.5 节)。对于本例子，用命令行进入 D:\2000\ch5_1，执行如下命令：

```
D:\2000>ch5_1>ant debug install
```

5.2　DatePicker 视图与 TimePicker 视图

DatePicker 与 TimePicker 提供为程序选择日期和时间的视图。使用 DatePicker 与 TimePicker 视图的好处是便于程序用统一的格式表示时间。

DatePicker 视图的继承关系如下：

```
android.view.ViewGroup
    └ android.widget.FrameLayout
        └ android.widget.DatePicker
```

使用 DatePicker 可以选择年、月、日。

TimePicker 视图的继承关系如下：

```
android.view.ViewGroup
    └ android.widget.FrameLayout
        └ android.widget.TimePicker
```

使用 TimePicker 可以选择时、分。

注：Android 也提供了更加方便的选择日期和时间的对话框（参见 6.8 节）。

1. DataPicker 视图的常用属性值以及作用

- android:calendarViewShown：确定是否用 CalendarView 子视图提供日期的选择。取值"true"或"false"。用户可以编辑下拉列表中的月和日。
- android:spinnersShown：确定是否用下拉列表视图提供日期的选择。取值"true"或"false"。用户可以编辑下拉列表中的年、月、日。
- android:startYear：确定可以选择的开始年。
- android:endYear：确定可以选择的结束年。

2. 日期与时间选择事件

用户选择日期(年、月、日)或时间(时、分)会触发日期选择事件或时间选择事件，Java 代码可以通过处理日期选择事件或时间选择事件，将有关日期和时间放入到程序中。处理日期选择事件或时间选择事件的接口分别是 DatePicker.OnDateChangedListener 接口和 TimePicker.OnTimeChangedListener 接口。

可以使用 public void init（int year, int monthOfYear, int dayOfMonth, DatePicker.OnDateChangedListener onDateChangedListener）方法为 DatePicker 视图注册监视器。DatePicker.OnDateChangedListener 接口中的方法是 public void onDateChanged（DatePicker view, int year, int month, int day）。

当处理日期选择事件时，该接口中的方法的参数 view 是当前 DatePicker 视图，year、month 和 day 分别是 DatePicker 视图上选择的年、月和日。

可以使用 public void setOnTimeChangedListener（TimePicker.OnTimeChangedListener onTimeChangedListener)方法为 TimePicker 视图注册监视器。TimePicker.OnTimeChangedListener 接口中的方法是 public void onTimeChanged(TimePicker view, int hour, int minute)。

当处理时间选择事件时，该接口中的方法的参数 view 是当前 TimePicker 视图，hour、minute 分别是 TimePicker 视图上选择的时和分。

3. 示例

例子 5-2 使用了 DatePicker 视图与 TimePicker 视图，让用户选择抗日战争的爆发日期，以及学校每天下午第一节课的上课时间。运行效果如图 5.2 所示。

例子 5-2

（1）创建名字为 ch5_2 的工程，主要 Activity 子类的名字为 Example5_2，使用的包名为 ch5.two。用命令行进入 D:\2000，创建工程 D:\2000＞android create project -t 3 -n ch5_2 -p ./ch5_2 -a Example5_2 -k ch5.two。

（2）增加值资源。修改值资源中的 strings.xml 文件，修改后的内容如下：

strings.xml

```
<?xml version = "1.0" encoding = "utf-8"?>
<resources>
    <string name = "app_name">Example5_2</string>
    <string name = "date">抗日战争的爆发日期:</string>
    <string name = "time">学校下午第一节课上课时间:</string>
</resources>
```

图 5.2　选择日期和时间的视图

（3）将下列和视图相关的 XML 文件保存到工程的\res\layout 目录中。

ch5_2.xml

```
<?xml version = "1.0" encoding = "utf-8"?>
<LinearLayout xmlns:android = "http://schemas.android.com/apk/res/android"
    android:layout_width = "match_parent"
    android:layout_height = "match_parent"
    android:background = "#87CEEB"
    android:orientation = "vertical">
    <TextView
        android:layout_width = "match_parent"
        android:layout_height = "wrap_content"
        android:gravity = "left|center"
        android:textColor = "#000000"
        android:textSize = "25dp"
        android:textStyle = "bold"
        android:background = "#FF0000"
        android:text = "@string/date" />
    <DatePicker
        android:id = "@+id/selectDate"
        android:layout_width = "match_parent"
        android:layout_height = "wrap_content"
        android:background = "#19FFAF"
        android:spinnersShown = "true"
        android:endYear = "1949"
        android:startYear = "1930" />
    <TextView
```

```xml
        android:layout_width = "match_parent"
        android:layout_height = "wrap_content"
        android:gravity = "left|center"
        android:textColor = "#0000FF"
        android:textSize = "25dp"
        android:background = "#55FFE9"
        android:text = "@string/time" />
    <TimePicker
        android:id = "@+id/selectTime"
        android:layout_width = "match_parent"
        android:layout_height = "wrap_content"
        android:background = "#00FFCC" />
    <TextView
        android:id = "@+id/textDate"
        android:layout_width = "match_parent"
        android:layout_height = "wrap_content"
        android:gravity = "left|center"
        android:textColor = "#FF0000"
        android:textStyle = "bold"
        android:textSize = "25dp"
        android:background = "#555555"  />
    <TextView
        android:id = "@+id/textTime"
        android:layout_width = "match_parent"
        android:layout_height = "wrap_content"
        android:gravity = "left|center"
        android:textSize = "25dp"
        android:textColor = "#0000FF"
        android:background = "#999999"  />
</LinearLayout>
```

(4) 修改工程\src\ch5\two 目录下的 Example5_2.java 文件，修改后的内容如下：

Example5_2.java

```java
package ch5.two;
import android.app.Activity;
import android.os.Bundle;
import android.widget.*;
import android.view.*;
public class Example5_2 extends Activity implements
DatePicker.OnDateChangedListener,TimePicker.OnTimeChangedListener{
    DatePicker datePicker;
    TimePicker timePicker;
    TextView textDate,textTime;
    public void onCreate(Bundle savedInstanceState){
        super.onCreate(savedInstanceState);
        setContentView(R.layout.ch5_2);
        datePicker = (DatePicker)findViewById(R.id.selectDate);
        timePicker = (TimePicker)findViewById(R.id.selectTime);
        textDate = (TextView)findViewById(R.id.textDate);
        textTime = (TextView)findViewById(R.id.textTime);
```

```
            datePicker.init(datePicker.getYear(),
                        datePicker.getMonth(),
                        datePicker.getDayOfMonth(),this);
            timePicker.setOnTimeChangedListener(this);
        }
        public void onDateChanged (DatePicker view,int year,int month,int day){
            String str = getResources().getString(R.string.date);
            month++;     //注意 month 是 0 表示一月,是 1 表示二月…是 11 表示十二月
            textDate.setText(str + "\n" + year + "/" + month + "/" + day);

        }
        public void onTimeChanged (TimePicker view,int hour,int minute){
            String str = getResources().getString(R.string.time);
            textTime.setText(str + "\n" + hour + ":" + minute);
        }
    }
```

（5）启动 AVD，进入工程的根目录，用快捷方式编译工程、安装应用程序到 AVD(有关知识点参见 1.5 节)。对于本例子,用命令行进入 D:\2000\ch5_2,执行如下命令：

```
D:\2000 > ch5_2 > ant debug install
```

5.3 ImageView 视图与 ImageButton 视图

ImageView 视图的继承关系如下：

android.view.View
　　└ android.widget.ImageView

ImageView 视图的特点是显示图像,并可以对图像进行缩放(scaling)处理。

ImageButton 视图的继承关系如下：

android.view.View
　　└ android.widget.ImageView
　　　　└ android.widget.ImageButton

ImageButton 是 ImageView 的子类,除了继承父类的属性以外,ImageButton 上可以触发 onClick 事件,这一点和普通的 Button 视图非常类似(但 ImageButton 不是 Button 的子类)。

1. ImageView 视图的常用属性值以及作用

- android：src：确定视图上显示的图像或颜色。取值是颜色或图像资源。
- android：scaleType：控制调整图像的大小和位置。取值,"matrix","fitXY","fitStart","fitCenter","fitEnd","center","centerCrop","centerInside"。该属性的默认值是"centerInside"。

2. 示例

我们曾在 3.13 节使用 ImageButton 视图浏览图像,以下例子 5-3 使用 ImageButton 视图显示四季的图像,程序处理了 ImageButton 视图上的 onClick 事件,用户点 ImageButton

视图更换其上的图像,效果如图 5.3 所示。

例子 5-3

(1) 创建名字为 ch5_3 的工程,主要 Activity 子类的名字为 Example5_3,使用的包名为 ch5.three。用命令行进入 D:\2000,创建工程 D:\2000>android create project -t 3 -n ch5_3 -p ./ch5_3 -a Example5_3 -k ch5.three。

图 5.3　ImageView 视图

(2) 增加值资源。修改值资源中的 strings.xml 文件(有关知识点参见 2.6 节),修改后的内容如下:

strings.xml

```
<?xml version = "1.0" encoding = "utf-8"?>
<resources>
    <string name = "app_name">Example5_3</string>
    <string name = "my_text">点击图片翻看下一张图片</string>
</resources>
```

(3) 增加图像资源。项目需要显示季节相关的图像,将名字为 spring.jpg,summer.jpg,autumn.jpg 和 winter.jpg 的图像保存到图像资源中(有关知识点参见 2.7 节)。

(4) 将下列和视图相关的 XML 文件保存到工程的 \res\layout 目录中。

ch5_3.xml

```
<?xml version = "1.0" encoding = "utf-8"?>
<LinearLayout xmlns:android = "http://schemas.android.com/apk/res/android"
    android:layout_width = "match_parent"
    android:layout_height = "match_parent"
    android:orientation = "vertical">
    <TextView  android:text = "@string/my_text"
               android:textSize = "28dp"
               android:layout_width = "match_parent"
               android:layout_height = "wrap_content" />
    <FrameLayout android:id = "@ + id/frame"
        android:layout_width = "match_parent"
        android:layout_height = "match_parent">
        <ImageButton android:src = "@drawable/spring"
               android:id = "@ + id/spring"
               android:visibility = "visible"
               android:onClick = "showNext"
               android:scaleType = "centerInside"
               android:layout_width = "match_parent"
               android:layout_height = "match_parent"/>
        <ImageButton android:src = "@drawable/summer"
               android:id = "@ + id/summer"
               android:visibility = "invisible"
               android:onClick = "showNext"
               android:scaleType = "fitStart"
               android:layout_width = "match_parent"
               android:layout_height = "match_parent"/>
        <ImageButton android:src = "@drawable/autumn"
```

```xml
            android:id = "@ + id/autumn"
            android:visibility = "invisible"
            android:onClick = "showNext"
            android:scaleType = "fitEnd"
            android:layout_width = "match_parent"
            android:layout_height = "match_parent"/>
    <ImageButton   android:src = "@drawable/winter"
            android:id = "@ + id/winter"
            android:visibility = "invisible"
            android:onClick = "showNext"
            android:scaleType = "fitCenter"
            android:layout_width = "match_parent"
            android:layout_height = "match_parent"/>
    </FrameLayout>
</LinearLayout>
```

(5) 修改工程\src\ch5\three 目录下的 Example5_3.java 文件,修改后的内容如下:

Example5_3.java

```java
package ch5.three;
import android.widget.*;
import android.view.*;
import android.app.*;
import android.os.Bundle;
public class Example5_3 extends Activity {
    ImageButton []p;
    FrameLayout frame;
    int count;
    public void onCreate(Bundle savedInstanceState) {
        super.onCreate(savedInstanceState);
        setContentView(R.layout.ch5_3);
        frame = (FrameLayout)findViewById(R.id.frame);
        count = frame.getChildCount();
        p    = new ImageButton[count];
        for(int i = 0;i < count;i++) {
            p[i] = (ImageButton)frame.getChildAt(i);
        }
    }
    public void showNext(View v) {
        ImageButton imageButton   = (ImageButton)v;
        int k = 0;
        for(int i = 0;i < count;i++) {
            if(imageButton == p[i])
                k = i;
        }
        int m = (k + 1) % count;
        p[m].setVisibility(View.VISIBLE);
        for(int j = 0;j < count;j++) {
            if(j!= m)
                p[j].setVisibility(View.INVISIBLE);
        }
    }
}
```

（6）启动 AVD，进入工程的根目录，用快捷方式编译工程、安装应用程序到 AVD（有关知识点参见 1.5 节）。对于本例子，用命令行进入 D:\2000\ch5_3，执行如下命令：

D:\2000＞ch5_3＞ant debug install

5.4　Chronometer 视图

Chronometer 视图的继承关系如下：

android.view.View
　　└ android.widget.TextView
　　　　└ android.widget.Chronometer

Chronometer 视图是文本样式的计时器。

1. 使用 Chronometer 视图配合周期操作

Chronometer 视图每隔 1 秒钟"计时"一次，如果程序需要周期地进行某个操作，就可以考虑使用 Chronometer 视图。比如，程序希望每隔 3 秒钟显示一个英文字母或一张图片，那么就可以使用 Chronometer 视图。Chronometer 视图上可以触发所谓的 Tick（嘀嗒）事件，即每隔 1 秒发生一次 Tick（嘀嗒）事件，处理事件的接口是 hronometer.OnChronometerTickListener 接口，该接口中的方法是 public void onChronometerTick（Chronometer chronometer），其中的参数 chronometer 就是当前发生 Tick（嘀嗒）事件的 Chronometer 视图。

Chronometer 视图调用 public void start()方法开始"计时"，调用 public void stop()方法停止"计时"。

2. 示例

以下例子 5-4 使用 Chronometer 视图每隔 3 秒钟显示一个英文字母，效果如图 5.4 所示。

例子 5-4

（1）创建名字为 ch5_4 的工程，主要 Activity 子类的名字为 Example5_4，使用的包名为 ch5.four。用命令行进入 D:\2000，创建工程 D:\2000＞android create project -t 3 -n ch5_4 -p ./ch5_4 -a Example5_4 -k ch5.four。

图 5.4　计时器视图

（2）将下列和视图相关的 XML 文件保存到工程的\res\layout 目录中。

ch5_4.xml

```
<?xml version = "1.0" encoding = "utf-8"?>
< LinearLayout xmlns:android = "http://schemas.android.com/apk/res/android"
    android:orientation = "vertical"
    android:layout_width = "match_parent"
    android:layout_height = "match_parent">
< Chronometer
    android:id = "@ + id/myTimer"
    android:layout_width = "match_parent"
    android:layout_height = "wrap_content"
```

```
            android:textSize = "30sp"
            android:background = "#777777" />
    <TextView
            android:id = "@+id/myText"
            android:layout_width = "match_parent"
            android:layout_height = "wrap_content"
            android:textSize = "50sp"
            android:textColor = "#0000FF"
            android:background = "#eeee00" />
</LinearLayout>
```

(3) 修改工程\src\ch5\four 目录下的 Example5_4.java 文件,修改后的内容如下:

Example5_4.java

```java
package ch5.four;
import android.app.Activity;
import android.os.Bundle;
import android.widget.*;
import android.view.*;
public class Example5_4 extends Activity implements Chronometer.OnChronometerTickListener {
    Chronometer timer;
    TextView text;
    int count = 0;
    char ch = 'A';
    public void onCreate(Bundle savedInstanceState) {
        super.onCreate(savedInstanceState);
        setContentView(R.layout.ch5_4);
        timer = (Chronometer)findViewById(R.id.myTimer);
        timer.setOnChronometerTickListener (this);
        text = (TextView)findViewById(R.id.myText);
        timer.start();
    }
    public void onChronometerTick (Chronometer chronometer) {
        count++;
        if(count == 3) {
            text.setText("" + ch);
            count = 0;
            ch++;
        }
        if(ch>'Z')
            ch = 'A';
    }
}
```

(4) 启动 AVD,进入工程的根目录,用快捷方式编译工程、安装应用程序到 AVD(有关知识点参见 1.5 节)。对于本例子,用命令行进入 D:\2000\ch5_4,执行如下命令:

```
D:\2000>ch5_4>ant debug install
```

5.5 Toast 视图

1. 视图的作用

Toast 并不是 View 类的子类,但也起着视图的作用,习惯称 Toast 为漂浮的提示条。Toast 类的继承关系如下:

```
java.lang.Object
    └ android.widget.Toast
```

Toast 漂浮条可以漂浮在 activity 上,起着提示信息的作用。Toast 漂浮条永远不能获得输入焦点,因此不接受输入操作。

可以用 Toast 的下述类方法返回一个 Toast 漂浮条:

public static Toast makeText (Context context, int resId, int duration);
public static Toast makeText (Context context, CharSequence text, int duration);

- 参数 context 是上下文对象,通常是应用程序或者 Activity 对象。
- 参数 resId 是值资源,例如 R. string. my_mess,是 Toast 视图显示的文本内容。
- 参数 text 就是要显示的文本内容。
- 参数 duration 是 Toast 视图持续漂浮的时间,可以取两个常量 Toast. LENGTH_SHORT 和 Toast. LENGTH_LONG。

Toast 漂浮条调用 show()方法将自己漂浮出来,调用 setGravity (int gravity, int xOffset, int yOffset)方法可以设置在 activity 上的漂浮位置。

2. 示例

下面的例子 5-5 中,用户在一个文本框中编辑信息,但需要经常输入一个容易打错的单词(economic)。程序使用 Toast 漂浮条提示这个单词,办法是每当用户单击按钮时 Toast 漂浮条就漂浮出来。运行效果如图 5.5 所示。

图 5.5 提示信息

例子 5-5

(1) 创建名字为 ch5_5 的工程,主要 Activity 子类的名字为 Example5_5,使用的包名为 ch5. five。用命令行进入 D:\2000,创建工程 D:\2000＞android create project -t 3 -n ch5_5 -p ./ch5_5 -a Example5_5 -k ch5. five。

(2) 增加和修改值资源。修改值资源中的 strings. xml 文件,修改后的内容见如下的 strings. xml。

strings. xml

```
<?xml version = "1.0" encoding = "utf-8"?>
<resources>
    <string name = "app_name">Example5_5</string>
    <string name = "word">economic(经济的)</string>
</resources>
```

(3) 增加视图资源。将下列和视图相关的 XML 文件保存到工程的\res\layout 目录中。

ch5_5.xml

```xml
<?xml version = "1.0" encoding = "utf-8"?>
<LinearLayout xmlns:android = "http://schemas.android.com/apk/res/android"
    android:orientation = "vertical"
    android:layout_width = "match_parent"
    android:layout_height = "match_parent"
    android:background = "#87CEEB"
    >
    <EditText
        android:layout_width = "match_parent"
        android:layout_height = "wrap_content" />
    <Button
        android:layout_width = "wrap_content"
        android:layout_height = "wrap_content"
        android:text = "提示一下吧"
        android:onClick = "showWord" />
</LinearLayout>
```

(4) 修改工程\src\ch5\five 目录下的 Example5_5.java 文件,修改后的内容如下。

Example5_5.java

```java
package ch5.five;
import android.app.Activity;
import android.os.Bundle;
import android.widget.*;
import android.view.*;
public class Example5_5 extends Activity {
    Toast toastOne, toastTwo;
    public void onCreate(Bundle savedInstanceState) {
        super.onCreate(savedInstanceState);
        setContentView(R.layout.ch5_5);
        toastOne = Toast.makeText (this, R.string.word, Toast.LENGTH_LONG);
        toastTwo = Toast.makeText (this, "inflation", Toast.LENGTH_SHORT);
        toastOne.setGravity(Gravity.TOP, 10, 30);
        toastTwo.setGravity(Gravity.TOP, -100, 30);
    }
    public void showWord(View v) {
        toastOne.show();
        toastTwo.show();
    }
}
```

(5) 启动 AVD,进入工程的根目录,用快捷方式编译工程、安装应用程序到 AVD(有关知识点参见 1.5 节)。对于本例子,用命令行进入 D:\2000\ch5_5,执行如下命令:

```
D:\2000>ch5_5>ant debug install
```

5.6 ProgressBar 视图

ProgressBar 视图继承关系如下：

```
android.view.View
    └ android.widget.ProgressBar
```

ProgressBar 提供显示某些操作执行进度的视图，即所谓的进度条。

1. ProgressBar 视图的常用属性

android:indeterminate：设置进度条的形状是否是圆形或水平条形，取值"false"是圆形，取值"true"是水平条形。水平条形通常用于可以给出进度情况的操作，圆形通常用于不能准确给出进度情况的操作。

android:max：该属性值确定进度条的最大值，取值正整数。对于水平进度条，其意义是将 ProgressBar 视图的水平空间平分为 max 份。

android:progress：属性值确定进度条的进度值，取值范围是 0 与进度条的最大值之间的整数。

2. ProgressBar 视图的样式

系统为 ProgressBar 视图提供了如下样式。

- Widget.ProgressBar.Horizontal：水平样式。
- Widget.ProgressBar.Small：小圆形样式。
- Widget.ProgressBar.Large：大圆形样式。
- Widget.ProgressBar.Inverse：反色样式（如果视图背景是浅颜色，则最好使用该样式）。
- Widget.ProgressBar.Small.Inverse：反色小圆形样式。
- Widget.ProgressBar.Large.Inverse：反色大圆形样式。

如果希望使用水平的 ProgressBar 视图，可以在视图的 XML 中包含如下设置：

```
Style = "@android:style/Widget.ProgressBar.Horizontal"
```

例如：

```
< ProgressBar
    android:layout_width = "wrap_content"
    android:layout_height = "wrap_content"
    style = "@android:style/Widget.ProgressBar.Horizontal" />
```

3. 示例

以下例子 5-6 使用了 ProgressBar 视图显示了得到随机数的进度，程序使用 Chronometer（计时器）每隔 1 秒得到一个 1~100 之间的随机数，如果该随机数小于 50，进度条就前进一个单位，进度条显示得到 10 个小于 50 的随机数的进度，效果如图 5.6 所示。

例子 5-6

（1）创建名字为 ch5_6 的工程，主要 Activity 子类的名字为 Example5_6，使用的包名为 ch5.six。用命令行进入 D:\

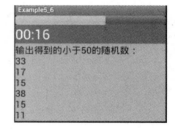

图 5.6　ProgressBar 视图

2000，创建工程 D：\2000＞android create project -t 3 -n ch5_6 -p ./ch5_6 -a Example5_6 -k ch5.six。

（2）将下列和视图相关的 XML 文件保存到工程的\res\layout 目录中。

ch5_6.xml

```xml
<LinearLayout  xmlns:android="http://schemas.android.com/apk/res/android"
    android:orientation="vertical"
    android:layout_width="match_parent"
    android:layout_height="match_parent">
    <ProgressBar
        android:id="@+id/my_progressBar"
        style="@android:style/Widget.ProgressBar.Horizontal"
        android:layout_width="match_parent"
        android:layout_height="wrap_content"
        android:max="10"/>
    <Chronometer
        android:id="@+id/myTimer"
        android:layout_width="match_parent"
        android:layout_height="wrap_content"
        android:textSize="30sp"
        android:background="#777777" />
    <ScrollView android:layout_width="match_parent"
        android:layout_height="match_parent"
        android:scrollbarStyle="outsideOverlay"  >
        <TextView
            android:id="@+id/my_text"
            android:layout_width="match_parent"
            android:layout_height="match_parent"
            android:background="#87CEEB"
            android:textSize="20sp"
            android:textColor="#000000"
            android:text="输出得到的小于50的随机数：" />
    </ScrollView>
</LinearLayout>
```

（3）修改工程\src\ch5\six 目录下的 Example5_6.java 文件，修改后的内容如下：

Example5_6.java

```java
package ch5.six;
import android.app.Activity;
import android.os.Bundle;
import android.widget.*;
import android.view.*;
public class Example5_6 extends Activity implements Chronometer.OnChronometerTickListener {
    TextView text;
    ProgressBar bar;
    Chronometer timer;
    int number=0,count=0;
    public void onCreate(Bundle savedInstanceState) {
        super.onCreate(savedInstanceState);
```

```
        setContentView(R.layout.ch5_6);
        text = (TextView)findViewById(R.id.my_text);
        bar = (ProgressBar)findViewById(R.id.my_progressBar);
        timer = (Chronometer)findViewById(R.id.myTimer);
        timer.setOnChronometerTickListener (this);
        timer.start();
    }
    public void onChronometerTick (Chronometer chronometer) {
        number = (int)(Math.random() * 100) + 1;
        if(number < 50) {
           count++;
           bar.setProgress (count) ;
           text.append("\n" + number);
        }
        if(count == 10)
            timer.stop();
    }
}
```

（4）启动 AVD,进入工程的根目录,用快捷方式编译工程、安装应用程序到 AVD(有关知识点参见 1.5 节)。对于本例子,用命令行进入 D:\2000\ch5_6,执行如下命令：

D:\2000 > ch5_6 > ant debug install

5.7　VideoView 视图

VideoView 视图继承关系如下：

android.view.View
　　↳ android.view.SurfaceView
　　　　↳ android.widget.VideoView

VideoView 提供播放视频或音频的视图。

1. VideoView 视图的几个重要方法

public void setVideoURI (Uri uri)：设置要播放的网络视频/音频的 Uri。

public void setVideoPath (String path)：设置要播放的本地视频/音频的位置。

public void setMediaController：设置播放器的控制条。用户单击 VideoView 视图可以看到播放器的控制条。

public void start()：开始播放。

public void stop()：停止播放。

public void pause()：暂停播放。

public void resume ()：恢复播放。

2. 播放网络、手机上的视频或 MP3

要播放网络上的视频的话,视频应该是渐进流式的,格式一般是 H.263 或者 H.264 格式、扩展名为 3gp 或者 mp4 的视频文件,或者是 MPEG4 SP 格式、扩展名为 3gp 的文件。例如,http://f3.3g.56.com/15/15/JGfMspPbHtzoqpzseFTPGUsKCEqMXFTW_smooth.

3gp 就是一个符合要求的视频。

假设在 Java 代码中获得了如下 VideoView 视图：

```
VideoView videoView = (VideoView)findViewById(R.id.videoView);
```

如果要播放网络上的视频/音频，首先要解析出一个 Uri，即使用 Uri 调用它的类方法 parse(String s) 返回一个 Uri 对象，例如：

```
String s = "http://f3.3g.56.com/15/15/JGfMspPbHtzoqpzseFTPGUsKCEqMXFTW_smooth.3gp";
Uri uri = Uri.parse(s);
```

然后 VideoView 视图调用 setVideoURI(Uri uri) 设置要播放的网络视频/音频的 Uri：

```
VideoView.setVideoURI(uri);
```

如果要播放手机上的视频/音频，VideoView 视图调用 setVideoPath(String path) 设置要播放的本地视频/音频的位置即可，比如播放手机 SD 卡上的视频 myVideo.mp4：

```
File f = new File("/sdcard/myVideo.mp4");
vedioView.setVideoPath(f.getAbsolutePath());
```

3. 应用程序自带视频/音频

由于播放网络上的视频/音频，需要开通 Internet 连接，这对调试使用 VideoView 视图程序不是很方便（播放手机本地的视频/音频参见 10.5 节）。Android 提供了应用程序可自带视频/音频的办法，然后用 AVD 或手机都可以播放应用程序自带的视频/音频。步骤如下。

首先在工程的资源目录 res 下建立名字为 raw 文件夹，将视频/音频文件保存在 raw 文件夹中，视频/音频文件名必须是有效的文件名（使用小写字母和数字，不要使用汉字）。

Uri 调用 parse 方法，用如下格式返回一个 Uri 对象：

```
Uri uri = Uri.parse("android.resource://项目的包名/" + R.raw.视频/MP3 文件名);
```

假设将视频文件 animal.mp4 保存到 raw 文件夹中，项目使用的包名为 ch5.six，那么返回 Uri 对象的代码是：

```
Uri uri = Uri.parse("android.resource://ch5.six/" + R.raw.animal);
```

由于程序中包含有视频/音频文件，安装程序到 AVD 或手机需要更长的时间，需要耐心等待。

4. 示例

以下例子 5-7 使用了 VideoView 视图播放 MP3 音乐，效果如图 5.7 所示。

例子 5-7

（1）创建名字为 ch5_7 的工程，主要 Activity 子类的名字为 Example5_7，使用的包名为 ch5.seven。用命令行进入 D:\2000，创建工程 D:\2000>android create project -t 3 -n ch5_7 -p ./ch5_7 -a Example5_7 -k ch5.seven。

图 5.7　VideoView 视图

（2）将名字为 pingan.mp3 的音频存放到工程的\res\raw 目录中。

（3）将下列和视图相关的 XML 文件保存到工程的\res\layout 目录中。
ch5_7.xml

```xml
<?xml version = "1.0" encoding = "utf-8"?>
<LinearLayout xmlns:android = "http://schemas.android.com/apk/res/android"
    android:orientation = "vertical"
    android:layout_width = "match_parent"
    android:layout_height = "wrap_content">
    <Button
        android:layout_width = "wrap_content"
        android:layout_height = "wrap_content"
        android:text = "开始播放平安夜音频"
        android:id = "@+id/my_button"
        android:onClick = "play"       />
    <!-- VideoView 视图: -->
    <VideoView
        android:id = "@+id/videoView"
        android:background = "#87CEEB"
        android:layout_gravity = "center"
        android:layout_width = "match_parent"
        android:layout_height = "60dp">
    </VideoView>
</LinearLayout>
```

（4）修改工程\src\ch5\seven 目录下的 Example5_7.java 文件，修改后的内容如下：
Example5_7.java

```java
package ch5.seven;
import android.app.Activity;
import android.net.Uri;
import android.os.Bundle;
import android.widget.*;
import android.view.*;
public class Example5_7 extends Activity {
    VideoView videoView;
    Uri uri;
    public void onCreate(Bundle savedInstanceState) {
        super.onCreate(savedInstanceState);
        this.setContentView(R.layout.ch5_7);
        videoView = (VideoView)findViewById(R.id.videoView);
    }
    public void play(View view) {
      try {
          uri = Uri.parse("android.resource://ch5.seven/" + R.raw.pingan);    //解析出 Uri
          videoView.setVideoURI(uri);                                         //指定需要播放的视频或音频的 Uri
          videoView.setMediaController(new MediaController(this));            //设置播放器的控制条
          videoView.start();                                                  //播放视频或音频
      }
      catch(NullPointerException exp){
      }
    }
}
```

(5) 启动 AVD,进入工程的根目录,用快捷方式编译工程、安装应用程序到 AVD(有关知识点参见 1.5 节)。对于本例子,用命令行进入 D:\2000\ch5_7,执行如下命令:

```
D:\2000 > ch5_7 > ant debug install
```

注:由于程序中含有视频/音频文件,安装程序到 AVD 或手机需要更长的时间,需要耐心等待。

5.8 WebView 视图

WebView 视图继承关系如下:

```
└ android.view.ViewGroup
        └ android.widget.AbsoluteLayout
                └ android.webkit.WebView
```

WebView 提供显示 Web 页的视图。如果程序中的 Activity 对象想在线显示某个 Web 页,就可以在 Activity 中使用 WebView 视图。WebView 视图不具备浏览器那样强大的功能,如果想使用浏览器显示 Web 页,需使用 Intent 对象启动 Android 系统内置的浏览器(参见 8.10 节)。

1. 修改配置文件

为了使得程序中的 WebView 视图能访问 Web 页,必须修改项目的配置文件 AndroidManifest.xml(在工程的根目录下),在 AndroidManifest.xml 加入如下内容:

```
< uses - permission android:name = "android.permission.INTERNET" />
```

例如:

```
< manifest ... >
    < uses - permission android:name = "android.permission.INTERNET" />
    ...
</manifest >
```

2. 加载 Web 页

WebView 视图使用 public void loadUrl (String url)方法加载参数指定的 Web 页,例如,假设视图中 WebView 视图的 id 是 webview,那么下列代码示意加载新浪网的 Web 页:

```
WebView myWebView = (WebView)findViewById(R.id.webview);
myWebView.loadUrl("http://www.sina.com");
```

3. 处理超链接

当 WebView 视图显示某个 Web 页之后,如果用户单击 Web 页中的超链接,将会导致打开系统的内置浏览器,并用内置浏览器访问超链接给出的 Web 页。如果不想用内置浏览器去访问超链接给出的 Web 页,而是继续使用 WebView 视图显示超链接给出的 Web 页,那么让 WebView 视图事先调用 void setWebViewClient(WebViewClient client)方法,并将 WebViewClient 类的实例传递给该方法的参数 client,例如可以执行如下的代码:

```
myWebView.setWebViewClient(new WebViewClient());
```

4. 支持 JavaScript

WebView 视图默认不支持 Web 页中的 JavaScript 脚本，如果准备让 WebView 视图支持 JavaScript 脚本，首先要获得 WebView 视图的 WebSettings 对象，代码如下所示：

```
WebSettings webSettings = myWebView.getSettings();
```

然后 WebSettings 对象开启对 JavaScript 脚本的支持，代码如下所示：

```
webSettings.setJavaScriptEnabled(true);
```

5. 支持 Web 页的缩放

WebView 视图默认不支持 Web 页中的缩放，如果准备让 WebView 视图支持缩放，首先要获得 WebView 视图的 WebSettings 对象，代码如下所示：

```
WebSettings webSettings = myWebView.getSettings();
```

然后 WebSettings 对象开启 WebView 视图的缩放功能，代码如下所示：

```
webSettings.setBuiltInZoomControls(true);
```

6. 示例

以下例子 5-8 使用了 WebView 视图显示了新浪网的 Web 页，效果如图 5.8 所示。

例子 5-8

（1）创建名字为 ch5_8 的工程，主要 Activity 的子类的名字为 Example5_8，使用的包名为 ch5.eight。用命令行进入 D:\2000，创建工程 D:\2000＞android create project -t 3 -n ch5_8 -p ./ch5_8 -a Example5_8 -k ch5.eight。

（2）将下列和视图相关的 XML 文件保存到工程的\res\layout 目录中。

ch5_8.xml

```
<?xml version = "1.0" encoding = "utf-8"?>
<LinearLayout xmlns:android = "http://schemas.android.com/apk/res/android"
    android:orientation = "vertical"
    android:layout_width = "match_parent"
    android:layout_height = "wrap_content">
    <LinearLayout
        android:layout_width = "match_parent"
        android:layout_height = "wrap_content">
        <EditText
            android:id = "@+id/edit"
            android:layout_width = "wrap_content"
            android:layout_height = "wrap_content"
            android:text = "http://www.sina.com" />
        <Button
            android:layout_width = "wrap_content"
            android:layout_height = "wrap_content"
```

图 5.8　WebView 视图

```xml
            android:text = "显示 Web 页"
            android:onClick = "showWeb"          />
    </LinearLayout>
    < LinearLayout
        android:layout_width = "match_parent"
        android:layout_height = "wrap_content">
     < Button
        android:layout_width = "wrap_content"
        android:layout_height = "wrap_content"
        android:text = "Back"
        android:onClick = "goBack"            />
     < Button
        android:layout_width = "wrap_content"
        android:layout_height = "wrap_content"
        android:text = "Forward"
        android:onClick = "goForward"          />
    </LinearLayout>
     < WebView
        android:id = "@ + id/webview"
        android:layout_width = "match_parent"
        android:layout_height = "match_parent" />
</LinearLayout>
```

(3) 修改工程\src\ch5\eight 目录下的 Example5_8.java 文件，修改后的内容如下：

Example5_8.java

```java
package ch5.eight;
import android.app.Activity;
import android.os.Bundle;
import android.widget. * ;
import android.view. * ;
import android.webkit.WebView;
import android.webkit.WebSettings;
import android.webkit.WebViewClient;
public class Example5_8 extends Activity {
    WebView myWebView;
    EditText edit;
    public void onCreate(Bundle savedInstanceState) {
        super.onCreate(savedInstanceState);
        setContentView(R.layout.ch5_8);
        myWebView = (WebView)findViewById(R.id.webview);
        myWebView.setWebViewClient(new WebViewClient());
        WebSettings webSettings = myWebView.getSettings();
        webSettings.setJavaScriptEnabled(true);
        webSettings.setBuiltInZoomControls(true);
        edit = (EditText)findViewById(R.id.edit);
    }
    public void showWeb(View view) {
        String url = edit.getText().toString();
        myWebView.loadUrl(url);
    }
```

```
        public void goBack(View view) {
            if(myWebView.canGoBack()) {
               myWebView.goBack () ;
            }
        }
        public void goForward(View view) {
            if(myWebView.canGoForward())
             myWebView.goForward()    ;
        }
    }
```

（4）修改项目的配置文件AndroidManifest.xml（在工程的根目录下）。在AndroidManifest.xml加入如下内容：＜uses-permission android：name＝"android.permission.INTERNET"/＞，使得程序中的WebView视图能访问Web页，如下所示：

AndroidManifest.xml

```
<?xml version = "1.0" encoding = "utf-8"?>
< manifest xmlns:android = "http://schemas.android.com/apk/res/android"
      ……>
   < uses - permission android:name = "android.permission.INTERNET" />
   < application ……
   </application>
</manifest>
```

（5）启动AVD，进入工程的根目录，用快捷方式编译工程、安装应用程序到AVD（有关知识点参见1.5节）。对于本例子，用命令行进入D:\2000\ch5_8，执行如下命令：

D:\2000＞ch5_8＞ant debug install

习　题　5

1．编写一个程序，用AnalogClock视图显示手机的当前时间。

2．编写一个程序，用TimePicker视图让用户选择银行的开门和关门时间。

3．编写一个程序，将多个ImageView视图作为LinearLayout视图的子视图，而LinearLayout又是ScrollView视图的子视图，每个ImageView视图负责显示一幅图像。

4．有6幅图像，编写程序，使用Chronometer视图周期地显示这6幅图像。

5．编写一个程序，使用VideoView视图显示一个你喜欢的网站的网页。

第6章 菜单、动作栏与对话框

主要内容：
- 菜单资源；
- 选项菜单；
- 上下文菜单；
- 弹出式菜单；
- 动作栏；
- 动作栏与选项菜单；
- AlertDialog 对话框图；
- DatePickerDialog 对话框与 TimePickerDialog 对话框；
- ProgressDialog 对话框；
- 使用 Dialog 创建对话框。

本章讲解供 Activity 使用的菜单（Menu），包括选项菜单（Options menus）、上下文菜单（Context menus）以及弹出式菜单（PopupMenu）。本章涉及的和菜单相关的 Menu 接口以及实现该接口的 ContextMenu，PopupMenu 等类均在 android.view 包中。

6.1 菜 单 资 源

本章介绍的选项菜单、弹出式菜单和上下文菜单都需要使用相关的 XML 文件来初始化菜单，因此单独用一节讲述怎样建立菜单有关的 XML 资源文件。

菜单有关的 XML 资源文件需要保存到工程的\res\menu 目录中（需要在 res 目录下建立名字为 menu 的子目录）。程序的 Java 代码使用系统提供的 R 类引用这个 XML 文件来构建菜单（具体代码参见后续内容）。

另外，需要再次强调的是，XML 文件和 Java 文件不同，默认的是 UTF-8 编码，因此在保存 XML 文件时必须将编码选择为"UTF-8"、保存类型选择为"所有文件"。菜单相关的 XML 文件的名字只能使用小写英文字母（a～z），数字（0～9），点（.）和下划线（_），不可以使用大写的英文字母。

和菜单相关的 XML 文件中包含有＜menu＞和＜item＞标记（标记也称元素），而且根标记必须是一个＜menu＞标记，名称空间的名字必须是 http://schemas.android.com/apk/res/android。

名称空间的引用名是 android。下列 my_first.xml 是一个简单的和菜单相关的 XML 文件。

my_first.xml

```xml
<?xml version = "1.0" encoding = "utf-8"?>
<menu xmlns:android = "http://schemas.android.com/apk/res/android">
    <item android:id = "@+id/cat"
          android:icon = "@drawable/cat"
          android:title = "小猫"  />
    <item android:id = "@+id/dog"
          android:icon = "@drawable/dog"
          android:title = "小狗"  />
</menu>
```

相关标记的作用

1. <menu>标记

<menu>标记负责定义菜单(Menu),<menu>可以有多个<item>子标记,即该菜单可以包含多个菜单项(MenuItem)。

2. <item>标记

<item>标记负责给出菜单中的菜单项,一个<item>标记负责给出菜单中的一个菜单项。另外,<item>标记可以把<menu>标记作为自己的一个子标记,形成子菜单。

3. <item>标记的属性

以下是<item>标记的几个常用属性。

- android:id

作用:确定菜单项的 ID 标识。在某些时候,Java 代码部分需要根据菜单项的 ID 寻找 XML 文件中给出的菜单项,以便做进一步的编码。属性取值是一个用字符串表示的整数,字符串由用户来指定,所代表的整数由系统的 R 类负责指定。给菜单项指定 ID 的格式是:

android:id = "@+id/ID 值"

- android:icon

作用:确定菜单项上的图标。属性取值是资源图像。

- android:title

作用:确定菜单项上的标题。属性取值是字符串。

6.2 选项菜单

每个 Activity 都自带一个选项菜单(Options menus),而且 Activity 只能有一个选项菜单,程序所要做的就是初始化 Activity 自带的这个选项菜单,即需要为这个选项菜单增加菜单项(MenuItem),并根据需要决定是否处理菜单项上触发的选择事件。

当用户按下手机或 ADV 上的 MENU 键后,可以在 Activity 的底部(Activity 容器的底端)看到菜单项。当菜单中有比较多的菜单项时,系统自动增加名字为"More"的菜单项,用户选"More"可以看到选项中其他的菜单项,如图 6.1(a),6.1(b)所示。

有些手机可能没有 MENU 键,这对于使用选项菜单是不合适的。Android 3.0(API level 11)之后,Android 提供了将选项菜单的菜单项放到动作栏(Action Bar)的机制,以方便地使用选项菜单。但是在默认情况下,Activity 不能使用动作栏,必须对项目的配置文件

(a) 提供More选项　　　　　　　　　　(b) 单击More显示的其余选项

图 6.1　More 选项的作用

AndroidManifest.xml（参见 2.2 节）进行必要的设置才能使用 Activity 的动作栏，因此我们将在 6.6 节讲解怎样将选项菜单的菜单项放到动作栏中。

1. Activity 加载菜单资源

Activity 的子类需要重写 public boolean onCreateOptionsMenu(Menu menu)方法，并使用菜单资源（工程\res\menu 下的 XML 文件）初始化参数 menu 给出的菜单，而参数 menu 就是 Activity 自带的选项菜单。在重写该方法时，首先要由 Activity 调用 getMenuInflater()获得一个用于初始化菜单的 MenuInflater（MenuInflater 类在 android.view 包中）对象，代码如下所示：

```
MenuInflater inflater = getMenuInflater();
```

MenuInflate 对象负责调用 void inflate(int menuRes,Menu menu)方法初始化参数 menu 指定的菜单，方法中的 menuRes 是负责初始化菜单的菜单资源（工程\res\menu 下的 XML 文件）。

假设菜单资源的 XML 文件的名字为 my_first.xml（和菜单相关的 XML 文件的有关知识参见 6.1 节），那么重写 onCreateOptionsMenu(Menu menu)方法的代码如下所示：

```
public boolean onCreateOptionsMenu(Menu menu) {
    MenuInflater inflater = getMenuInflater();
    inflater.inflate(R.menu.my_first,menu);
    return true;
}
```

用户首次按下 ADV 上（或手机）的 MENU 键，Activity 就会调用重写的 onCreateOptionsMenu(Menu menu)方法对选项菜单进行初始化，并让菜单项出现在 Activity 的底部，以后用户再次按下手机或 ADV 上的 MENU 键，Activity 不再调用重写的 onCreateOptionsMenu(Menu menu)方法，只是让菜单项出现在 Activity 对象底部。用户一旦选中某个菜单项，选项菜单的所有菜单项就都自动隐藏起来，用户想再次看见菜单项，就必须再次按下手机或 ADV 上的 MENU 键。

注：如果让 onCreateOptionsMenu(Menu menu)方法返回 false，用户按下手机或 ADV 上的 MENU 键，选项菜单就不出现在 Activity 的底部。

2. 处理菜单项上的选择事件

重写 onOptionsItemSelected 方法处理菜单项上的选择事件，可以在 Activity 的子类中重写父类的 public boolean onOptionsItemSelected（MenuItem item）方法。系统认为菜单项上的选择事件的监视器是当前的 Activity 对象。当菜单项上触发选择事件后，Activity 对象调用子类重写的 onOptionsItemSelected（MenuItem item）方法，该方法中的参数 item 就是当前触发事件的选项菜单中的菜单项。例如：

```
public boolean onOptionsItemSelected(MenuItem item) {
    //Handle item selection
    switch (item.getItemId()) {
        case R.id.cat:
            //需要处理和菜单 ID 是 cat 有关的代码
            return true;
        case R.id.dog:
            //需要处理和菜单 ID 是 dog 有关的代码
            return true;
        default:
            return super.onOptionsItemSelected(item);
    }
}
```

3. 示例

例子 6-1 使用了菜单，菜单中有 4 个菜单项：猫、狗、虎和狮，用户选择一个菜单项，程序在一个 TextView 视图显示菜单项的有关信息，在 ImageButton 视图中显示菜单项上的图标，用户按 ADV 上的 MENU 键，在 Activity 视图的底部就会出现选项菜单。运行效果如图 6.2(a)、6.2(b)所示。

(a) 按MENU键之前

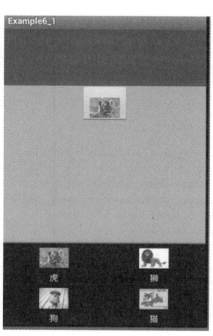
(b) 按MENU键之后

图 6.2 运行效果

例子 6-1

(1) 创建名字为 ch6_1 的工程，主要 Activity 子类的名字为 Example6_1，使用的包名为 ch6.one。用命令行进入 D:\2000，创建工程 D:\2000＞android create project -t 3 -n ch6_1 -p ./ch6_1 -a Example6_1 -k ch6.one。

(2) 增加图像资源。准备让 4 个菜单项都带有图标，将名字为 cat.jpg、dog.jpg、tiger.jpg 和 lion.jpg 的图像保存到图像资源中，建议每个图像的大小控制在 70*50 以内（有关知识点参见 2.7 节）。

(3) 将菜单资源文件 animal_menu.xml 保存到工程的 \res\menu 目录中（需要在工程的 \res 目录下建立名字为 menu 的子目录）。

animal_menu.xml

```xml
<?xml version = "1.0" encoding = "utf-8"?>
<menu xmlns:android = "http://schemas.android.com/apk/res/android">
    <item android:id = "@+id/cat"
        android:icon = "@drawable/cat"
        android:title = "猫"
        android:orderInCategory = "4" />
    <item android:id = "@+id/dog"
        android:icon = "@drawable/dog"
        android:title = "狗"
        android:orderInCategory = "3" />
    <item android:id = "@+id/tiger"
        android:icon = "@drawable/tiger"
        android:title = "虎"
        android:orderInCategory = "1" />
    <item android:id = "@+id/lion"
        android:icon = "@drawable/lion"
        android:title = "狮"
        android:orderInCategory = "2" />
</menu>
```

(4) 将下列和视图相关的 XML 文件保存到工程的 \res\layout 目录中。

ch6_1.xml

```xml
<?xml version = "1.0" encoding = "utf-8"?>
<LinearLayout xmlns:android = "http://schemas.android.com/apk/res/android"
    android:layout_width = "match_parent"
    android:layout_height = "match_parent"
    android:background = "#EFAA99"
    android:orientation = "vertical">
    <TextView
        android:id = "@+id/my_text"
        android:textColor = "#806400"
        android:layout_width = "match_parent"
        android:layout_height = "wrap_content"
        android:textSize = "25sp"
        android:background = "#AA00EE"
        android:text = "按手机或 ADV 上的 MENU 键，就会出现选择菜单"/>
```

```xml
< ImageButton
    android:id = "@ + id/imageButton"
    android:scaleType = "centerInside"
    android:layout_gravity = "center"
    android:layout_width = "75dp"
    android:layout_height = "60dp"/>
</LinearLayout >
```

(5) 修改工程\src\ch6\one 目录下的 Example6_1.java 文件,修改后的内容如下：

Example6_1.java

```java
package ch6.one;
import android.widget. * ;
import android.view. * ;
import android.app. * ;
import android.os.Bundle;
import android.graphics.drawable.Drawable;
public class Example6_1 extends Activity {
    ImageButton imageButton;
    TextView text;
    public void onCreate(Bundle savedInstanceState) {
        super.onCreate(savedInstanceState);
        setContentView(R.layout.ch6_1);
        imageButton = (ImageButton)findViewById(R.id.imageButton);
        text = (TextView)findViewById(R.id.my_text);
    }
    public boolean onCreateOptionsMenu(Menu menu) {
        MenuInflater inflater = getMenuInflater();
        inflater.inflate(R.menu.animal_menu,menu);
        return true;
    }
    public boolean onOptionsItemSelected(MenuItem item) {
        Drawable drawable;
        drawable = item.getIcon();
        imageButton.setImageDrawable(drawable);
        int index = item.getOrder() ;
        CharSequence title = item.getTitle();
        text.setText(null);
        text.append("The item's order is " + index + "\n");
        text.append("The item's title is " + title + "\n");
        return true;
    }
}
```

(6) 启动 AVD,进入工程的根目录,用快捷方式编译工程、安装应用程序到 AVD(有关知识点参见 1.5 节)。对于本例子,用命令行进入 D:\2000\ch6_1,执行如下命令：

D:\2000 > ch6_1 > ant debug install

6.3 上下文菜单

程序可以为 Activity 对象中的任何视图注册（register）一个上下文菜单（ContextMenu）。当用户在注册了上下文菜单的视图上实施了"长按"（a long-click）操作，那么相应的上下文菜单将显示自己的全部菜单项。

1. 为视图注册上下文菜单

Activity 对象使用 public void registerForContextMenu(View view)方法为视图 view 注册一个上下文菜单。例如，假设 button 是 Activity 对象中的一个按钮视图，那么为 button 注册一个上下文菜单的代码如下：

```
registerForContextMenu(button);
```

2. Activity 对象加载菜单资源

Activity 的子类需要重写 public void onCreateContextMenu(ContextMenu menu, View view, ContextMenuInfo menuInfo)方法，并使用菜单资源（工程\res\menu 下的 XML 文件）对上下文菜单 menu 进行初始化。

用户在注册了上下文菜单的视图上首次实施"长按"操作时，Activity 对象就会调用重写的 onCreateContextMenu()方法对该视图上注册的上下文菜单 menu 进行初始化，并让上下文菜单出现在屏幕上，其中参数 view 就是当前用户实施"长按"操作的视图。需要注意的是，用户再次实施"长按"操作时，Activity 对象会再次调用重写的 onCreateContextMenu()方法（这一点和选项菜单的机制不同），并让上下文菜单出现在屏幕上。用户一旦选中某个菜单项，上下文菜单就自动隐藏起来，用户想再次看见选项菜单，就必须再次实施"长按"操作。在重写 onCreateContextMenu(ContextMenu menu, View view, ContextMenuInfo menuInfo)方法时，Activity 对象调用 getMenuInflater()获得一个负责初始化菜单的 MenuInflater 对象。假设菜单资源的 XML 文件的名字为 context_menu.xml（和菜单相关的 XML 文件的有关知识参见 6.1 节），那么重写 onCreateContextMenu()方法的代码如下：

```
public void onCreateContextMenu(ContextMenu menu,View view,
                                ContextMenuInfo menuInfo) {
    MenuInflater inflater = getMenuInflater();
    inflater.inflate(R.menu.context_menu,menu);
}
```

3. 处理菜单项上的选择事件

程序通过重写 onContextItemSelected 方法来处理菜单项上的选择事件，需要在 Activity 的子类中重写父类的 public boolean onContextItemSelected(MenuItem item)方法。系统认为菜单项上的选择事件的监视器是当前的 Activity 对象。当菜单项上触发选择事件后，Activity 对象调用子类重写的 onContextItemSelected(MenuItem item)方法，该方法中的参数 item 就是当前触发事件的上下文菜单中的菜单项。例如：

```
public boolean onContextItemSelected MenuItem item) {
    switch (item.getItemId()) {
        case R.id.cat:
```

```
            //需要处理和菜单 ID 是 cat 有关的代码
                return true;
            case R.id.dog:
                //需要处理和菜单 ID 是 dog 有关的代码
                return true;
            default:
                return super.onContextItemSelected(item);
        }
    }
}
```

注：上下文菜单的菜单项上无法显示图标（菜单资源中设置的图标无效）。

4. 示例

例子 6-2 在两个 ViewText 视图分别注册了上下文菜单，用户通过一个 TextView 注册的上下文菜单可以更改两个 ViewText 视图中文字大小，通过另一个 TextView 注册的上下文菜单可以更改两个 ViewText 视图中文字颜色。运行效果如图 6.3 所示。

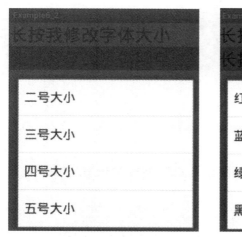

(a) 字体大小上下文菜单　　　　　(b) 字体颜色上下文菜单

图 6.3　运行效果

例子 6-2

(1) 创建名字为 ch6_2 的工程，主要 Activity 子类的名字为 Example6_2，使用的包名为 ch6.two。用命令行进入 D:\2000，创建工程 D:\2000＞android create project -t 3 -n ch6_2 -p ./ch6_2 -a Example6_2 -k ch6.two。

(2) 将下列菜单资源文件 font_size.xml 和 font_color.xml 保存到工程的\res\menu 目录中（需要在工程的\res 目录下建立名字为 menu 的子目录）。

font_size.xml

```xml
<?xml version = "1.0" encoding = "utf-8"?>
<menu xmlns:android = "http://schemas.android.com/apk/res/android">
    <item android:id = "@+id/size_5"
        android:title = "五号大小"
        android:orderInCategory = "4" />
    <item android:id = "@+id/size_4"
```

```xml
            android:title = "四号大小"
            android:orderInCategory = "3"/>
    <item android:id = "@ + id/size_3"
            android:title = "三号大小"
            android:orderInCategory = "2" />
    <item android:id = "@ + id/size_2"
            android:title = "二号大小"
            android:orderInCategory = "1" />
</menu>
```

font_color.xml

```xml
<?xml version = "1.0" encoding = "utf - 8"?>
<menu xmlns:android = "http://schemas.android.com/apk/res/android">
    <item android:id = "@ + id/red"
            android:title = "红色"
            android:orderInCategory = "1" />
    <item android:id = "@ + id/blue"
            android:title = "蓝色"
            android:orderInCategory = "2"/>
    <item android:id = "@ + id/green"
            android:title = "绿色"
            android:orderInCategory = "3" />
    <item android:id = "@ + id/black"
            android:title = "黑色"
            android:orderInCategory = "4" />
</menu>
```

（3）将下列和视图相关的 XML 文件保存到工程的\res\layout 目录中。

ch6_2.xml

```xml
<?xml version = "1.0" encoding = "utf - 8"?>
<LinearLayout xmlns:android = "http://schemas.android.com/apk/res/android"
    android:layout_width = "match_parent"
    android:layout_height = "match_parent"
    android:background = "#EFAA99"
    android:orientation = "vertical">
    <TextView
        android:id = "@ + id/textOne"
        android:layout_width = "match_parent"
        android:layout_height = "wrap_content"
        android:background = "#AABB11"
        android:textColor = "#000000"
        android:fontFamily = "Arial"
        android:text = "长按我修改字体大小" />
    <TextView
        android:id = "@ + id/textTwo"
        android:layout_width = "match_parent"
        android:layout_height = "wrap_content"
        android:background = "#EE0099"
        android:textColor = "#000000"
        android:text = "长按我修改字体颜色" />
```

</LinearLayout>

(4) 修改工程\src\ch6\two 目录下的 Example6_2.java 文件,修改后的内容如下:

Example6_2.java

```java
package ch6.two;
import android.widget.*;
import android.view.*;
import android.app.*;
import android.os.Bundle;
import android.view.ContextMenu.ContextMenuInfo;
import android.graphics.Color;
public class Example6_2 extends Activity {
    TextView textOne,textTwo;
    public void onCreate(Bundle savedInstanceState) {
        super.onCreate(savedInstanceState);
        setContentView(R.layout.ch6_2);
        textOne = (TextView)findViewById(R.id.textOne);
        textTwo = (TextView)findViewById(R.id.textTwo);
        registerForContextMenu(textOne);
        registerForContextMenu(textTwo);
    }
    public void onCreateContextMenu(ContextMenu menu, View view, ContextMenuInfo menuInfo) {
        if(view.getId() == R.id.textOne){
            if(menu.size() == 0) {
                MenuInflater inflater = getMenuInflater();
                inflater.inflate(R.menu.font_size,menu);
            }
        }
        if(view.getId() == R.id.textTwo){
            if(menu.size() == 0) {
                MenuInflater inflater = getMenuInflater();
                inflater.inflate(R.menu.font_color,menu);
            }
        }
    }
    public boolean onContextItemSelected(MenuItem item) {
        switch (item.getItemId()) {
            case R.id.size_5:
                setFontSize(16);
                return true;
            case R.id.size_4:
                setFontSize(25);
                return true;
            case R.id.size_3:
                setFontSize(30);
                return true;
            case R.id.size_2:
                setFontSize(35);
                return true;
            case R.id.red:
```

```
                setFontColor(Color.RED);
                return true;
            case R.id.blue:
                setFontColor(Color.BLUE);
                return true;
            case R.id.green:
                setFontColor(Color.GREEN);
                return true;
            case R.id.black:
                setFontColor(Color.BLACK);
                return true;
            default:
                return super.onContextItemSelected(item);
        }
    }
    void setFontSize(int size) {
        textOne.setTextSize(size);
        textTwo.setTextSize(size);
    }
    void setFontColor(int c) {
        textOne.setTextColor(c);
        textTwo.setTextColor(c);
    }
}
```

(5) 启动 AVD,进入工程的根目录,用快捷方式编译工程、安装应用程序到 AVD(有关知识点参见 1.5 节)。对于本例子,用命令行进入 D:\2000\ch6_2,执行如下命令:

```
D:\2000>ch6_2>ant debug install
```

6.4 弹出式菜单

Activity 对象中的弹出式菜单(PopupMenu)需要依托(anchored)于 Activity 对象中的某个视图。如果弹出式菜单所依托的视图的下方有足够的空间,弹出式菜单在依托的视图的下方弹出自己的全部菜单项,否则弹出式菜单在依托的视图的正前方弹出自己的全部菜单项。程序需要在代码部分通过处理某种事件让弹出式菜单弹出自己的菜单项。

1. 创建弹出式菜单

需要使用 PopupMenu 类的构造方法 PopupMenu(Context context,View anchor)创建弹出式菜单,其中参数 context 是弹出式菜单所在的上下文对象,anchor 是弹出式菜单所依托的视图。例如:

```
PopupMenu popup = new PopupMenu(this,new Button(this));
```

2. 初始化弹出式菜单

弹出式菜单调用 void inflate(int menuRes)方法,用参数 menuRes 指定的菜单资源,即与菜单相关的 XML 文件初始化弹出式菜单,例如:

```
popup.inflate(R.menu.view_color);
```

如果使用的 Android 版本低于 Android 3.0（API level 11），可以使用 MenuInflater 对象初始化菜单，例如：

```
MenuInflater inflater = popup.getMenuInflater();
inflater.inflate(R.menu.actions,popup.getMenu());
```

其中 view_color 是保存在工程的\res\menu 目录中的 XML 文件的名字（和菜单相关的 XML 文件的有关知识参见 6.1 节）。

3. 弹出菜单项

弹出式菜单调用 void show()方法弹出自己的菜单项，例如：

```
popup.show();
```

4. 处理菜单项上的选择事件

弹出式菜单的菜单项上可以触发单击菜单项事件（MenuItemClicked），处理事件的接口是 PopupMenu.OnMenuItemClickListener，该接口中的方法是 public boolean onMenuItemClick (MenuItem item)。

弹出式菜单使用 public void setOnMenuItemClickListener (PopupMenu.OnMenuItemClickListener listener)方法注册事件监视器。

注：和选项菜单不同，在一个 Activity 对象中用户可以根据需要创建多个不同的弹出式菜单。

5. 示例

例子 6-3 使用了弹出式菜单，该弹出式菜单依托的视图是一个 Button 视图，用户单击 Button 视图弹出菜单项，这些菜单项分别是"红色"、"绿色"、"蓝色"和"机器人"，用户选择某个菜单项可以改变 Button 视图的背景颜色或背景图案。运行效果如图 6.4 所示。

例子 6-3

（1）创建名字为 ch6_3 的工程，主要 Activity 子类的名字为 Example6_3，使用的包名为 ch6.three。用命令行进入 D:\2000，创建工程 D:\2000＞android create project -t 3 -n ch6_3 -p ./ch6_3 -a Example6_3 -k ch6.three。

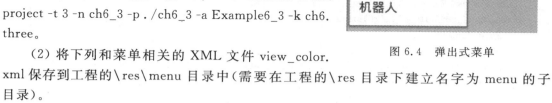

图 6.4　弹出式菜单

（2）将下列和菜单相关的 XML 文件 view_color.xml 保存到工程的\res\menu 目录中（需要在工程的\res 目录下建立名字为 menu 的子目录）。

view_color.xml

```
<?xml version = "1.0" encoding = "utf-8"?>
<menu xmlns:android = "http://schemas.android.com/apk/res/android">
    <item android:id = "@ + id/red"
        android:title = "红色"
        android:orderInCategory = "1" />
    <item android:id = "@ + id/blue"
```

```
            android:title = "蓝色"
            android:orderInCategory = "2"/>
    < item android:id = "@ + id/green"
            android:title = "绿色"
            android:orderInCategory = "3" />
    < item android:id = "@ + id/robot"
            android:title = "机器人"
            android:orderInCategory = "4" />
</menu >
```

(3) 将下列和视图相关的 XML 文件保存到工程的 \res\layout 目录中。

ch6_3.xml

```
<?xml version = "1.0" encoding = "utf - 8"?>
< LinearLayout xmlns:android = "http://schemas.android.com/apk/res/android"
    android:layout_width = "match_parent"
    android:layout_height = "match_parent"
    android:background = "#EFAA99"
    android:orientation = "vertical">
    < Button
        android:id = "@ + id/button"
        android:layout_width = "wrap_content"
        android:layout_height = "wrap_content"
        android:text = "修改背景颜色"
        android:onClick = "changeColor"/>
</LinearLayout >
```

(4) 修改工程\src\ch6\three 目录下的 Example6_3.java 文件,修改后的内容如下:

Example6_3.java

```
package ch6.three;
import android.app.*;
import android.os.Bundle;
import android.widget.*;
import android.view.*;
import android.graphics.Color;
public class Example6_3 extends Activity implements
PopupMenu.OnMenuItemClickListener {
    PopupMenu popup;
    Button button;
    public void onCreate(Bundle savedInstanceState) {
        super.onCreate(savedInstanceState);
        setContentView(R.layout.ch6_3);
        button = (Button)findViewById(R.id.button);
        popup = new PopupMenu(this,button);
        popup.inflate(R.menu.view_color);
        popup.setOnMenuItemClickListener(this);
    }
    public boolean onMenuItemClick (MenuItem item) {
        switch (item.getItemId()) {
            case R.id.red:
```

```
                button.setBackgroundColor(Color.RED);
                return true;
            case R.id.blue:
                button.setBackgroundColor(Color.BLUE);
                return true;
            case R.id.green:
                button.setBackgroundColor(Color.GREEN);
                return true;
            case R.id.robot:
                button.setBackgroundResource(R.drawable.ic_launcher);
                return true;
            default:
                return true;
        }
    }
    public void changeColor(View view) {
        popup.show();
    }
}
```

（5）启动 AVD，进入工程的根目录，用快捷方式编译工程、安装应用程序到 AVD（有关知识点参见 1.5 节）。对于本例子，用命令行进入 D:\2000\ch6_3，执行如下命令：

D:\2000>ch6_3>ant debug install

6.5 动 作 栏

Android 3.0（API level 11）版本之后，系统为 Activity 对象提供了动作栏（ActionBar），主要目的有两个，一是使用动作栏制作导航条，即将起导航作用的 Tab 选项卡放在动作栏中，方便程序在各个模块之间进行切换，二是将选项菜单的菜单项放置在动作栏中，方便用户使用选项菜单中的菜单项（参见 6.1 节和 6.6 节）。

当 Activity 对象开启使用动作栏后，动作栏将出现在 Activity 对象视图区的上方。动作栏有图标和标题（在动作栏的左侧或上方），如果不特意设置它们的话，动作栏的图标/标题与应用程序的图标/标题相同。动作栏中可以放置 Tab 选项卡或选项菜单的菜单项（参见 6.6 节）。Android 4.1（API level 16）版本之后，当动作栏中的 Tab 选项卡较多时，允许用户左、右拖移动作栏。

1. 开启动作栏

默认情况下，Activity 对象不能使用动作栏。对项目的配置文件（AndroidManifest.xml，参见 2.2 节）进行额外的设置后，就能使用 Activity 对象的动作栏。

在 AndroidManifest.xml 配置文件的＜Activity＞标记中给出 android:theme 属性的值，该值是"@android:style/Theme.Holo"，表示开启该 Activity 对象的动作栏。例如：

AndroidManifest.xml

```
<?xml version = "1.0" encoding = "utf - 8"?>
<manifest xmlns:android = "http://schemas.android.com/apk/res/android"
```

```
    ...
    <application   ... >
        <!--在 activity 标记中给出 android:theme 的值,启用动作栏:-->
        <activity
            android:theme = "@android:style/Theme.Holo"

        >
        </activity>
        ...
    </application>
</manifest>
```

如果希望应用程序中的所有 Activity 对象都可以使用动作栏,可以在 AndroidManifest.xml 配置文件的＜application＞标记中给出 android:theme 属性的值,例如:

AndroidManifest.xml

```
<?xml version = "1.0" encoding = "utf-8"?>
<manifest xmlns:android = "http://schemas.android.com/apk/res/android"
    ... >
    <application   ...
        <!-- 在 application 标记中给出 android:theme 的值,启用所有 activity 的动作栏:-->
        android:theme = "@android:style/Theme.Holo"   >
        <activity >
        </activity>
        ...
    </application>
</manifest>
```

2. 向动作栏增加 Tab 选项卡

1) 获得动作栏

ActionBar 类在 android.app 包中,Activity 对象可以使用 getActionBar()得到系统提供的 ActionBar 对象,即得到动作栏。得到动作栏之后,动作栏默认可见,如果想隐藏动作栏,可以让动作栏调用 hide()方法,想再显示动作栏可以让动作栏调用 show()方法。

动作栏主要用于放置 Tab 选项卡,因此,获得动作栏之后,需要将动作栏的模式设置成 NAVIGATION_MODE_TABS 模式,如下所示:

```
ActionBar actionBar = getActionBar();
actionBar.setNavigationMode(ActionBar.NAVIGATION_MODE_TABS);
```

2) 获得 Tab 选项卡

ActionBar.Tab 类(Tab 类是 ActionBar 的静态内部类)的实例称为 Tab 选项卡。动作栏调用 newTab()方法可以返回一个 Tab 选项卡,如下所示:

```
ActionBar.Tab tab = actionBar.newTab();
```

要求 newTab()方法返回的 Tab 选项卡必须注册有 TabListener 监视器,即要求 Tab 选项卡能对用户选择它作出某种响应(比如能担当导航作用)。

Tab 选项卡注册监视器的方法是:

```
setTabListener(ActionBar.TabListener listener);
```

ActionBar.TabListener 接口(TabListener 接口是 ActionBar 类中的静态接口)中有如下三个方法。

- public void onTabSelected(ActionBar.Tab tab,FragmentTransaction ft),用户将未选中的 Tab 选项卡变为选中状态时,监视器调用该方法,参数 tab 是当前用户操作的 Tab 选项卡。
- public void onTabReselected(ActionBar.Tab tab,FragmentTransaction ft),用户释放选中的 Tab 选项卡时,监视器调用该方法,参数 tab 是当前用户操作的 Tab 选项卡。
- public void onTabUnselected(ActionBar.Tab tab,FragmentTransaction ft),用户将选中的 Tab 选项卡变为未选中状态时,监视器调用该方法,参数 tab 是当前用户操作的 Tab 选项卡。

3) 向动作栏添加 Tab 选项卡

动作栏可以调用 addTab(ActionBar.Tab tab)方法添加参数指定的 Tab 选项卡,例如:

```
actionBar.addTab(tab);
```

需要特别注意的是,如果动作栏添加了一个没有注册 TabListener 监视器的 Tab 选项卡,那么项目可以通过编译(Debug),但会发生运行异常,系统将终止程序的运行。

3. ActionBar 的常用方法

void addTab(ActionBar.Tab tab,boolean setSelected):添加 Tab 选项卡,并指定状态。

void addTab(ActionBar.Tab tab,int position):添加 Tab 选项卡,并指定位置,位置索引从 0 开始。

int getTabCount():返回其中的 Tab 选项卡的数目。

void setIcon(int resId) 与 void setIcon(Drawable icon):设置动作栏的图标。

void setTitle(int resId) 与 void setTitle(CharSequence title):设置动作栏的标题。

void setDisplayShowTitleEnabled(boolean showTitle):设置是否显示动作栏的标题。

void setDisplayShowHomeEnabled(boolean showHome):设置是否显示动作栏的图标。当取值 false 时,动作栏将出现在最顶端。

4. Tab 的常用方法

ActionBar.Tab setIcon(Drawable icon) 与 setIcon(int resId):设置选项卡上的图标。

ActionBar.Tab setTabListener(ActionBar.TabListener listener):为选项卡注册监视器。

ActionBar.Tab setTag(Object obj):为选项卡设置一个备用对象。

ActionBar.Tab setText(int resId) 与 setText(CharSequence text):设置选项卡上的文本。

5. 示例

例子 6-4 使用动作栏,用户选中动作栏上的 Tab 选项卡,程序显示 Tab 选项卡的有关信息。运行效果如图 6.5 所示。

例子 6-4

(1) 创建名字为 ch6_4 的工程,主要 Activity 子类的名字为 Example6_4,使用的包名为 ch6.four。用命令行进入 D:\2000,创建工程 D:\2000>android create project -t 3 -n ch6_4 -p ./ch6_4 -a Example6_4 -k ch6.four。

图 6.5 动作栏

（2）修改 AndroidManifest.xml 配置文件。打开工程根目录（对于本例子，就是\ch6_4 目录）下的 AndroidManifest.xml 配置文件，在 AndroidManifest.xml 配置文件的＜activity＞标记中给出 android:theme 属性的值 android:theme＝"@android:style/Theme.Holo"，然后用另存方式替换原有的 AndroidManifest.xml（注意，另存时，编码要选择成"UTF-8"）。修改后的 AndroidManifest.xml 内容如下：

AndroidManifest.xml

```xml
<?xml version = "1.0" encoding = "utf-8"?>
<manifest xmlns:android = "http://schemas.android.com/apk/res/android"
    package = "ch6.four"
    android:versionCode = "1"
    android:versionName = "1.0">
    <application android:label = "@string/app_name" android:icon = "@drawable/ic_launcher">
        <activity android:name = "Example6_4"
            android:theme = "@android:style/Theme.Holo"
            android:label = "@string/app_name">
            <intent-filter>
                <action android:name = "android.intent.action.MAIN" />
                <category android:name = "android.intent.category.LAUNCHER" />
            </intent-filter>
        </activity>
    </application>
</manifest>
```

（3）将下列和视图相关的 XML 文件保存到工程的\res\layout 目录中。

ch6_4.xml

```xml
<?xml version = "1.0" encoding = "utf-8"?>
<LinearLayout xmlns:android = "http://schemas.android.com/apk/res/android"
    android:layout_width = "match_parent"
    android:layout_height = "match_parent"
    android:background = "#EFAA99"
    android:orientation = "vertical">
    <TextView
        android:id = "@+id/text"
        android:layout_width = "match_parent"
        android:layout_height = "wrap_content"
        android:textSize = "20sp"
        android:background = "#E910AA"
        android:textColor = "#000000"/>
</LinearLayout>
```

（4）修改工程\src\ch6\four 目录下的 Example6_4.java 文件，修改后的内容如下：

Example6_4.java

```java
package ch6.four;
import android.app.*;
import android.os.Bundle;
import android.widget.*;
```

```java
import android.view.*;
public class Example6_4 extends Activity implements ActionBar.TabListener {
    ActionBar actionBar;
    ActionBar.Tab [] tab;
    TextView text;
    public void onCreate(Bundle savedInstanceState) {
        super.onCreate(savedInstanceState);
        actionBar = getActionBar();
        actionBar.setNavigationMode(ActionBar.NAVIGATION_MODE_TABS);
        setContentView(R.layout.ch6_4);
        text = (TextView)findViewById(R.id.text);
        tab = new ActionBar.Tab[5];
        String [] tabName = {"cat","dog","tiger","monkey","lion"};
        for(int i = 0;i < tab.length;i++) {
            tab[i] = actionBar.newTab();
            tab[i].setText(tabName[i]);
            tab[i].setTabListener(this);
            actionBar.addTab(tab[i],false);
        }
    }
    public  void onTabSelected (ActionBar.Tab tab,FragmentTransaction ft) {
        int index = tab.getPosition();
        String str = tab.getText().toString();
        text.append("index:" + index + "\tname:" + str + "\n");
    }
    public void onTabReselected (ActionBar.Tab tab,FragmentTransaction ft){}
    public void onTabUnselected (ActionBar.Tab tab,FragmentTransaction ft) {}
}
```

（5）启动 AVD，进入工程的根目录，用快捷方式编译工程、安装应用程序到 AVD（有关知识点参见 1.5 节）。对于本例子，用命令行进入 D:\2000\ch6_4，执行如下命令：

D:\2000 > ch6_4 > ant debug install

6.6　动作栏与选项菜单

在 6.1 节我们讲述了选项菜单，其特点是当用户按下手机或 ADV 上的 MENU 键后，可以在 Activity 对象的底部（activity 容器的底端）看到菜单项。但是，在某些情况下可能对使用选项菜单很不利，比如有些手机可能没有 MENU 键。为了克服这个不利因素，可以开启动作栏，把选项菜单的菜单项放到动作栏中。

1. 菜单项与 Tab 选项卡的位置关系

动作栏自动分成互不干扰的两行，第一行用于放置选项菜单的菜单项，第二行放置 Tab 选项卡。需要注意的是，当动作栏中放置了选项菜单的菜单项后，程序将不再显示动作栏的标题，只显示动作栏的图标（主要是为菜单项节省空间），而且如果向动作栏放置了较多的菜单项后，使得菜单项的宽度之和超出了 Activity 对象的视图的宽度，动作栏的第一行不会自动水平滚动，用户将无法看到全部的菜单项。

注：需要特别注意的是，如果菜单项带有图标，动作栏仅显示菜单项的图标。

2. android:showAsAction 属性

默认情况下，选项菜单不使用动作栏，即不将自己的菜单项放到动作栏中。如果想让选项菜单把自己的菜单项放到动作栏中，可以在菜单资源的 XML 文件中为菜单项（item）增加 android:showAsAction 属性，其属性值给出将菜单项放到动作栏中的方式，例如，下列菜单资源文件中的菜单项＜item＞使用 android:showAsAction 属性把自己放到动作栏中：

```
<?xml version = "1.0" encoding = "utf-8"?>
< menu xmlns:android = "http://schemas.android.com/apk/res/android">
    < item android:id = "@ + id/cat"
        android:title = "我在动作栏中了"
        android:showAsAction = "always" />   <!--菜单项放到动作栏中-->
    < item android:id = "@ + id/dog"
        android:icon = "@drawable/dog"
        android:title = "I am on ActonBar"
        android:showAsAction = "always"/>    <!--菜单项放到动作栏中-->
</menu>
```

android:showAsAction 属性可以取下列值。

- ifRoom：如果动作栏有剩余空间，就将菜单项放到动作栏中。当用这个属性值时，如果目前动作栏中放置的菜单项的宽度之和已经超出了 Activity 对象的视图的宽度，Activity 对象就不会把该菜单项放到动作栏中，用户仍需要按手机或 ADV 上的 MENU 键后，才能看到这个没有放到动作栏中的菜单项。
- always：一定要把菜单项放到动作栏中。需要注意的是，和放置 Tab 选项卡不同，如果动作栏中放置的菜单项的宽度之和超出了 Activity 对象的视图的宽度，用户将无法看到全部的菜单项，因此要慎用 always 属性值。
- never：不把菜单项放到动作栏中。

3. ActionBar 图标上的选择事件

当动作栏可见后，如果不特意设置动作栏的图标，那么动作栏的图标与应用程序的图标相同。系统默认动作栏的图标可以发生 ItemSeleted 事件，当用户选择动作栏的图标，系统自动调用 Activity 对象重写的 public boolean onOptionsItemSelected（MenuItem item）方法，即系统将动作栏的图标当成选项菜单中的一个选项（其索引位置是 0），并指定动作栏的图标的 id 是 android.R.id.home。

4. 示例

例子 6-5

在例子 6-1（参见 6.2 节）中的菜单资源文件 animal_menu.xml 中为菜单项＜item＞增加 android:showAsAction 属性，并取值"always"。开启动作栏，即修改 AndroidManifest.xml 配置文件，在＜activity＞标记中给出 android:theme 属性的值是"@android:style/Theme.Holo"。

重新编译、运行 6.2 节中的例子 6-1，效果如图 6.6 所示。

图 6.6 动作栏中的菜单项

6.7　AlertDialog 对话框

AlertDialog 类的继承关系如下：

```
java.lang.Object
  └ android.app.Dialog
      └ android.app.AlertDialog
```

使用 AlertDialog 类可以得到带按钮的对话框，即警示对话框，或带列选表的对话框，即选择对话框，前者经常用于提示警示信息，后者经常用于提供选择信息。

1. 警示或选择对话框

警示对话框是程序设计经常使用的对话框之一，其外观区域按垂直方向从上到下分成三部分（如图 6.7 所示）。

标题区域（Title area）：放置标题的文本和图标。

内容区域（Content area）：主要用于放置提示的信息。

按钮区域（Button area）：放置 NegativeButton、NeutralButton 或 PositiveButton 按钮。

当程序进行一个重要的操作之前，用户的操作会对程序的继续运行产生重要的影响或需要提示用户某些重要的信息时，弹出一个警示对话框是非常适宜的。

选择对话框是程序设计经常使用的对话框之一，主要是为用户提供重要的选项，往往是程序要求用户必须选择的选项，否则程序将无法继续执行。选择对话框的外观区域按垂直方向从上到下分成两部分（如图 6.8 所示）。

图 6.7　警示对话框的结构

图 6.8　选择对话框的结构

标题区域（Title area）：放置标题的文本和图标。

内容区域（Content area）：用于放置列表选项。

2. 使用生成器构建警示或选择对话框

构建警示或选择对话框采用了设计模式中的生成器模式，其主要思想是用一个称作生成器（Builder）的对象负责构建警示或选择对话框，而不是使用 AlertDialog 类的构造方法，具体内容如下：

1）Builder 对象

AlertDialog.Builder 类的实例负责构建警示或选择对话框（Builder 是 AlertDialog 的静态内部类）。

可以使用构造方法 AlertDialog.Builder(Context context)创建一个 AlertDialog.Builder 类

的对象，例如在 Activity 的子类中创建生成器 builder：

```
AlertDialog.Builder builder = new AlertDialog.Builder(this);
```

也可以使用构造方法 AlertDialog.Builder（Context context，int theme）创建一个 AlertDialog.Builder 类的对象，其中参数 theme 可以取 AlertDialog 类中的下列整数值，以决定所构造的对话框的背景颜色：

AlertDialog.THEME_DEVICE_DEFAULT_DARK

AlertDialog.THEME_DEVICE_DEFAULT_LIGHT

AlertDialog.THEME_HOLO_DARK

AlertDialog.THEME_HOLO_LIGHT

2）设置警示或选择对话框的标题和文本信息

生成器 builder 负责设置警示对话框上的标题以及文本信息，主要方法如下。

AlertDialog.Builder setTitle(CharSequence title)和 setTitle(int titleId)：设置标题上的文本。

AlertDialog.Builder setIcon(Drawable icon)和 setIcon(int iconId)：设置标题上的图标。

AlertDialog.Builder setMessage(int messageId)和 setMessage(CharSequence message)：设置包含的文本信息。

3）设置警示对话框中的按钮

生成器 builder 负责设置警示对话框上的按钮，并为按钮注册 DialogInterface.OnClickListener 监视器，DialogInterface.OnClickListener 接口在 android.content 包中，接口中的方法是 abstract void onClick(DialogInterface dialog, int which)。

对话框上可以有确定（PositiveButton）、否定（NegativeButton）和取消（NeutralButton）按钮。无论用户单击哪个按钮，对话框都立刻消失，并导致接口中的 onClick（DialogInterface dialog，int which）方法被执行，其中 dialog 就是当前消失的对话框，which 的值依赖于用户所单击的按钮，即 which 的值是下列整数之一：

DialogInterface.BUTTON_NEGATIVE

DialogInterface.BUTTON_NEUTRAL

DialogInterface.BUTTON_POSITIVE

程序可以根据需要设置对话框上是否有确定（PositiveButton）、否定（NegativeButton）和取消（NeutralButton）按钮。具体方法如下。

setPositiveButton(int textId, DialogInterface.OnClickListener listener)：设置对话框上有"确定"按钮，该"确定"按钮的名字由参数 textId 指定，监视器由 listener 指定。

setPositiveButton(CharSequence text, DialogInterface.OnClickListener listener)：设置对话框上有"确定"按钮，该"确定"按钮的名字由参数 text 指定，监视器由 listener 指定。

setNegativeButton(int textId, DialogInterface.OnClickListener listener)：设置对话框上有"否定"按钮，该"否定"按钮的名字由参数 textId 指定，监视器由 listener 指定。

setNegativeButton(CharSequence text, DialogInterface.OnClickListener listener)：设置对话框上有"否定"按钮，该"否定"按钮的名字由参数 text 指定，监视器由 listener 指定。

setNeutralButton(int textId, DialogInterface.OnClickListener listener)：设置对话框

上有"取消"按钮,该"取消"按钮的名字由参数 textId 指定,监视器由 listener 指定。

setNeutralButton(CharSequence text,DialogInterface.OnClickListener listener):设置对话框上有"取消"按钮,该"取消"按钮的名字由参数 text 指定,监视器由 listener 指定。

4) 设置选择对话框中的选项

setItems(CharSequence[] items,DialogInterface.OnClickListener listener):设置选择对话框的内容区域中的单选列表,列表的选项由参数 items 指定,监视器由 listener 指定。

setItems(int items,DialogInterface.OnClickListener listener):设置选择对话框的内容区域中的单选列表,列表的选项由参数 items 指定,监视器由 listener 指定。该方法的参数 items 要求使用值资源,因此需要建立值资源,比如,将下列 array.xml 保存到值资源中,即保存到工程的\res\values 目录中(有关知识点参见 2.6 节)。

```xml
<?xml version = "1.0" encoding = "utf - 8"?>
<resources>
    <array name = "car_list">
        <item>audi</item>
        <item>jeep</item>
        <item>ford</item>
    </array>
</resources>
```

那么,方法的参数 items 可以是 R.array.carlist,即对话框中的列表中包含有 audi、jeep 和 ford 三个选项。

对于选择对话框,无论用户选择哪个选项,对话框都即刻消失,并导致接口中的 onClick(DialogInterface dialog,int which)方法被执行,其中的 dialog 就是当前消失的对话框,which 是被选中的选项在列表中的位置索引(索引从 0 开始)。

5) 返回所构建的警示或选择对话框

最后,生成器 builder 调用 create()方法返回所构建的警示或选择对话框,例如:

```
AlertDialog dialog = builder.create();
```

3. 弹出警示对话框或选择对话框

对话框调用 show()方法使得对话框可见,即弹出对话框,调用 hide()方法可以隐藏对话框(show()和 hide()都是父类 Dialog 类中的方法)。如果不希望用户使用手机上的回复键让对话框消失,即希望用户必须对弹出的对话框作出响应,可以让生成器在构造对话框时,调用 setCancelable(boolean cancelable)方法,参数取值 false,例如:

```
builder.setCancelable(false);
```

或让对话框自身调用 setCancelable(boolean cancelable)方法(Builder 和 AlertDialog 类都有这个方法),例如:

```
dialog.setCancelable(false);
```

那么用户必须单击对话框上的某个按钮让对话框消失,否则程序将无法继续执行。

4. 示例

下面的例子 6-6 中,用户在 EditText 输入一个正整数,程序计算这个正整数的全部因子,如果用户的输入不符合要求,程序弹出警示对话框(见图 6.9),提示用户输入错误。运行效果如图 6.9 所示。

例子 6-6

(1) 创建名字为 ch6_6 的工程,主要 Activity 子类的名字为 Example6_6,使用的包名为 ch6.six。用命令行进入 D:\2000,创建工程 D:\2000>android create project -t 3 -n ch6_6 -p ./ch6_6 -a Example6_6 -k ch6.six。

(2) 将下列和视图相关的 XML 文件保存到工程的 \res\layout 目录中。

图 6.9 警示对话框

ch6_6.xml

```
<?xml version = "1.0" encoding = "utf-8"?>
<LinearLayout xmlns:android = "http://schemas.android.com/apk/res/android"
    android:layout_width = "match_parent"
    android:layout_height = "match_parent"
    android:background = "#EFAA99"
    android:orientation = "vertical">
    <EditText
        android:layout_width = "match_parent"
        android:layout_height = "wrap_content"
        android:inputType = "numberDecimal|textMultiLine"
        android:id = "@+id/edit" />
    <Button
        android:layout_width = "wrap_content"
        android:layout_height = "wrap_content"
        android:text = "计算因子"
        android:id = "@+id/button"
        android:onClick = "computer"        />
    <TextView
        android:id = "@+id/text"
        android:layout_width = "match_parent"
        android:layout_height = "wrap_content"
        android:textSize = "20sp"
        android:background = "#E910AA"
        android:textColor = "#000000"/>
</LinearLayout>
```

(3) 修改工程\src\ch6\six 目录下的 Example6_6.java 文件,修改后的内容如下:

Example6_6.java

```
package ch6.six;
import android.app.*;
import android.os.Bundle;
import android.widget.*;
import android.view.*;
import android.content.DialogInterface;
public class Example6_6 extends Activity implements DialogInterface.OnClickListener{
    TextView text;
```

```java
        EditText edit;
        public void onCreate(Bundle savedInstanceState) {
            super.onCreate(savedInstanceState);
            setContentView(R.layout.ch6_6);
            text = (TextView)findViewById(R.id.text);
            edit = (EditText)findViewById(R.id.edit);
        }
        AlertDialog getAlertDialogWithButton(String mess) {
            AlertDialog.Builder builder = new AlertDialog.Builder(this,AlertDialog.THEME_HOLO_LIGHT);
            builder.setTitle("input Error");
            builder.setMessage("" + mess + " is not positive integer");
            builder.setIcon(R.drawable.ic_launcher);
            builder.setPositiveButton("agree",this);
            builder.setNeutralButton("cancel",this);
            builder.setCancelable(false);
            AlertDialog dialog = builder.create();
            return dialog;
        }
        public void onClick(DialogInterface dialog,int which) {
            if(which == dialog.BUTTON_POSITIVE) {
                edit.setText(null);
            }
            if(which == dialog.BUTTON_NEGATIVE) {}
            if(which == dialog.BUTTON_NEUTRAL) {}
        }
        public void computer(View view) {
            String str = edit.getText().toString();
            int number = 0;
            try {   number = Integer.parseInt(str);
                    if(number <= 0) {
                        AlertDialog dialog = getAlertDialogWithButton(str);
                        dialog.show();
                    }
                    else {
                        for(int i = 1;i < number;i++) {
                            if(number % i == 0)
                                text.append("" + i + "\n");
                        }
                    }
            }
            catch(Exception exp) {
                AlertDialog dialog = getAlertDialogWithButton(str);
                dialog.show();
            }
        }
    }
```

（4）启动 AVD,进入工程的根目录,用快捷方式编译工程、安装应用程序到 AVD(有关知识点参见 1.5 节)。对于本例子,用命令行进入 D:\2000\ch6_6,执行如下命令：

D:\2000 > ch6_6 > ant debug install

下面的例子 6-7 中，使用选择对话框把 Button 视图的颜色设置成红色、黄色或绿色。程序一开始就弹出一个选择对话框，用户在对话框的列选表中可以选择红色、黄色或绿色选项，程序根据其选择设置按钮的颜色，如果想继续更改按钮的颜色，可以单击该按钮，再次弹出选择对话框，选择红色、黄色或绿色选项。运行效果如图 6.10 所示。

图 6.10　选择对话框

例子 6-7

（1）创建名字为 ch6_7 的工程，主要 Activity 子类的名字为 Example6_7，使用的包名为 ch6.seven。用命令行进入 D:\2000，创建工程 D:\2000＞android create project -t 3 -n ch6_7 -p ./ch6_7 -a Example6_7 -k ch6.seven。

（2）将下列值资源文件 color.xml 保存在值资源中，即保存到工程的\res\values 目录中（有关知识点参见 2.6 节），并修改值资源中的 strings.xml 文件，修改后的内容见如下的 strings.xml。

color.xml

```xml
<?xml version = "1.0" encoding = "utf-8"?>
<resources>
    <array name = "color_list">
        <item>红色</item>
        <item>黄色</item>
        <item>绿色</item>
    </array>
</resources>
```

strings.xml

```xml
<?xml version = "1.0" encoding = "utf-8"?>
<resources>
    <string name = "app_name">Example6_7</string>
    <string name = "dialog_title">选择下列颜色之一</string>
    <string name = "button_color">改变我的颜色</string>
</resources>
```

（3）将下列和视图相关的 XML 文件保存到工程的\res\layout 目录中。

ch6_7.xml

```xml
<?xml version = "1.0" encoding = "utf-8"?>
<LinearLayout xmlns:android = "http://schemas.android.com/apk/res/android"
    android:layout_width = "match_parent"
    android:layout_height = "match_parent"
    android:background = "#EFAA99"
    android:orientation = "vertical">
    <Button
        android:layout_width = "match_parent"
        android:layout_height = "wrap_content"
        android:text = "@string/button_color"
        android:id = "@+id/button"
```

```
            android:onClick = "changeColor"        />
</LinearLayout >
```

(4) 修改工程\src\ch6\seven 目录下的 Example6_7.java 文件,修改后的内容如下:

Example6_7.java

```java
package ch6.seven;
import android.app.*;
import android.os.Bundle;
import android.widget.*;
import android.view.*;
import android.content.DialogInterface;
import android.graphics.Color;
public class Example6_7 extends Activity implements DialogInterface.OnClickListener{
    Button button;
    public void onCreate(Bundle savedInstanceState) {
        super.onCreate(savedInstanceState);
        setContentView(R.layout.ch6_7);
        button = (Button)findViewById(R.id.button);
        AlertDialog dialog = getAlertDialogWithList(R.array.color_list);
        dialog.show();
    }
    AlertDialog getAlertDialogWithList(int itemId) {
        AlertDialog.Builder builder = new AlertDialog.Builder(this,AlertDialog.THEME_HOLO_LIGHT);
        builder.setTitle(R.string.dialog_title);
        builder.setIcon(R.drawable.ic_launcher);
        builder.setItems(itemId,this);
        builder.setCancelable(false);
        AlertDialog dialog = builder.create();
        return dialog;
    }
    public void onClick(DialogInterface dialog,int which) {
        if(which == 0) {
            button.setBackgroundColor(Color.RED);
        }
        if(which == 1) {
            button.setBackgroundColor(Color.YELLOW);
        }
        if(which == 2) {
            button.setBackgroundColor(Color.GREEN);
        }
    }
    public void changeColor(View view) {
        AlertDialog dialog = getAlertDialogWithList(R.array.color_list);
        dialog.show();
    }
}
```

(5) 启动 AVD,进入工程的根目录,用快捷方式编译工程、安装应用程序到 AVD(有关知识点参见 1.5 节)。对于本例子,用命令行进入 D:\2000\ch6_7,执行如下命令:

```
D:\2000 > ch6_7 > ant debug install
```

6.8　DatePickerDialog 对话框与 TimePickerDialog 对话框

DatePickerDialog 类和 TimePickerDialog 类都是 AlertDialog 的子类。

DatePickerDialog 类和 TimePickerDialog 类分别用来创建日期对话框和时间对话框，其目的是方便程序选择日期和时间。

1. 日期/时间对话框

当程序需要选择日期(年、月、日)时，弹出一个日期对话框是非常适宜的。日期对话框的外观区域按垂直方向从上到下分成三部分(如图 6.11 所示)。

标题区域(Title area)：放置标题的文本和图标，需要注意的是，一旦用户使用日期区域选择了日期，标题的文本将被更改为用户当前选择的日期。

内容区域(Content area)：主要用于放置提示的信息。

日期区域(Date area)：该区域内置了一个 DatePicker 视图。

当程序需要选择时间(时、分)时，弹出一个时间对话框是非常适宜的。时间对话框的外观区域按垂直方向从上到下分成三部分(如图 6.12 所示)。

图 6.11　日期对话框的结构

图 6.12　时间对话框的结构

标题区域(Title area)：放置标题的文本和图标。

内容区域(Content area)：主要用于放置提示的信息。

时间区域(Time area)：该区域内置了一个 TimePicker 视图。

2. 构造日期/时间对话框

1) 构造日期对话框

可以使用 DatePickerDialog 类的下述构造方法创建日期对话框：

```
DatePickerDialog(Context context, DatePickerDialog.OnDateSetListener callBack, int year, int monthOfYear, int dayOfMonth);
DatePickerDialog(Context context, int theme, DatePickerDialog.OnDateSetListener callBack, int year, int monthOfYear, int dayOfMonth);
```

参数意义如下所示。

context：日期对话框所在的某个上下文对象，一般取当前 Activity 对象即可。

callBack：实现 DatePickerDialog.OnDateSetListener 接口的类的实例，DatePickerDialog.OnDateSetListener 接口中的方法是 public void onDateSet(DatePicker view, int year, int monthOfYear, int dayOfMonth)。当日期对话框消失，或用户单击日期对话框上的 Done 按钮，都会导致 callBack 调用类实现的接口中的 onDateSet 方法，此时，onDateSet 方法中的 view 就是日期对话框中的 DatePicker 视图，year 就是用户在 DatePicker 视图中的选择的年份，monthOfYear 就是用户在 DatePicker 视图中选择的月份，dayOfMont 就是用户在 DatePicker 视图中选择的日期。需要特别注意的是，如果选择的是一月（Jan），monthOfYear 的值是 0，选择的是二月（Feb），monthOfYear 的值是 1，以此类推，选择的是十二月（Dec），monthOfYear 的值是 11。

year：用于给出日期对话框上显示的初始年份。

monthOfYear：用于给出日期对话框上显示的初始月份（特别注意取值 0 表示一月，11 表示十二月）。

dayOfMonth：用于给出日期对话框上显示的初始日期。

theme：用于给出日期对话框的背景颜色，取下列值之一，AlertDialog.THEME_DEVICE_DEFAULT_DARK，AlertDialog.THEME_DEVICE_DEFAULT_LIGHT，AlertDialog.THEME_HOLO_DARK，AlertDialog.THEME_HOLO_LIGHT。

2）构造时间对话框

可以使用 TimePickerDialog 类的下述构造方法创建时间对话框：

```
TimePickerDialog(Context context,TimePickerDialog.OnTimeSetListener callBack, int hourOfDay,
int minute,boolean is24HourView);
TimePickerDialog(Context context, int theme, TimePickerDialog.OnTimeSetListener callBack, int
hourOfDay, int minute,boolean is24HourView);
```

参数意义如下所示。

context：时间对话框所在的某个上下文对象，一般取当前 Activity 对象即可。

callBack：实现 TimePickerDialog.OnTimeSetListener 接口的类的实例，TimePickerDialog.OnTimeSetListener 接口中的方法是 public void onTimeSet（TimePicker view, int hourOfDay, int minute）。当时间对话框消失或用户单击时间对话框上的 Done 按钮，都会导致 callBack 调用类实现的接口中的 onTimeSet 方法，此时，onTimeSet 方法中的 view 就是时间对话框中的 TimePicker 视图，hourOfDay 就是用户在 TimePicker 视图中选择的小时，minute 就是用户在 TimePicker 视图中选择的分钟。

hourOfDay：用于给出时间对话框上显示的小时。

minute：用于给出时间对话框上显示的分钟。

is24HourView：确定是否使用 24 时制。

3. 弹出日期/时间对话框

对话框调用 show() 方法使得对话框可见，即弹出对话框，调用 hide() 方法可以隐藏对话框（show() 和 hide() 都是父类 Dialog 类中的方法）。如果不希望用户使用手机上的回复键让对话框消失，即希望用户必须对弹出的对话框作出响应，可以让对话框 dialog 调用 setCancelable(boolean cancelable) 方法，参数取值 false，例如：

```
dialog.setCancelable(false);
```
那么用户必须单击对话框上的 Done 按钮让对话框消失,否则程序将无法继续执行。

4. 示例

例子 6-8 让用户使用 DatePickerDialog 对话框与 TimePickerDialog 对话框给出抗日战争的爆发日期,以及学校每天下午第一节课的上课时间。运行效果如图 6.13(a),6.13(b)所示。

(a) 选出抗日战争爆发的日期

(b) 选出下午上课时间

图 6.13 运行效果

例子 6-8

(1) 创建名字为 ch6_8 的工程,主要 Activity 子类的名字为 Example6_8,使用的包名为 ch6.eight。用命令行进入 D:\2000,创建工程 D:\2000>android create project -t 3 -n ch6_8 -p ./ch6_8 -a Example6_8 -k ch6.eight。

(2) 修改值资源中的 strings.xml 文件,修改后的内容见如下的 strings.xml。

strings.xml

```
<?xml version = "1.0" encoding = "utf-8"?>
<resources>
    <string name = "app_name">Example6_8</string>
    <string name = "date">抗日战争爆发日期</string>
    <string name = "time">下午上课时间</string>
</resources>
```

(3) 将下列和视图相关的 XML 文件保存到工程的\res\layout 目录中。

ch6_8.xml

```
<?xml version = "1.0" encoding = "utf-8"?>
<RelativeLayout xmlns:android = "http://schemas.android.com/apk/res/android"
    android:layout_width = "match_parent"
    android:layout_height = "match_parent"
    android:background = "#EFAA99"
    android:orientation = "vertical">
    <Button
```

```xml
        android:layout_width = "wrap_content"
        android:layout_height = "wrap_content"
        android:layout_alignParentLeft = "true"
        android:layout_alignParentTop = "true"
        android:textSize = "20sp"
        android:text = "抗日战争日期"
        android:id = "@ + id/buttonOne"
        android:onClick = "openDateDialog"    />
<Button
        android:layout_width = "wrap_content"
        android:layout_height = "wrap_content"
        android:layout_alignParentRight = "true"
        android:layout_alignParentTop = "true"
        android:textSize = "20sp"
        android:text = "下午上课时间"
        android:id = "@ + id/buttonTwo"
        android:onClick = "openTimeDialog"    />
<TextView
    android:id = "@ + id/textDate"
    android:layout_width = "wrap_content"
    android:layout_height = "wrap_content"
    android:layout_below = "@ + id/buttonOne"
    android:layout_alignParentLeft = "true"
    android:textSize = "20sp"
    android:background = "#E910AA"
    android:textColor = "#000000"/>
<TextView
    android:id = "@ + id/textTime"
    android:layout_width = "wrap_content"
    android:layout_height = "wrap_content"
      android:layout_below = "@ + id/buttonTwo"
    android:layout_alignParentRight = "true"
    android:textSize = "20sp"
    android:background = "#AA98E8"
    android:textColor = "#000000"/>
</RelativeLayout>
```

(4) 修改工程\src\ch6\eight 目录下的 Example6_8.java 文件,修改后的内容如下:

Example6_8.java

```java
package ch6.eight;
import android.app.*;
import android.os.Bundle;
import android.widget.*;
import android.view.*;
public class Example6_8 extends Activity implements
DatePickerDialog.OnDateSetListener,TimePickerDialog.OnTimeSetListener {
    TextView textDate,textTime;
    int year,monthOfYear,dayOfMonth;
    int hour,minute;
    DatePickerDialog dateDialog;
```

```java
        TimePickerDialog timeDialog;
        public void onCreate(Bundle savedInstanceState) {
            super.onCreate(savedInstanceState);
            setContentView(R.layout.ch6_8);
            textDate = (TextView)findViewById(R.id.textDate);
            textTime = (TextView)findViewById(R.id.textTime);
            dateDialog = new DatePickerDialog(this,this,1930,0,1);
            dateDialog.setTitle("Picker Date");
            dateDialog.setIcon(R.drawable.ic_launcher);
            dateDialog.setMessage("you can select date");
            dateDialog.setCancelable(false);
            timeDialog = new TimePickerDialog(this,this,11,01,false);
            timeDialog.setTitle("Picker Time");
            timeDialog.setIcon(R.drawable.ic_launcher);
            timeDialog.setMessage("you can select time");
        }
        public void  onDateSet(DatePicker view,int year,int monthOfYear,int dayOfMonth) {
            this.year = year;
            this.monthOfYear = monthOfYear;
            this.dayOfMonth = dayOfMonth;
            setDateMess();
        }
        public void  onTimeSet(TimePicker view,int hourOfDay,int minute) {
            hour = hourOfDay;
            this.minute = minute;
            setTimeMess();
        }
        void setDateMess() {
            textDate.setText(R.string.date);
            textDate.append("\n" + year + " - " + monthOfYear + " - " + dayOfMonth);
        }
        void setTimeMess() {
            textTime.setText(R.string.time);
            textTime.append("\n" + hour + ":" + minute);
        }
        public void openTimeDialog(View v) {
            timeDialog.show();
        }
        public void openDateDialog(View v) {
            dateDialog.show();
        }
    }
```

(5) 启动 AVD，进入工程的根目录，用快捷方式编译工程、安装应用程序到 AVD（有关知识点参见 1.5 节）。对于本例子，用命令行进入 D:\2000\ch6_8，执行如下命令：

```
D:\2000>ch6_8>ant debug install
```

6.9　ProgressDialog 对话框

ProgressDialog 是 AlertDialog 的子类，用来创建进度条对话框。当进行一个耗时的操作时，弹出一个进度条对话框是非常适宜的。

1. 进度条对话框的外观

进度条的外观区域由三部分构成(如图 6.14(a),6.14(b)所示)。

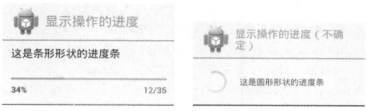

(a) 条形形状的进度条　　　　(b) 图形形状的进度条

图 6.14　进度条的外观区域

标题区域(Title area)：放置标题的文本和图标。

内容区域(Content area)：主要用于放置提示的信息。

进度区域(Progress area)：该区域内置了一个表示进度的视图。进度视图是一个水平条形状(通常用于可以给出进度情况的操作)或圆形形状(通常用于不能准确给出进度情况的操作)。

2. 使用构造方法创建进度条对话框

可以使用 ProgressDialog 类的下述构造方法创建进度条对话框：

ProgressDialog(Context context);
ProgressDialog(Context context,int theme);

参数意义如下所示。

context：进度条对话框所在的某个上下文对象,一般取当前 Activity 对象即可。

theme：用于给出进度条对话框的背景颜色,取下列值之一,AlertDialog. THEME_DEVICE_DEFAULT_DARK, AlertDialog. THEME_DEVICE_DEFAULT_LIGHT, AlertDialog. THEME_HOLO_DARK, AlertDialog. THEME_HOLO_LIGHT。

3. 使用静态方法返回进度条对话框

可以使用 ProgressDialog 类的下述静态方法返回一个进度条对话框,其特点是,这些静态方法不仅返回一个进度条对话框,而且立刻弹出该进度条。

static ProgressDialog　show(Context context,CharSequence title,CharSequence message);
static ProgressDialog　show(Context context,CharSequence title,CharSequence message,boolean indeterminate,boolean cancelable);
static ProgressDialog　show(Context context,CharSequence title,CharSequence message,boolean indeterminate);

参数意义如下所示。

context：进度条对话框所在的某个上下文对象,一般取当前 Activity 对象即可。

title：标题区域的标题的文本。

message：内容区域中的文本信息。

Indeterminate：取值 false 或 true,决定进度条是水平条形状或圆形形状。

cancelable：取值 true 或 false,决定用户能否使用手机上的回复键让对话框消失,取值

true是可以用手机上的回复键让对话框消失。

4. 进度条的常用方法

void setProgressStyle(int style)：设置进度条是水平条形或圆形进度条，参数 style 取值是 ProgressDialog 类中的静态常量，STYLE_HORIZONTAL 或 STYLE_SPINNER。

void setMax(int max)：设置水平进度条的最大值，即进度条被平分为 max 份。

void setProgress(int value)：设置进度值，取值在 0~max 之间。

void setIndeterminate(boolean indeterminate)：设置进度视图中的视图是否是圆形的进度条。

5. 弹出进度对话框

对话框调用 show()方法使得对话框可见，即弹出对话框，调用 hide()方法可以隐藏对话框（show()和 hide()都是父类 Dialog 类中的方法）。如果不希望用户使用手机上的回复键让对话框消失，可以让对话框 dialog 调用 setCancelable(boolean cancelable)方法，参数取值 false，例如：

```
dialog.setCancelable(false);
```

6. 示例

以下例子 6-9 使用了进度条对话框显示了查找随机数的进度，程序使用 Chronometer（计时器）每隔 1 秒得到一个 1~100 之间的随机数，如果该随机数小于 50，进度条就前进一个单位，进度条显示得到 10 个小于 50 的随机数的进度。运行效果如图 6.15 所示。

例子 6-9

（1）创建名字为 ch6_9 的工程，主要 Activity 子类的名字为 Example6_9，使用的包名为 ch6.nine。用命令行进入 D:\2000，创建工程 D:\2000＞android create project -t 3 -n ch6_9 -p ./ch6_9 -a Example6_9 -k ch6.nine。

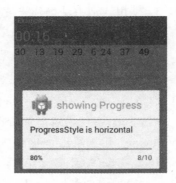

图 6.15　进度条对话框显示得到随机数的进度

（2）将下列和视图相关的 XML 文件保存到工程的\res\layout 目录中。

ch6_9.xml

```
<?xml version = "1.0" encoding = "utf - 8"?>
< LinearLayout xmlns:android = "http://schemas.android.com/apk/res/android"
    android:layout_width = "match_parent"
    android:layout_height = "match_parent"
    android:background = "#EFAA99"
    android:orientation = "vertical">
    < Chronometer
      android:id = "@ + id/myTimer"
      android:layout_width = "match_parent"
      android:layout_height = "wrap_content"
      android:textSize = "30sp"
      android:background = "#777777" />
    < TextView
```

```
        android:id = "@ + id/text"
        android:layout_width = "match_parent"
        android:layout_height = "match_parent"
        android:textSize = "20sp"
        android:background = "#FFFFAA"
        android:textColor = "#000000"/>
</RelativeLayout>
```

(3) 修改工程\src\ch6\nine 目录下的 Example6_9.java 文件,修改后的内容如下:

Example6_9.java

```java
package ch6.nine;
import android.app.*;
import android.os.Bundle;
import android.widget.*;
import android.view.*;
public class Example6_9 extends Activity implements Chronometer.OnChronometerTickListener{
    ProgressDialog dialog;
    TextView text;
    Chronometer timer;
    int count = 0, N = 10, MAX = 50;
    public void onCreate(Bundle savedInstanceState) {
        super.onCreate(savedInstanceState);
        setContentView(R.layout.ch6_9);
        text = (TextView)findViewById(R.id.text);
        timer = (Chronometer)findViewById(R.id.myTimer);
        timer.setOnChronometerTickListener (this);
        timer.start();
        dialog = new ProgressDialog(this,ProgressDialog.THEME_HOLO_LIGHT);
        dialog.setProgressStyle(ProgressDialog.STYLE_HORIZONTAL);
        dialog.setIndeterminate(false);
        dialog.setCancelable(false);
        dialog.setMax(N);
        dialog.setTitle("showing Progress");
        dialog.setIcon(R.drawable.ic_launcher);
        dialog.setMessage("ProgressStyle is horizontal");
        dialog.show();
    }
    public void onChronometerTick (Chronometer chronometer) {
        int number = (int)(Math.random() * 100) + 1;
        if(number < MAX) {
           count++;
           dialog.setProgress (count) ;
           text.append(number + "\t");
        }
        if(count == N) {
            timer.stop();
            dialog.hide();
        }
    }
}
```

(4) 启动 AVD,进入工程的根目录,用快捷方式编译工程、安装应用程序到 AVD(有关知识点参见 1.5 节)。对于本例子,用命令行进入 D:\2000\ch6_9,执行如下命令:

D:\2000>ch6_9>ant debug install

6.10 使用 Dialog 创建对话框

可以直接用 Dialog 或自己编写 Dialog 的子类创建对象,即创建对话框。

1. 对话框的外观

Dialog 对话框的外观区域由两部分构成。

标题区域(Title area):放置标题的文本。用户需使用 Dialog 类的 setTitle(int titleId) 或 setTitle(CharSequence title)方法指定对话框上标题的文本。

视图区域(View area):用户需使用 Dialog 类的 setContentView(View view)或 setContentView(int layoutResID)方法指定对话框上视图区域中的视图。

2. 为视图区域提供视图资源

尽管可以使用 setContentView(View view)为对话框的视图区设置视图,但更提倡使用 setContentView(int layoutResID)方法为对话框的视图区设置视图。因此,需要编写相应的视图资源文件,并存放到视图资源中,即存放到工程的\res\layout 目录中。

3. 处理视图区域中的事件

对话框的主要目的是提供和用户进行交互的视图,所以对话框常常在视图区域中添加诸如 Button、CheckBox 的视图,并根据需要编写处理事件的代码,以便决定怎样处理有关的数据、怎样关闭对话框等。比如,用户在一个登录对话框中输入自己的 login,按登录按钮,那么对话框就需要处理用户单击登录按钮事件,判断用户输入的 login 是否可以登录。再比如,用户希望使用对话框为其计算若干个数的平均值,那么对话框的视图区应该包含为用户提供输入数据的 EditText 视图,同时也应包含负责关闭对话框的视图,比如 Button。

应当由对话框的拥有者,比如 Activity,监视对话框中的视图上的事件,比如监视对话框中的 Button 视图上的 onClick 事件,以便 Activity 能方便地控制对话框。

4. 示例

以下例子 6-10 使用了自定义的登录对话框,如果用户在登录对话框时输入的登录名是 geng,密码是 java,程序将显示 Android 的 Logo(绿色小机器人),否则将显示登录名不正确。运行效果如图 6.16 所示。

例子 6-10

(1) 创建名字为 ch6_10 的工程,主要 Activity 子类的名字为 Example6_10,使用的包名为 ch6.ten。用命令行进入 D:\2000,创建工程 D:\2000>android create project -t 3 -n ch6_10 -p ./ch6_10 -a Example6_10 -k ch6.ten。

(2) 修改值资源中的 strings.xml 文件,修改后的内容如下:

图 6.16 登录对话框

strings.xml

```xml
<?xml version = "1.0" encoding = "utf-8"?>
<resources>
    <string name = "app_name">Example6_10</string>
    <string name = "input_login_name">登录名称</string>
    <string name = "login_password">登录密码</string>
    <string name = "positive_button">登录</string>
    <string name = "cancel_button">取消</string>
</resources>
```

（3）将下列和视图相关的 XML 文件保存到工程的\res\layout 目录中。其中 ch6_10.xml 是 activity 使用的视图，dialog.xml 是登录对话框使用的视图。

dialog.xml

```xml
<TableLayout xmlns:android = "http://schemas.android.com/apk/res/android"
    android:layout_width = "match_parent"
    android:layout_height = "match_parent"
    android:background = "#87CEEB" android:stretchColumns = "0,1">
  <TableRow>
    <TextView
      android:layout_width = "wrap_content"
      android:layout_height = "wrap_content"
      android:textColor = "#0000FF"
      android:text = "@string/input_login_name" />
    <EditText
      android:id = "@+id/loginName"
      android:layout_width = "wrap_content"
      android:layout_height = "wrap_content"
      android:singleLine = "true" />
  </TableRow>
  <TableRow>
    <TextView
      android:layout_width = "wrap_content"
      android:layout_height = "wrap_content"
      android:textColor = "#000000"
      android:text = "@string/login_password" />
    <EditText
      android:id = "@+id/password"
      android:layout_width = "wrap_content"
      android:layout_height = "wrap_content"
      android:singleLine = "true" />
  </TableRow>
  <TableRow>
    <Button
      android:id = "@+id/positive_button"
      android:layout_width = "wrap_content"
      android:layout_height = "wrap_content"
      android:text = "@string/positive_button" />
    <Button
      android:id = "@+id/cancel_button"
```

```
            android:layout_width = "wrap_content"
            android:layout_height = "wrap_content"
            android:text = "@string/cancel_button" />
    </TableRow>
</TableLayout>
```

ch6_10.xml

```
<?xml version = "1.0" encoding = "utf-8"?>
<LinearLayout xmlns:android = "http://schemas.android.com/apk/res/android"
    android:orientation = "vertical"
    android:background = "#EEEEEE"
    android:layout_width = "match_parent"
    android:layout_height = "match_parent">
    <TextView
        android:id = "@+id/text"
        android:textSize = "18sp"
        android:gravity = "center"
        android:background = "#555555"
        android:layout_width = "match_parent"
        android:layout_height = "wrap_content"/>
    <ImageButton
        android:id = "@+id/image"
        android:src = "@drawable/ic_launcher"
        android:visibility = "invisible"
        android:scaleType = "centerInside"
        android:layout_width = "match_parent"
        android:layout_height = "wrap_content"/>
    <Button
        android:layout_width = "wrap_content"
        android:layout_height = "wrap_content"
        android:text = "我要登录"
        android:onClick = "login" />
</LinearLayout>
```

(4) 修改工程\src\ch6\ten 目录下的 Example6_10.java 文件,修改后的内容如下:

Example6_10.java

```
package ch6.ten;
import android.app.*;
import android.os.Bundle;
import android.widget.*;
import android.view.*;
public class Example6_10 extends Activity implements View.OnClickListener{
    Dialog dialog;
    TextView text;
    EditText inputName, inputPassword;
    ImageButton image;
    Button positiveButton, cancelButton;
    public void onCreate(Bundle savedInstanceState) {
        super.onCreate(savedInstanceState);
        setContentView(R.layout.ch6_10);
```

```java
        dialog = new Dialog(this);
        dialog.setTitle("Login");
        dialog.setContentView(R.layout.dialog);
        text = (TextView)findViewById(R.id.text);
        image = (ImageButton)findViewById(R.id.image);
        positiveButton = (Button)dialog.findViewById(R.id.positive_button);
        cancelButton = (Button)dialog.findViewById(R.id.cancel_button);
        positiveButton.setOnClickListener(this);
        cancelButton.setOnClickListener(this);
        inputName = (EditText)dialog.findViewById(R.id.loginName);
        inputPassword = (EditText)dialog.findViewById(R.id.password);
    }
    public void onClick(View v) {
        if(v == positiveButton) {
            String name = inputName.getText().toString();
            String pass = inputPassword.getText().toString();
            if(name.equals("geng")&&pass.equals("java")) {
                image.setVisibility(View.VISIBLE);
                text.setText("login name and password are right");
            }
            else {
                text.setText("login name or password is error");
                image.setVisibility(View.INVISIBLE);
            }
            dialog.hide();
        }
        if(v == cancelButton) {
            dialog.hide();
        }
    }
    public void login(View v) {
        dialog.show();
    }
}
```

（5）启动 AVD，进入工程的根目录，用快捷方式编译工程、安装应用程序到 AVD（有关知识点参见 1.5 节）。对于本例子，用命令行进入 D:\2000\ch6_10，执行如下命令：

D:\2000>ch6_10>ant debug install

6.11 长按事件与对话框

在 6.4 节介绍了上下文菜单，其特点是用户通过在视图上实施"长按"（a long-click）操作显示上下文菜单，本节介绍，怎样通过"长按"事件弹出一个对话框。

1. 注册长按事件监视器

程序可以在任何视图上注册"长按"事件的监视器，当用户在视图上实施"长按"操作后，监视器将会调用相应的方法进行有关的操作。

为视图注册"长按"事件监视器的方法是：

```
void setOnLongClickListener(View.OnLongClickListener listener);
```

监视器由实现 View.OnLongClickListener 接口的类负责创建,View.OnLongClickListener 接口有如下的方法:

```
public boolean onLongClick(View v);
```

当用户在注册了"长按"事件监视器的视图上实施"长按"操作后,监视器将调用 onLongClick(View v)方法,其中的参数 v 就是当前被实施了"长按"的视图。

2. 示例

以下例子 6-11 中,用户在 TextView 视图上实施"长按"操作后,程序弹出一个可以改变 TextView 颜色的对话框。运行效果如图 6.17 所示。

例子 6-11

(1) 创建名字为 ch6_11 的工程,主要 Activity 子类的名字为 Example6_11,使用的包名为 ch1.eleven。用命令行进入 D:\2000,创建工程 D:\2000＞android create project -t 3 -n ch6_11 -p ./ch6_11 -a Example6_11 -k ch6.eleven。

(2) 将下列和视图相关的 XML 文件保存到工程的 \res\layout 目录中。

图 6.17 长按弹出对话框

ch6_11.xml

```
<?xml version = "1.0" encoding = "utf-8"?>
<LinearLayout xmlns:android = "http://schemas.android.com/apk/res/android"
    android:orientation = "vertical"
    android:background = "#EEEEEE"
    android:layout_width = "match_parent"
    android:layout_height = "match_parent">
  <TextView
    android:id = "@+id/text"
    android:textSize = "18sp"
    android:text = "长按 TextView 视图,可弹出一个对话框"
    android:background = "#555555"
    android:layout_width = "match_parent"
    android:layout_height = "200dp"/>
</RelativeLayout>
```

(3) 修改工程\src\ch6\eleven 目录下的 Example6_11.java 文件,修改后的内容如下:

Example6_11.java

```
package ch6.eleven;
import android.app.*;
import android.os.Bundle;
import android.widget.*;
import android.view.*;
import android.content.DialogInterface;
import android.graphics.Color;
public class Example6_11 extends Activity
```

```java
    implements View.OnLongClickListener,DialogInterface.OnClickListener{
        TextView text;
        public void onCreate(Bundle savedInstanceState) {
            super.onCreate(savedInstanceState);
            setContentView(R.layout.ch6_11);
            text = (TextView)findViewById(R.id.text);
            text.setOnLongClickListener(this);
        }
        public boolean onLongClick(View v) {
          getAlertDialogWithButton().show();
          return true;
        }
        public void  onClick(DialogInterface dialog,int which) {
            if(which == dialog.BUTTON_POSITIVE) {
                text.setBackgroundColor(Color.RED);
            }
            if(which == dialog.BUTTON_NEGATIVE) {
                text.setBackgroundColor(Color.YELLOW);
            }
            if(which == dialog.BUTTON_NEUTRAL) {
                text.setBackgroundColor(Color.GREEN);
            }
        }
        AlertDialog getAlertDialogWithButton() {
            AlertDialog.Builder builder = new AlertDialog.Builder(this);
            builder.setTitle("Red Yellow,Green");
            builder.setMessage("Change TextView Color");
            builder.setIcon(R.drawable.ic_launcher);
            builder.setPositiveButton("Red",this);
            builder.setNegativeButton("Yellow",this);
            builder.setNeutralButton("Green",this);
            builder.setCancelable(false);
            AlertDialog dialog = builder.create();
            return dialog;
        }
    }
```

(4) 启动 AVD，进入工程的根目录，用快捷方式编译工程、安装应用程序到 AVD（有关知识点参见 1.5 节）。对于本例子，用命令行进入 D:\2000\ch6_11，执行如下命令：

D:\2000>ch6_11>ant debug install

习 题 6

1. 每个 Activity 对象都自带一个选项菜单（Options menus）吗？
2. 编写程序使用选项菜单（Options menus），要求选项菜单（Options menus）有三个菜单项：红、蓝、绿，用户选择一个菜单项，程序将一个 TextView 视图的背景设置成相应的颜色。

3. Activity 对象中只能有一个选项菜单(Options menus)吗？最多只能有一个上下文菜单(ContextMenu)吗？

4. 编写程序使用上下文菜单(ContextMenu)，当用户在一个 EditView 视图上实施"长按"操作后，上下文菜单将显示自己的全部菜单项，用户选择一个菜单项，程序将菜单项的名字放入到 EditView 视图中。

5. 编写程序使用弹出式菜单(PopupMenu)，用户单击 Button 视图可以弹出弹出式菜单的菜单项，要求菜单项分别是 3 种字体的名称，用户选择某个菜单项可以改变 TextView 视图中的文字的字体。

6. 编写一个程序，程序的 Activity 对象中有一个 FrameLayout 视图，FrameLayout 视图中又有 4 个 ImageView 子视图，这 4 个 ImageView 子视图分别显示春、夏、秋、冬的图像。让 Activity 对象开启动作栏，动作栏中有名字是春、夏、秋、冬的 Tab 选项卡，用户单击某个选项卡，FrameLayout 视图显示相应的 ImageView 子视图(可参见本章的例子 6-4 和第 4 章的例子 4-6)。

7. 编写一个程序，用户在一个 EditText 输入一个小于等于 100、大于等于 0 的代表考试成绩的整数，然后单击 Button 视图。如果用户输入合乎要求，程序在 TextView 视图中显示成绩是否是"优秀"、"良好"、"及格"或"不及格"，如果用户的输入不符合要求，程序弹出警示对话框，提示用户输入错误。

8. 编写一个程序，程序的 Activity 对象中有一个 TextView 视图和一个 Button 视图，用户在 TextView 视图上实施"长按"操作后，Button 视图上的文字变成"Hello"；用户在 Button 视图上实施"长按"操作后，TextView 视图中的文字变成"How are you"。

第 7 章　2D 绘图

主要内容：
- Drawable 类；
- Canvas 类；
- SurfaceView 类；
- 使用画布绘制位图。

应用程序可能需要绘制图形或图像，特别是编写游戏类的应用程序，因此，本章介绍 2D 绘图有关的类。Android 也支持 3D 绘图，但 3D 绘图需要一定的 3D 知识，因此本书没有讲解 3D 绘图。

7.1　Drawable 类

Drawable 类的继承关系如下：

```
java.lang.Object
    └ android.graphics.drawable.Drawable
```

当需要在一个视图上绘制一幅图像时，就可以考虑 Drawable 类。

1. 图像资源

创建项目后，可以把程序要绘制的图像存放到图像资源中（有关知识点参见 2.7 节），比如可以把名字相同，仅仅大小（像素）不同的 4 幅图像，分别存放在工程的下列目录中：

```
\res\drawable-hdpi
\res\drawable-ldpi
\res\drawable-mdpi
\res\drawable-xhdpi
```

图像文件的名字只能使用小写英文字母（a~z），数字（0~9），下划线（_），不可以使用大写的英文字母。如果不希望应用程序根据图像的大小不同选择其中之一，那么也可以在工程的\res 目录下新建子目录 drawable，然后把图像存放到该目录下。

2. 直接引用图像资源的方法

任何视图都可以调用 setBackgroundResource(int resid)方法把背景设置成资源下的一幅图像，例如，假设图像资源中有名字为 gamePic 的图像，view 是程序代码中的某个 View 视图，那么 view 设置背景的代码如下：

```
view.setBackgroundResource(R.drawable.gamePic);
```

ImageView 视图(有关知识点参见 5.3 节)调用 void setImageResource(int resId)方法绘制图像,假设 image 是程序代码中的一个 ImageView 视图,那么 image 绘制图像的代码如下:

```
image.setImageResource(R.drawable.gamePic);
```

3. 引用图像资源创建 Drawable 对象

程序代码中可以使用图像资源得到一个 Drawable 对象。Context 类提供了获取资源的方法 public Resources getResources(),该方法返回一个 Resources 类的实例(有关知识点参见 2.8 节),使用该实例可以访问已有的资源。例如,Resources 类的实例使用 Drawable getDrawable(int id) 方法可以使用图像资源 id 得到一个 Drawable 对象,例如:

```
Drawable drawable = getResources().getDrawable(R.drawable.gamePic);
```

视图也可以使用 void setBackground (Drawable background)设置背景,代码如下:

```
view.setBackground(drawable);
```

7.2 Canvas 类

Canvas 类的继承关系如下:

```
java.lang.Object
    └ android.graphics.Canvas
```

当需要动态绘制图形或图像时就可以使用 Canvas 类。Canvas 类的对象调用方法可以在自身上绘制图形或图像,因此当需要动态绘制图形或图像时,就可以使用 Canvas 类的实例。

1. 使用依托于视图的 Canvas 对象

在编写代码时,程序可以编写 View 的子类,并重写 View 类的 public void onDraw (Canvas canvas)方法。当程序编写的 View 子类的实例可见时(视图可见时),该实例会自动调用 onDraw(Canvas canvas)方法(View 子类的实例调用 invalidate()方法,也会导致调用 onDraw 方法),参数 canvas 由系统自动创建(该 canvas 是依托于视图的画布),因此,用户需要编写的主要代码是让 canvas 调用绘制图形或图像的方法在画布上绘制图形或图像。

下列是重写 onDraw(Canvas canvas)方法的代码:

```
class MyView extends View {
    MyView(Context c) {
        super(c);
        setMinimumHeight(300);
        setMinimumWidth(100);
    }
    public void onDraw(Canvas canvas) {   //系统传递给参数 canvas 一个 Canvas 的实例
        super.onDraw(canvas);
        Paint paint = new Paint();        //创建一个画笔
        paint.setColor(Color.RED);
        canvas.drawColor(Color.BLUE);     //设置画布的颜色
```

```
        canvas.clipRect(0,0,getWidth(),getHeight());        //设置画布的可见区域
        canvas.drawRect(new Rect(0,0,100,100),paint);       //在画布上绘制一个矩形
    }
}
```

由于画布需要设置裁剪区,因此在 View 子类的构造方法中应当给出视图的最小宽度和最小高度。

2. Canvas 类的绘制方法

Canvas 的可见区域的大小取决于它依托的视图的大小,视图坐标系的左上角是坐标系的原点,向右的方向是 x-轴,向下的方向是 y-轴。当 canvas 调用绘制方法(诸如 draw 方法)时,需要向绘制方法传递一个 Paint 类的实例(该实例就好比画笔),该实例会完成具体的绘制工作。

Canvas 类提供丰富的绘制图形的方法,这里不打算一一列出,只列出例子 7-1 将要使用的一些方法,如果需要绘制复杂图形,只需查看 Canvas 类的帮助文档即可。

- boolean clipRect(float left,float top,float right,float bottom):设置画布上的裁剪区,裁剪区的左上角的坐标是(left,top),右下角的坐标是(right,bottom)。只有在裁剪区的图形是可见的。只可以设置一个裁剪区(裁剪区也称图形的可见区),如果不设置任何裁剪区,裁剪区默认是整个画布。
- public boolean clipRect(RectF rect,Region.Op op):将当前裁剪区与参数 rect 指定的矩形进行 op 运算得出新的裁剪区,其中 op 取值是 Region.Op 类的静态常量,DIFFERENCE,INTERSECT,REPLACE,REVERSE_DIFFERENCE,UNION,XOR。
- void drawColor(int color):设置裁剪区的颜色(底色)。
- void drawCircle(float cx,float cy,float radius,Paint paint):绘制一个圆心在(cx,cy),半径是 radius 的圆。
- void drawLine(float startX,float startY,float stopX,float stopY,Paint paint):绘制一条直线。
- void drawRect(float left,float top,float right,float bottom,Paint paint):绘制一个矩形,矩形的左上角坐标是(left,top),右下角坐标是(right,bottom)。
- void drawOval(RectF oval,Paint paint):绘制一个椭圆,椭圆的外接矩形是 oval。
- void drawPoints(float[] pts,Paint paint):绘制参数 pts 给出的点。
- void drawText(String text,float x,float y,Paint paint):绘制文本,文本的起点坐标是(x,y)。
- void drawBitmap(Bitmap bitmap,float left,float top,Paint paint):绘制位图图像。

3. Paint 类的常用方法

画布在绘制图形时,需要一个 Paint 类的实例来帮助它完成绘制工作。Paint 类的常用方法如下。

- void setStyle(Paint.Style style):设置绘制方式,当参数取值 Paint.Style.FILL 时,绘制方式是填充式,也是 Paint 实例的默认方式,参数取值 Paint.Style.STROKE 时,绘制方式是线条方式。
- void setTextSize(float textSize):设置所绘制的文本的大小。

- void setStrokeWidth(float width)：当绘制方式是线条方式时,设置线条的粗细。
- void setColor(int color)：设置画笔的颜色。

4. 旋转画布上的裁剪区

由于只有裁剪区的图形是可见的,因此,当需要旋转画布上某个图形时(不是全部的图形),可以通过调整裁剪区(见 clipRect(RectF rect,Region.Op op)方法),并旋转调整后的裁剪区来旋转画布上一个图形,旋转裁剪区需要以下三个步骤。

1) 保存原有的裁剪区

int save()：应当首先保存好当前裁剪区,即调用 save 方法保存好当前的裁剪区。

2) 旋转裁剪区

rotate(float degrees)：画布调用该方法以原点为轴点旋转裁剪区。旋转的角度是 degrees(单位是度,不是弧度)。如果 degree 为正值是逆时针旋转,否则是顺时针旋转。

rotate(float degrees,float px,float py)：画布调用该方法以(px,py)为轴点旋转裁剪区。

3) 绘制图形

如果此时画布使用绘制方法绘制的图形,相对于原始的裁剪区,就是旋转的图形。

4) 回复原始的裁剪区

void restore()：画布调用该方法将画布中的裁剪区回复成 save 保存的裁剪区。

在画布上缩放和移动图形的步骤与旋转图形类似,平移和缩放裁剪区的方法分别是：

```
void translate(float dx,float dy);
void scale(float sx,float sy);
```

5. 示例

以下例子 7-1 中,一个画布上绘制了一些基本图形,一个画布上绘制了若干个旋转的椭圆和旋转的直线,一个画布上绘制了 Android 的 Logo(小机器人)。效果如图 7.1 所示。

例子 7-1

(1) 创建名字为 ch7_1 的工程,主要 Activity 子类的名字为 Example7_1,使用的包名为 ch7.one。用命令行进入 D:\2000,创建工程 D:\2000＞ android create project -t 3 -n ch7_1 -p ./ch7_1 -a Example7_1 -k ch7.one。

(2) 将下列和视图相关的 XML 文件保存到工程的 \res\layout 目录中。

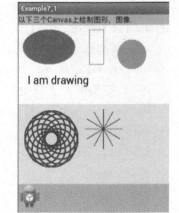

图 7.1 使用 Canvas 绘制图形、图像

ch7_1.xml

```
<?xml version = "1.0" encoding = "utf-8"?>
<ScrollView xmlns:android = "http://schemas.android.com/apk/res/android"
    android:layout_width = "match_parent"
    android:layout_height = "match_parent"
    android:scrollbarStyle = "outsideOverlay"
```

```xml
        android:background = "#87CEEB">
    <LinearLayout
        android:id = "@+id/layout"
        android:orientation = "vertical"
        android:layout_width = "match_parent"
        android:layout_height = "match_parent"
        android:background = "#87CEEB">
        <TextView
            android:background = "#AAEE00"
            android:textColor = "#000000"
            android:layout_width = "match_parent"
            android:layout_height = "wrap_content"
            android:text = "以下三个Canvas上绘制图形、图像."
            android:textSize = "15sp" />
    </LinearLayout>
</ScrollView>
```

（3）将下列画布依托的View视图的源文件MyView.java保存到工程的\src\ch7\one目录下，并修改工程\src\ch7\one目录下的Example7_1.java文件。MyView.java和修改后的Example7_1.java内容如下。

MyView.java

```java
package ch7.one;
import android.graphics.*;
import android.view.*;
import android.content.*;
public class MyView extends View {
    int n;
    MyView(Context c, int n) {
        super(c);
        this.n = n;
        setMinimumHeight(160);
        setMinimumWidth(200);
    }
    public void onDraw(Canvas canvas) {
        super.onDraw(canvas);
        canvas.clipRect(0, 0, getWidth(), getHeight());
        Paint paint = new Paint();
        if(n == 1) {
            paint.setColor(Color.RED);
            canvas.drawColor(Color.WHITE);
            canvas.drawOval(new RectF(10, 10, 120, 80), paint);
            paint.setStyle(Paint.Style.STROKE);
            canvas.drawRect(150, 10, 180, 80, paint);
            paint.setColor(Color.GREEN);
            paint.setStyle(Paint.Style.FILL);
            canvas.drawCircle(240, 60, 30, paint);
            paint.setTextSize(22);
            paint.setColor(Color.BLACK);
            canvas.drawText("I am drawing", 20, 120, paint);
```

```java
        }
        else if(n == 2) {
            paint.setColor(Color.BLUE);
            canvas.drawColor(Color.YELLOW);
            paint.setStyle(Paint.Style.STROKE);
            paint.setStrokeWidth(2.6F);
            canvas.save();
            for(int i = 1; i <= 12; i++){
                float degree = 30;
                canvas.rotate(degree,70,70);
                canvas.drawOval(new RectF(50,50,120,120),paint);
            }
            canvas.restore();
            paint.setStrokeWidth(1);
            canvas.save();
            for(int i = 1; i <= 12; i++){
                float degree = 30;
                canvas.rotate(degree,180,50);
                canvas.drawLine(180,50,220,50,paint);
            }
            canvas.restore();
        }
        if(n == 3) {
            Bitmap bm = BitmapFactory.decodeResource(getResources(),R.drawable.ic_launcher);
            canvas.drawBitmap(bm,0,0,paint);
        }
    }
}
```

Example7_1.java

```java
package ch7.one;
import android.app.*;
import android.os.Bundle;
import android.widget.*;
import android.view.*;
public class Example7_1 extends Activity {
    public void onCreate(Bundle savedInstanceState) {
        super.onCreate(savedInstanceState);
        setContentView (R.layout.ch7_1);
        LinearLayout layout = (LinearLayout)findViewById(R.id.layout);
        View view_one = new MyView(this,1);
        layout.addView(view_one);
        view_one.invalidate();
        View view_two = new MyView(this,2);
        layout.addView(view_two);
        View view_three = new MyView(this,3);
        layout.addView(view_three);
    }
}
```

(4) 启动 AVD，进入工程的根目录，用快捷方式编译工程、安装应用程序到 AVD(有关知识点参见 1.5 节)。对于本例子，用命令行进入 D:\2000\ch7_1，执行如下命令：

```
D:\2000> ch7_1> ant debug install
```

7.3　SurfaceView 类

SurfaceView 类的继承关系如下：

```
└ android.view.View
    └ android.view.SurfaceView
```

SurfaceView 类是 View 类的子类，允许用户在 SurfaceView 视图的画布上绘制图形(SurfaceView 里内嵌了一个 Canvas 的实例，用于绘制图形)，其最大的特点是，允许在单独的工作线程中完成画布上的绘制任务，这样一来，当绘制任务需要较长的时间时，也不影响应用程序中的 UI 线程(有关应用程序的线程问题的详细讨论参见第 9 章)。

1. 编写 SurfaceView 类的子类

为了能在 SurfaceView 视图上使用画布并绘制图形，系统要求必须编写 SurfaceView 类的子类，该子类要实现 SurfaceHolder.Callback 接口。对于 SurfaceView 视图，有三件事情，系统要求 SurfaceView 类的子类必须给予处理：

- SurfaceView 视图被创建，并可见。
- SurfaceView 视图的大小被改变。
- SurfaceView 视图被摧毁。

SurfaceView 类的子类通过实现 SurfaceHolder.Callback 接口来处理上述三件事情，SurfaceHolder.Callback 接口中有三个方法：

- void surfaceChanged(SurfaceHolder holder, int format, int width, int height)　当在 SurfaceView 视图可见时(内嵌的画布就会可见)，该方法被执行。
- void surfaceCreated(SurfaceHolder holder)　SurfaceView 视图内嵌的画布的大小被改变时，该方法被执行
- void surfaceDestroyed(SurfaceHolder holder)　SurfaceView 视图被摧毁时，该方法被执行。

SurfaceView 类的子类通过上述三个方法决定什么时候开始自己的绘制工作或结束绘制工作，比如，在 surfaceCreate 方法中启动负责绘制的线程，在 surfaceDestroyed 方法中保存一些重要的结果等。

2. 在程序的 Activity 中创建 SurfaceView 对象

在应用程序的 Activity 的子类中使用 SurfaceView 类的子类，比如使用 UserSurfaceView 子类创建了一个名字为 surface 的对象。surface 的对象中有一个 SurfaceHolder 类型的对象，首先让 surface 的对象调用 getHolder( ) 方法返回它的 SurfaceHolder 对象，比如返回的对象为 holder，再让 holder 对象调用 addCallback()方法注册 surface 对象，即 addCallback()方法应当在 Activity 中进行(必须在应用程序的 UI 线程中进行)。

经过上述步骤后，当 surface 对象可见时、大小被调整时或销毁时，SurfaceView 类的子类实现的 SurfaceHolder.Callback 接口的相应方法就会被调用执行。代码如下：

```
public void onCreate(Bundle savedInstanceState) {
    UserSurfaceView   surface = new UserSurfaceView()
    SurfaceHolder holder = surface.getHolder();
    holder.addCallback(surface);
}
```

3. 在工作线程中进行绘制工作

SurfaceView 视图中的 SurfaceHolder 对象维护着一个画布，即该画布依托于 SurfaceView 视图。SurfaceView 视图其最大的特点是，允许在一个单独的工作线程中完成画布上的绘制任务。因此可以在编写的 SurfaceView 的子类中创建一个线程（Thread 类创建的对象），并在画布上绘制图形。

需要特别注意的是，每次绘制之前，必须先让 SurfaceHolder 对象调用 Canvas lockCanvas() 锁住它维护的画布，只有这样才能在画布上进行绘制，即才可以编辑画布。绘制完毕后，再让 SurfaceHolder 对象调用 void unlockCanvasAndPost（Canvas canvas）方法解锁 SurfaceHolder 对象曾锁住的画布，其目的是让应用程序中的 UI 线程立刻显示 Thread 线程在画布上绘制的图形（Thread 线程无法完成显示图形任务）。代码如下：

```
Canvas canvas = getHolder().lockCanvas();
//canvas 调用方法绘制图形
    getHolder().unlockCanvasAndPost(Canvas canvas);
```

4. 子类的构造方法

由于是在 Activity 的子类中用 SurfaceView 的子类创建对象，因此也可以在 SurfaceView 的子类的构造方法中让 SurfaceHolder 对象调用 addCallback() 方法完成注册任务，即注册 SurfaceView 的子类的对象到 SurfaceHolder 对象。

以下是一个 SurfaceView 类的子类 UserSurfaceView。

```
class UserSurfaceView extends SurfaceView
implements SurfaceHolder.Callback,Runnable{
   SurfaceHolder holder ;           //SurfaceHolder 对象
   Thread thread;                    //负责绘制的线程
   public UserSurfaceView() {
      holder = getHolder();          //得到 SurfaceHolder 对象
      holder.addCallback(this);      //将当前 SurfaceView 视图注册到 holder
      thread = new Thread(this);
   }
   //SurfaceHolder.Callback 接口中的方法：
   public void surfaceChanged(SurfaceHolder holder,int format,int width,int height)  {
   }
   public void  surfaceCreated(SurfaceHolder holder) {
       thread.start();
   }
   public void  surfaceDestroyed(SurfaceHolder holder) {}
   public void run() {
      Canvas canvas = holder.lockCanvas();
```

```
        ///canvas 调用绘制图形的方法
        holder.unlockCanvasAndPost(canvas);
    }
}
```

5．在子类中重写 draw(Canvas canvas)方法

在编写 SurfaceView 的子类时,也可以重写 void draw(Canvas canvas)方法,该方法不会被系统自动调用(与前一节中 View 子类重写的 onDraw 方法不同)。可以在 Thread 线程中调用 draw 方法绘制图形。需要注意的是,如果调用重写 draw 方法进行绘制,建议在一个同步块中调用 draw 方法,以防止应用程序在线程的外边调用 draw 方法引起不必要的麻烦。

6．示例

以下例子 7-2 中,使用 SurfaceView 视图实现小动画,在 SurfaceView 提供的画布上实现平行移动矩形的动画。效果如图 7.2 所示。

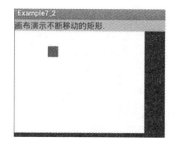

图 7.2　使用 SurfaceView 绘制图形

例子 7-2

(1) 创建名字为 ch7_2 的工程,主要 Activity 子类的名字为 Example7_2,使用的包名为 ch7.two。用命令行进入 D:\2000,创建工程 D:\2000＞android create project -t 3 -n ch7_2 -p ./ch7_2 -a Example7_2 -k ch7.two。

(2) 将下列和视图相关的 XML 文件保存到工程的\res\layout 目录中。

ch7_2.xml

```xml
<?xml version = "1.0" encoding = "utf - 8"?>
<ScrollView xmlns:android = "http://schemas.android.com/apk/res/android"
    android:layout_width = "match_parent"
    android:layout_height = "match_parent"
    android:scrollbarStyle = "outsideOverlay"
    android:background = " # 87CEEB">
    <LinearLayout
        android:id = "@ + id/layout"
        android:orientation = "vertical"
        android:layout_width = "match_parent"
        android:layout_height = "match_parent"
        android:background = " # 87CEEB">
        <TextView
            android:background = " # AAEE00"
            android:textColor = " # 000000"
            android:layout_width = "match_parent"
            android:layout_height = "wrap_content"
            android:text = "画布演示不断移动的矩形."
            android:textSize = "15sp" />
    </LinearLayout>
</ScrollView>
```

(3) 将下列 UserSurfaceView.java 保存到工程的 \src\ch7\two 目录下，并修改工程 \src\ch7\two 目录下的 Example7_2.java 文件。UserSurfaceView.java 和修改后的 Example7_2.java 内容如下。

UserSurfaceView.java

```java
package ch7.two;
import android.graphics.*;
import android.view.*;
import android.content.*;
import android.widget.*;
public class UserSurfaceView extends SurfaceView
implements SurfaceHolder.Callback, Runnable{
    Thread thread;
    Paint paint;
    boolean die;
    SurfaceHolder holder;
    public UserSurfaceView (Context c){
        super(c);
        setMinimumHeight(160);
        setMinimumWidth(100);
        paint = new Paint();
        holder = getHolder();
        holder.addCallback(this);
        thread = new Thread(this);
    }
    public void surfaceChanged(SurfaceHolder holder,int format,int width,int height){}
    public void  surfaceCreated(SurfaceHolder holder) {
        thread.start();
    }
    public void  surfaceDestroyed(SurfaceHolder holder) {
        die = true;
    }
    public void run() {
        int left = 10, top = 30, right = 30, bottom = 50;
        while(true) {
            if(die == true) return;
            RectF rect = new RectF(left,top,right,bottom);
            Canvas canvas = holder.lockCanvas();
            drawRect(canvas,rect);
            holder.unlockCanvasAndPost(canvas);
            left++;
            right++;
            if(right >= 260) {
              left = 10;
              right = 30;
            }
            try { Thread.sleep(100);
            }
            catch(Exception exp){}
        }
```

```java
    }
    private void drawRect(Canvas canvas,RectF rect) {
        canvas.clipRect(0,0,260,200);
        paint.setColor(Color.BLUE);
        canvas.drawColor(Color.WHITE);
        canvas.drawRect(rect,paint);
    }
}
```

Example7_2.java

```java
package ch7.two;
import android.app.*;
import android.os.Bundle;
import android.widget.*;
import android.view.*;
public class Example7_2 extends Activity {
    public void onCreate(Bundle savedInstanceState) {
        super.onCreate(savedInstanceState);
        setContentView(R.layout.ch7_2);
        LinearLayout layout = (LinearLayout)findViewById(R.id.layout);
        UserSurfaceView view = new UserSurfaceView(this);
        layout.addView(view,300,300);

    }
}
```

（4）启动 AVD，进入工程的根目录，用快捷方式编译工程、安装应用程序到 AVD（有关知识点参见 1.5 节）。对于本例子，用命令行进入 D:\2000\ch7_2，执行如下命令：

D:\2000>ch7_2>ant debug install

7.4 使用画布绘制位图

前面的 7.2 节和 7.3 节学习了怎样使用依托于 View 和 SurfaceView 的画布绘制图形或图像。有时候程序可能需要动态绘制一个位图图像（Bitmap），然后再将绘制的位图图像绘制在画布中。本节学习怎样借助程序中的画布对象绘制位图图像。

1. 为 Canvas 对象指定 Bitmap 对象

当我们使用 Canvas 创建对象时，即创建画布时，可以为其指定一个 Bitmap 对象。可以使用 Bitmap 类重载的 static Bitmap createBitmap()方法，例如，调用 createBitmap(int width,int height,Bitmap.Config config)方法创建一个位图图像，创建一个没有任何像素点的位图图像的代码如下：

```java
Bitmap bp = Bitmap.createBitmap(100,100,Bitmap.Config.ARGB_8888);
Canvas c = new Canvas(b);    //为画布指定一个 Bitmap 对象
```

其中 100,100 是位图图像的宽度，常量 Bitmap.Config.ARGB_8888 表示位图图像中每个像素占有 4 个字节。

一旦为画布指定了一个位图图像（只能指定一个），那么画布调用绘制方法绘制的全部图形或图像就都绘制在这个位图图像中。

2. 在依托于 View 或 SurfaceView 的画布上绘制位图图像

然后程序可以使用 7.2 节和 7.3 节介绍的绘制方法，将位图图像 bp 绘制到依托于 View 或 SurfaceView 的画布上。

3. 示例

以下例子 7-3 中，绘制了一幅位图图像，该位图图像上有一个红色的矩形和一个字符串"drawing a bitmap"。效果如图 7.3 所示。

图 7.3　绘制位图图像

例子 7-3

（1）创建名字为 ch7_3 的工程，主要 Activity 子类的名字为 Example7_3，使用的包名为 ch7.three。用命令行进入 D:\2000，创建工程 D:\2000> android create project -t 3 -n ch7_3 -p ./ch7_3 -a Example7_3 -k ch7.three。

（2）将下列和视图相关的 XML 文件保存到工程的\res\layout 目录中。

ch7_3.xml

```xml
<?xml version = "1.0" encoding = "utf-8"?>
<LinearLayout xmlns:android = "http://schemas.android.com/apk/res/android"
    android:id = "@+id/layout"
    android:orientation = "vertical"
    android:layout_width = "match_parent"
    android:layout_height = "match_parent"
    android:background = "#87CEEB">
</LinearLayout>
```

（3）修改工程\src\ch7\three 目录下的 Example7_3.java 文件。修改后的内容如下：

Example7_3.java

```java
package ch7.three;
import android.app.*;
import android.os.Bundle;
import android.widget.*;
import android.view.*;
import android.graphics.*;
import android.content.*;
public class Example7_3 extends Activity {
    Canvas canvas;
    public void onCreate(Bundle savedInstanceState) {
        super.onCreate(savedInstanceState);
        setContentView(R.layout.ch7_3);
        LinearLayout layout = (LinearLayout)findViewById(R.id.layout);
        Bitmap bp = Bitmap.createBitmap(150,150,Bitmap.Config.ARGB_8888);
        canvas = new Canvas(bp);
        Paint paint = new Paint();
        paint.setColor(Color.RED);
        canvas.clipRect(0,0,200,200);
```

```
            canvas.drawRect(new RectF(10,10,80,80),paint);
            canvas.drawText("drawing a bitmap",12,100,paint);
            View view = new CanvasView(this,bp);
            layout.addView(view);
        }
    }
    class CanvasView extends View {
        Bitmap bp;
        CanvasView(Context c,Bitmap bp) {
            super(c);
            this.bp = bp;
            setMinimumHeight(200);
            setMinimumWidth(200);
        }
        public void onDraw(Canvas canvas) {
            Paint paint = new Paint();
            canvas.drawBitmap(bp,0,0,paint);
        }
    }
```

(4) 启动 AVD，进入工程的根目录，用快捷方式编译工程、安装应用程序到 AVD(有关知识点参见 1.5 节)。对于本例子，用命令行进入 D:\2000\ch7_3，执行如下命令：

D:\2000 > ch7_3 > ant debug install

习 题 7

1. 参考例子 7-1，使用依托 View 的画布绘制一个直角三角形。
2. 参考例子 7-2，编写一个小动画程序，模拟龟兔赛跑，用圆形表示小乌龟，用矩形表示小兔子。

第 8 章　Intent 对象与 Activity 对象

主要内容：
- Intent 对象及使用步骤；
- Intent 对象与 AndroidManifest 配置文件；
- 内置范畴与自定义范畴；
- 内置动作与自定义动作；
- Intent 对象的附加数据；
- 启动拨号的 Activity 对象；
- 启动发送短信的 Activity 对象；
- 启动播放视频的 Activity 对象；
- 启动使用 Google 地图的 Activity 对象；
- 启动使用浏览器的 Activity 对象；
- 具有多个 Activity 对象的程序；
- 让 Activity 对象返回数据；
- 启动使用照相机的 Activity 对象。

一个应用程序可以包含若干个 Activity 对象（参见 2.1 节），那么一个应用程序中的一个 Activity 对象可能需要启动当前应用程序中的其他的 Activity 对象，比如一个负责注册的 Activity 对象在接收了用户的注册信息后，可能就需要启动一个负责登录的 Activity 对象，以便用户登录。另一方面，一个应用程序中的一个 Activity 对象可能需要启动其他应用程序中的 Activity 对象为自己服务，比如其他某个应用程序中已经有一个负责拨号的 Activity 对象，当前应用程序想进行拨号时，就可以想办法启动这个负责拨号的 Activity 对象。

应用程序可以使用 Intent 对象帮助当前应用程序启动其他的 Activity 对象，Intent 对象是 Activity 对象进行交往、互通信息的"桥梁"。另外，Intent 对象也能帮助当前应用程序启动 Service 对象和 BroadcastReceiver 对象（有关内容参见第 9 章）。

8.1　Intent 对象及使用步骤

Intent 类的继承关系如下：

```
java.lang.Object
   └ android.content.Intent
```

1. Intent 对象的目的

Intent 对象的字面意思是意图（打算），体现了 Intent 的真正作用。Intent 对象用一个动作(action)或一个动作和数据来描述它的意图，即 Intent 对象给出了一个意图。Intent 对象给出一个意图，然后程序去寻找能完成意图的 Activity 对象。如果找到一个这样的 Activity 对象，就激活这个 Activity 对象，如果找到多个这样的 Activity 对象，就给出一个选择列表，让用户选择其中一个 Activity 对象，并激活所选择的 Activity 对象。

2. 使用 Intent 对象的关键步骤

当前应用程序使用 Intent 对象启动一个或多个 Activity 对象的关键步骤如下。

1）创建 Intent 对象

可以使用 Intent 的构造方法 Intent(String action)创建一个 Intent 对象（有关细节将陆续讲解），比如，Intent 类中的静态常量 ACTION_DIAL 表示这样一个动作(action)：拨打电话，因此，可如下创建一个 Intent 对象：

```
Intent intent = new Intent(Intent.ACTION_DIAL);
```

2）当前应用程序调用 startActivity(Intent intent)方法

当前应用程序调用 startActivity(Intent intent)方法，将相应的 Intent 对象传给方法的参数，那么该方法就会按着参数 intent 中表示的意图（拨打电话）去寻找 Activity 对象。startActivity（Intent intent）方法不能找到 Activity 对象时就会抛出 ActivityNotFoundException 异常，因此建议在 try-catch 语句中使用 startActivity(Intent intent)方法，例如：

```
try {
        startActivity(intent);
}
catch(ActivityNotFoundException exp) {}
```

注：被 startActivity(Intent intent)方法启动的 Activity 对象会进入自己的生命周期，即首先调用 onCreate()方法，然后调用 onStart()方法等（参见 2.4 节）。

3. Intent 的常用构造方法

1）Intent()

创建一个没有任何意图的 Intent 对象，该对象可以使用 setAction(String action)方法添加相应的动作，以便体现 Intent 对象的意图，例如：

```
Intent intent = new Intent();
intent.setAction(Intent.ACTION_DIAL);
```

2）Intent(String action)

创建一个 Intent 对象，该对象中的动作由参数 action 来描述，例如：

```
Intent intent = new Intent(Intent.ACTION_DIAL);
```

也就是说 Intent(String action)构造方法的关键是通过一个动作来构造一个 Intent 对象，即通过一个动作来体现一个意图。

3) Intent(String action, Uri uri)

Intent(String action, Uri uri)构造方法的关键是通过一个动作 action 和一个 Uri 类型的数据 uri 来体现一个意图,即构造一个 Intent 对象,例如:

```
Uri uri = Uri.parse("tel:13887698765");
Intent intent = new Intent(Intent.ACTION_DIAL,uri);
```

那么,Intent 对象体现的意图就是"拨打电话 13887698765",即该意图是通过一个动作"拨打电话"和一个数据"13887698765"来体现的。

Uri 的 parse 方法会从"tel:13887698765"中解析出一个电话号码"13887698765"放到 Uri 对象中,比如,对于 Uri uri = Uri.parse("tel:-1police10");Uri 对象中的电话号码是 110。

上述三个构造方法创建的 Intent 对象所体现的意图被习惯地称为隐式意图(Implicit Intents),即没有明确给出启用哪个 Activity 对象来完成意图,因此使用上述构造方法创建的 Intent 对象,可能启动多个 Activity 对象。比如,Intent 类的静态常量 VIEW 表示的动作是将数据显示给用户(甚至没有说显示怎样类型的数据),那么程序执行下列代码后:

```
Intent intent = new Intent(Intent.ACTION_VIEW);
startActivity(intent);
```

startActivity(Intent intent)方法在手机(或 AVD)的 Android 系统内找到有 5 个 Activity 对象能完成这个意图(这 5 个 Activity 主要是显示系统内部的一些数据),因此列出了这 5 个 Activity 对象,让用户选择其中一个 Activity 对象,并激活所选择的 Activity 对象,效果如图 8.1 所示。

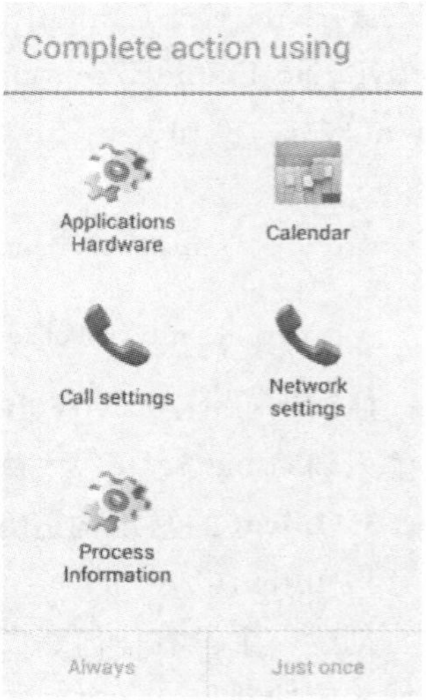

4) Intent(Context packageContext, Class<?> cls)

该构造方法创建的 Intent 对象体现的意图被习惯地称为显式意图(Explicit Intents),即非常准确地给出了要启动的 Activity 对象。该构造方法的参数 packageContext 是当前应用程序所在的上下文,参数 cls 是打算启动的 Activity 对象的类的名字(该类负责创建要启动的 Activity 对象),比如,当前应用程序中有名字为 Hello 的 Activity 的子类,该类负责创建某个 Activity 对象,那么可如下创建一个 Intent 对象:

```
Intent intent = new Intent(this,Hello.class);
```

图 8.1 找到的能完成意图的 Activity 对象

4. 示例

在下面的例子 8-1 中,我们的程序中的 Activity 对象提供了两个按钮,用户单击一个按钮,程序启动拨号的 Activity 对象,单击另一个按钮,程序启动发短信的 Activity 对象。运行效果如图 8.2(a),8.2(b),8.2(c)所示。

例子 8-1

(1) 创建名字为 ch8_1 的工程,主要 Activity 子类的名字为 Example8_1,使用的包名为 ch8.one。用命令行进入 D:\2000,创建工程 D:\2000>android create project -t 3 -n ch8_1 -p ./ch8_1 -a Example8_1 -k ch8.one。

(a) 程序中的Activity　　(b) 拨打电话的Activity　　(c) 发短信的Activity

图 8.2　Activity

(2) 将下列和视图相关的 XML 文件 ch8_1.xml 保存到工程的\res\layout 目录中。
ch8_1.xml

```xml
<?xml version = "1.0" encoding = "utf-8"?>
<LinearLayout xmlns:android = "http://schemas.android.com/apk/res/android"
    android:orientation = "vertical"
    android:layout_width = "match_parent"
    android:layout_height = "match_parent"
    android:background = "#87CEEB"    >
<Button
    android:layout_width = "wrap_content"
    android:layout_height = "wrap_content"
    android:text = "我要拨打电话"
    android:onClick = "phone"       />
<Button
    android:layout_width = "wrap_content"
    android:layout_height = "wrap_content"
    android:text = "我要发短信"
    android:onClick = "sentMess"    />
</LinearLayout>
```

(3) 修改工程\src\ch8\one 目录下的 Example8_1.java 文件,修改后的内容如下:
Example8_1.java

```java
package ch8.one;
import android.content.*;
import android.app.*;
import android.os.Bundle;
import android.net.*;
import android.widget.*;
import android.view.*;
public class Example8_1 extends Activity {
    public void onCreate(Bundle savedInstanceState) {
        super.onCreate(savedInstanceState);
        setContentView(R.layout.ch8_1);
    }
```

```java
    public void phone(View v) {
        Uri uri = Uri.parse("tel:1390000987");
        Intent intent = new Intent(Intent.ACTION_DIAL,uri);
        try {
            startActivity(intent);
        }
        catch(ActivityNotFoundException exp) {
            AlertDialog.Builder build = new AlertDialog.Builder(this);
            AlertDialog dialog = build.create();
            dialog.setTitle("can not find activity!");
            dialog.show();
        }
    }
    public void sentMess(View v) {
        Uri uri = Uri.parse("smsto:1390000987");
        Intent intent = new Intent(Intent.ACTION_SENDTO,uri);
        try {
            startActivity(intent);
        }
        catch(ActivityNotFoundException exp) {
            AlertDialog.Builder build = new AlertDialog.Builder(this);
            AlertDialog dialog = build.create();
            dialog.setTitle("can not find activity!");
            dialog.show();
        }
    }
}
```

（4）启动 AVD，进入工程的根目录，用快捷方式编译工程、安装应用程序到 AVD（有关知识点参见 1.5 节）。对于本例子，用命令行进入 D:\2000\ch8_1，执行如下命令：

D:\2000 > ch8_1 > ant debug install

8.2 Intent 对象与 AndroidManifest 配置文件

1. Intent 对象的动作与范畴

Intent 对象在寻找 Activity 对象时通常需要包含以下两部分内容。

1）action（动作）

该内容给出的信息是一个字符串，该字符串被习惯地称作 Intent 对象中的 action，Intent 对象用 action 要求的所寻找的 Activity 对象应当能进行的动作。

一个 Intent 对象中可以有至多一个 action，Intent 对象可以使用方法 setAction（String action）设置新的 action（如果没有的话）或替换它已有的 action。可以使用方法 String getAction()返回它的 action，如果 Intent 对象没有 action，该方法返回 null。

2）category（范畴）

该内容给出的信息是一个字符串，被称作 Intent 对象的中一个 category，Intent 对象用 category 要求的所寻找的 Activity 对象应在的范畴。Intent 对象可以使用 addCategory

(String category)方法为自己增加新的 category。如果 Intent 对象中有 n 个 category,那么就是要求的所寻找的 Activity 对象必须同时在这 n 个 category 要求的范畴内。Intent 对象使用一个 Set＜String＞集合存放它的全部 category,Intent 对象可以调用 Set＜String＞ getCategories()方法返回它的全部 category。Intent 对象可以调用 removeCategory(String category)方法移除一个 category。

一个 Intent 对象要通过与 Activity 对象相关的 AndroidManifest.xml 配置文件(该文件在工程的根目录下)来确定这个 Activity 是否是它要启动的 Activity 对象,以下介绍 AndroidManifest.xml 文件中的相关内容。

2. AndroidManifest.xml 中的＜action＞与＜category＞标记

Activity 对象一定与项目的配置文件 AndroidManifest.xml 中的一个＜activity＞标记相对应(参见 2.1 节)。一个 Activity 对象对应的＜activity＞标记使用＜intent-filter＞子标记给出自己能被 Intent 对象找到的条件。

＜intent-filter＞标记可有 0 个或多个＜action＞子标记,0 个或多个＜category＞标记。

- ＜action＞标记给出 Activity 对象的动作(但不强迫实际的代码能完成这个动作)。一个＜action＞标记表明 Activty 对象能进行的动作。一个 Activity 对象可以能进行多个动作,就像一个人能进行"跑"、"走"和"跳"等多个动作。
- ＜category＞标记给出 Activity 对象所在的范畴。一个＜category＞标记表明 Activty 对象所在的一个范畴。一个 Activity 对象可以同时在多个范畴内,就像一个人可以同时在"男人"、"大学生"和"公务员"等多个范畴内。

需要注意的是,startActivity(Intent intent)方法在寻找 Activity 对象时,一定会给参数 intent 多增加一个范畴(如果没有就增加这个范畴),增加的这个范畴的值是"android. intent.category.DEFAULT",因此,如果 Activity 对象想被 Intent 对象找到,应当包含一个 ＜category＞标记,并让这个＜category＞标记 android:name 属性的值是"android.intent. category.DEFAULT"。

例如,某个应用程序在配置文件中＜activity＞标记的一个＜intent-filter＞子标记包含的内容如下所示(注意加重的文字)。

AndroidManifest.xml

```
<?xml version = "1.0" encoding = "utf-8"?>
<manifest xmlns:android = "http://schemas.android.com/apk/res/android"
    package = "ch5.six"
    android:versionCode = "1"
    android:versionName = "1.0">
    <application android:label = "@string/app_name" android:icon = "@drawable/ic_launcher">
        <activity android:name = "Example5_6"
            android:label = "@string/app_name">
            <intent-filter>
                <action android:name = "android.intent.action.MAIN" />
                <action android:name = "android.intent.action.VIEW" />
                <category android:name = "android.intent.category.LAUNCHER" />
                <category android:name = "android.intent.category.DEFAULT" />
                <type android:mimeType = "video/mpeg" />
```

```
            </intent-filter>
        </activity>
    </application>
</manifest>
```

上述配置文件中＜activity＞标记注册的 Activity 对象能进行的动作有"android. intent. action. MAIN","android. intent. action. VIEW"。

Activity 对象同时在"android. intent. category. LAUNCHER"和"android. intent. category. DEFAULT"范畴内。

Intent 对象需要对照 AndroidManifest. xml 中＜action＞和＜category＞标记给出的信息,来确定该 Activity 对象是否是它所寻找的 Activity 对象。

例如,Intent 的静态常量 ACTION_MAIN 是一个动作,该常量是一个字符串,这个字符串就是"android. intent. action. MAIN"。

如果如下创建一个 intent 对象:

```
Intent intent = new Intent(Intent.ACTION_MAIN);
```

等价写法是:

```
Intent intent = new Intent("android.intent.action.MAIN");
```

那么该对象中的 action 就是"android. intent. action. MAIN"。

如果执行如下代码:

```
startActivity(intent);
```

那么任何一个 Activity 对象,如果在其 AndroidManifest. xml 配置文件中给出的多个 action 中有一个是"android. intent. action. MAIN",多个 category 中有一个是"android. intent. category. DEFAULT",那么这个 Activity 对象就会被 startActivity(intent)方法找到,比如,前面配置文件中＜activity＞对应的 Activity 对象就会被找到。

也就是说,如果一个 Activity 对象的多个 action 中有一个与 Intent 对象的 action 一致,Activity 的多个 category 包含了 Intent 对象的全部 category,那么这个 Activity 对象就会被 startActivity(intent)方法找到,如果找到多个 Activity 对象,就提示用户选择其中的某一个。

比如,生活中的如下信息,
张三(模拟一个 Activity 对象)给出的应聘信息(模拟 AndroidManifest 配置文件):
能"打字"(模拟一个 action),
能"游泳"(模拟一个 action),
是"男人"(模拟一个 category),
是"中国人"(模拟一个 category),
是"本科学历"(模拟一个 category)
一个公司(模拟另一个 Activity 对象)用招聘广告(模拟 Intent 对象)给出如下条件:
能"游泳"(模拟一个 action),
是"男人"(模拟一个 category),
是"本科学历"(模拟一个 category)

那么该公司只要贴出广告(模拟执行 startActivity(Intent intent)方法),就可以找到张三。

＜activity＞标记可以有多个＜intent-filter＞子标记,每个＜intent-filter＞子标记都是独立给出 Activity 对象能被 Intent 对象找到的一组条件。例如,一个 Activity 对象的配置文件中对应的＜activity＞标记中有如下内容(给出两组可以被找到的条件):

```
< intent - filter >    <!—给出一组能被找到的条件>
    < action android:name = "android.intent.action.MAIN" />
    < action android:name = "android.intent.action.VIEW" />
    < category android:name = "android.intent.category.LAUNCHER" />
    < category android:name = "android.intent.category.DEFAULT" />
</intent - filter >
< intent - filter >    <!—给出另一组能被找到的条件>
    < action android:name = "android.intent.action.PICK" />
    < category android:name = "android.intent.category.DEFAULT" />
</intent - filter >
```

如果如下创建一个 Intent 对象:

Intent intent＝new Intent(Intent.ACTION_VIEW);

那么 intent 就能找到这个 Activity 对象。如果如下创建一个 Intent 对象:

Intent intent＝new Intent(Intent.ACTION_PICK);

那么 intent 也能找到这个 Activity 对象。

3. Intent 对象中的 data

Intent 中可以有一种特殊的数据,即 Uri 类型的数据,称作 Intent 对象中的 data,data 是 Intent 对象要求寻找的 Activity 对象有能力处理的一个数据。

一个 Intent 对象中可以有至多一个 data,Intent 对象可以使用方法 setData(Uri data) 设置新的 data(如果没有的话)或替换它已有的 data。

如果 Intent 包含了这种特殊的数据,Intent 对象要从找到的 Activity 对象中进一步筛选出能处理这种数据的 Activity 对象(系统会调用这些 Activity 对象在 ContentProvider 定义的一些信息,并与该 Activity 对象在配置文件中使用 data 标记定义的相应信息进行匹配,以判断这个 Activity 对象是否是 Intent 对象要寻找的 Activity 对象。深入讨论这些问题已经超出了本书的范围)。

4. 在 Intent 对象中指定应用程序

我们已经知道,创建 Intent 对象 intent 后,startActivity(intent)方法将按着意图 intent 在所有的应用程序中寻找 Activity 对象。如果希望在某个特定应用程序中寻找 Activity 对象,Intent 对象 intent 可以事先使用下列方法给出应用程序的包名(应用程序使用包名来标识自己,参见 2.2 节):

Intent setPackage(String packageName);

例如,假设某个应用程序的包名是 ch8.two,代码如下:

intent.setPackage("ch8.two");

有关代码详细说明见下面例子 8-2 之后给出的注释。

5. 示例

我们已经知道,我们创建工程时指定的 Activity 对象一定能进行"android.intent.action.MAIN"动作,并且在"android.intent.category.LAUNCHER"范畴内,即是可以被操作系统直接加载的 Activity 对象(参见 2.3 节)。

在例子 8-2 中,用户单击一个按钮,程序使用 Intent 对象寻找进行"android.intent.action.MAIN"动作,并且在"android.intent.category.LAUNCHER"范畴内的 Activity 对象(找到了多个,包括本例子 8-2 提供的 Activity 对象,如图 8.2(b)所示)。用户单击另一个按钮,程序使用 Intent 对象列出了能进行"android.intent.action.VIEW"动作,并且在"android.intent.category.DEFAULT"范畴内的 Activity 对象(找到了多个,如图 8.3(c)所示)。运行效果如图 8.3(a),8.3(b),8.3(c)所示(具体运行效果依赖手机中或 AVD 中应用程序的数量)。

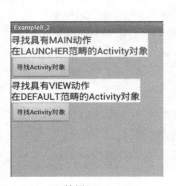
(a) 使用Intent

(b) 找到的Activity

(c) 找到的Activity

图 8.3 运行效果

例子 8-2

(1) 创建名字为 ch8_2 的工程,主要 Activity 子类的名字为 Example8_2,使用的包名为 ch8.two。用命令行进入 D:\2000,创建工程 D:\2000>android create project -t 3 -n ch8_2 -p ./ch8_2 -a Example8_2 -k ch8.two。

(2) 让 Example8_2 类创建的 Activity 对象能进行"android.intent.action.VIEW"动作,并且在"android.intent.category.DEFAULT"范畴内,因此需要修改 AndroidManifest.xml 配置文件,修改后的内容如下:

AndroidManifest.xml

```
<?xml version = "1.0" encoding = "utf-8"?>
< manifest xmlns:android = "http://schemas.android.com/apk/res/android"
        package = "ch8.two"
        android:versionCode = "1"
        android:versionName = "1.0">
```

```xml
<application android:label = "@string/app_name" android:icon = "@drawable/ic_launcher">
    <activity android:name = "Example8_2"
              android:label = "@string/app_name">
        <intent-filter>
            <action android:name = "android.intent.action.MAIN" />
            <action android:name = "android.intent.action.VIEW" />        <!--新增加的动作-->
            <category android:name = "android.intent.category.LAUNCHER" />
            <category android:name = "android.intent.category.DEFAULT" /> <!--新增加的范畴-->
        </intent-filter>
    </activity>
</application>
</manifest>
```

(3) 将下列和视图相关的 XML 文件保存到工程的\res\layout 目录中。

ch8_2.xml

```xml
<?xml version = "1.0" encoding = "utf-8"?>
<LinearLayout xmlns:android = "http://schemas.android.com/apk/res/android"
    android:orientation = "vertical"
    android:layout_width = "match_parent"
    android:layout_height = "match_parent"
    android:background = "#87CEEB"  >
    <TextView
        android:layout_width = "wrap_content"
        android:layout_height = "wrap_content"
        android:background = "#FFFFFF"
        android:textColor = "#0000FF"
        android:textSize = "20sp"
        android:text = "寻找具有 MAIN 动作\n 在 LAUNCHER 范畴的 Activity 对象"/>
    <Button
        android:layout_width = "wrap_content"
        android:layout_height = "wrap_content"
        android:text = "寻找 Activity 对象"
        android:onClick = "findOne"        />
    <TextView
        android:layout_width = "wrap_content"
        android:layout_height = "wrap_content"
        android:background = "#FFFFFF"
        android:textColor = "#000000"
        android:textSize = "20sp"
        android:text = "寻找具有 VIEW 动作\n 在 DEFAULT 范畴的 Activity 对象"/>
    <Button
        android:layout_width = "wrap_content"
        android:layout_height = "wrap_content"
        android:text = "寻找 Activity 对象"
        android:onClick = "findTwo"        />
</LinearLayout>
```

(4) 修改工程\src\ch8\two 目录下的 Example8_2.java 文件,修改后的内容如下:

Example8_2.java

```java
package ch8.two;
import android.content.*;
```

```
    import android.app.Activity;
    import android.os.Bundle;
    import android.widget.*;
    import android.view.*;
    public class Example8_2 extends Activity {
        public void onCreate(Bundle savedInstanceState) {
            super.onCreate(savedInstanceState);
            setContentView(R.layout.ch8_2);
        }
        public void findOne(View v) {
            Intent intent = new Intent("android.intent.action.MAIN");
            intent.addCategory("android.intent.category.LAUNCHER") ;//增加一个 category
            startActivity(intent);
        }
        public void findTwo(View v) {
            Intent intent = new Intent(Intent.ACTION_VIEW);
            startActivity(intent);
        }
    }
```

（5）启动 AVD，进入工程的根目录，用快捷方式编译工程、安装应用程序到 AVD（有关知识点参见 1.5 节）。对于本例子，用命令行进入 D:\2000\ch8_2，执行如下命令：

```
D:\2000>ch8_2>ant debug install
```

注：如果在 Example8_2.java 的 findTwo(View v)方法中增加如下代码：

```
intent.setPackage("ch8.two");
```

那么单击按钮后，startActivity(intent)方法只能找到 ch8.two 应用程序中的 Activity 对象，并启动该 Activity 对象。

8.3　内置范畴与自定义范畴

1. 内置范畴

Intent 类用一些静态的 String 常量表示一些范畴，这些范畴称作内置范畴，如果希望寻找系统内置的一些特殊 Activity 对象，就可以在 Intent 对象中使用这些内置范畴（参见 8.6 节至 8.11 节）。提倡 Activity 对象在其配置文件中让自己属于内置的 CATEGORY_DEFAULT（"android.intent.category.DEFAULT"）范畴，以便自己能被 Intent 对象找到，其原因是 startActivity(Intent intent)方法在寻找 Activity 对象时，一定会给参数 intent 多增加一个范畴（如果没有就增加这个范畴），增加的这个范畴的值正是"android.intent.category.DEFAULT"，因此，如果 Activity 对象想被 Intent 对象找到，应当包含一个＜category＞标记，并让这个＜category＞标记 android:name 属性的值是"android.intent.category.DEFAULT"。Android 系统一共有 16 个内置范畴，读者可以查看帮助文档了解这些内置范畴。

2. 自定义范畴

自定义 category（范畴）的目的不仅是让自己属于的范畴不同于内置范畴，而且也便于用户程序准确寻找到用户自己编写的 Activity 对象。category 是一个字符串，例如"android.intent.category.GENG"，其中的 GENG 习惯上用大写。category 的包名可任意给定，例如，包名是 sohu.com.cn 的范畴"sohu.com.cn.BIRD"。

3. 示例

例子 8-3 需要创建两个工程，其中一个工程中的 Activity 对象可以使用 Intent 对象启动另外一个工程中的 Activity 对象。第一个将自己属于"android.intent.category.DEFAULT"范畴和"android.intent.category.GENG"，第二个工程都将自己属于"android.intent.category.DEFAULT"范畴和"android.intent.category.ZHANG"范畴。运行效果如图 8.4(a)，8.4(b)所示。

(a) Example8_3_1中的Activity对象　　(b) Example8_3_2中的Activity对象

图 8.4　运行效果

例子 8-3

1) 工程一

(1) 创建名字为 ch8_3_1 的工程，主要 Activity 子类的名字为 Example8_3_1，使用的包名为 ch8.three_1。用命令行进入 D:\2000，创建工程 D:\2000＞android create project -t 3 -n ch8_3_1 -p ./ch8_3_1 -a Example8_3_1 -k ch8.three_1。

(2) 修改工程 ch8_3_1 中的 AndroidManifest.xml 配置文件(在工程的根目录中，本例就是 ch8_3_1 中)，新增两个范畴，修改后的内容如下：

AndroidManifest.xml

```xml
<?xml version="1.0" encoding="utf-8"?>
<manifest xmlns:android="http://schemas.android.com/apk/res/android"
      package="ch8.three_1"
      android:versionCode="1"
      android:versionName="1.0">
  <application android:label="@string/app_name" android:icon="@drawable/ic_launcher">
      <activity android:name="Example8_3_1"
                android:label="@string/app_name">
          <intent-filter>
              <action android:name="android.intent.action.MAIN" />
              <category android:name="android.intent.category.LAUNCHER" />
              <category android:name="android.intent.category.DEFAULT" />
              <category android:name="sina.com.cn.GENG" />
          </intent-filter>
      </activity>
  </application>
</manifest>
```

(3) 将下列和视图相关的 XML 文件保存到工程的\res\layout 目录中。
ch8_3_1.xml

```xml
<?xml version = "1.0" encoding = "utf-8"?>
<LinearLayout xmlns:android = "http://schemas.android.com/apk/res/android"
    android:orientation = "vertical"
    android:layout_width = "match_parent"
    android:layout_height = "match_parent"
    android:background = "#87CEEB"    >
    <TextView
        android:layout_width = "wrap_content"
        android:layout_height = "wrap_content"
        android:background = "#FFFFFF"
        android:textColor = "#000000"
        android:textSize = "20sp"
        android:text = "我是 Exampe8_3_1\n 寻找 Example8_3_2 的 Activity 对象\n"/>
    <Button
        android:layout_width = "wrap_content"
        android:layout_height = "wrap_content"
        android:text = "寻找 Activity 对象"
        android:onClick = "find"    />
</LinearLayout>
```

(4) 修改工程\src\ch8\three_1 目录下的 Example8_3_1.java 文件,修改后的内容如下:

Example8_3_1.java

```java
package ch8.three_1;
import android.content.*;
import android.app.*;
import android.os.Bundle;
import android.widget.*;
import android.view.*;
public class Example8_3_1 extends Activity {
    public void onCreate(Bundle savedInstanceState) {
        super.onCreate(savedInstanceState);
        setContentView(R.layout.ch8_3_1);
    }
    public void find(View v) {
        Intent intent = new Intent("android.intent.action.MAIN");
        intent.addCategory("sohu.com.cn.ZHANG");  //增加一个 category
        try {
            startActivity(intent);
        }
        catch(ActivityNotFoundException exp) {
            AlertDialog.Builder build = new AlertDialog.Builder(this);
            AlertDialog dialog = build.create();
            dialog.setTitle("can not find activity!");
            dialog.show();
        }
    }
}
```

（5）启动 AVD，进入工程的根目录，用快捷方式编译工程、安装应用程序到 AVD（有关知识点参见 1.5 节）。对于本例子，用命令行进入 D:\2000\ch8_3_1，执行如下命令：

D:\2000＞ch8_3_1＞ant debug install

2）工程二

（1）创建名字为 ch8_3_2 的工程，主要 Activity 子类的名字为 Example8_3_2，使用的包名为 ch8.three_2。用命令行进入 D:\2000，创建工程 D:\2000＞android create project -t 3 -n ch8_3_2 -p ./ch8_3_2 -a Example8_3_2 -k ch8.three_2。

（2）修改工程 ch8_3_2 中的 AndroidManifest.xml 配置文件（在工程的根目录中，本例就是 ch8_3_2 中），新增两个范畴，修改后的内容如下：

AndroidManifest.xml

```
<?xml version = "1.0" encoding = "utf-8"?>
<manifest xmlns:android = "http://schemas.android.com/apk/res/android"
      package = "ch8.three_2"
      android:versionCode = "1"
      android:versionName = "1.0">
   <application android:label = "@string/app_name" android:icon = "@drawable/ic_launcher">
         <activity android:name = "Example8_3_2"
                android:label = "@string/app_name">
            <intent-filter>
               <action android:name = "android.intent.action.MAIN" />
               <category android:name = "android.intent.category.LAUNCHER" />
               <category android:name = "android.intent.category.DEFAULT" />
               <category android:name = "sohu.com.cn.ZHANG" />
            </intent-filter>
         </activity>
   </application>
</manifest>
```

（3）将下列和视图相关的 XML 文件 ch8_3_2.xml 保存到工程的\res\layout 目录中。

ch8_3_2.xml

```
<?xml version = "1.0" encoding = "utf-8"?>
<LinearLayout xmlns:android = "http://schemas.android.com/apk/res/android"
      android:orientation = "vertical"
      android:layout_width = "match_parent"
      android:layout_height = "match_parent"
      android:background = "#87CEEB"     >
      <TextView
         android:layout_width = "wrap_content"
         android:layout_height = "wrap_content"
         android:background = "#000000"
         android:textColor = "#FFFFFF"
         android:textSize = "20sp"
         android:text = "我是 Exampe8_3_2\n 寻找 Example8_3_1 的 Activity 对象\n"/>
      <Button
         android:layout_width = "wrap_content"
```

```
            android:layout_height = "wrap_content"
            android:text = "寻找 Activity 对象"
            android:onClick = "find"        />
</LinearLayout>
```

(4) 修改工程\src\ch8\three_2 目录下的 Example8_3_2.java 文件，修改后的内容如下：

Example8_3_2.java

```
package ch8.three_2;
import android.content.*;
import android.app.*;
import android.os.Bundle;
import android.widget.*;
import android.view.*;
public class Example8_3_2 extends Activity {
    public void onCreate(Bundle savedInstanceState) {
        super.onCreate(savedInstanceState);
        setContentView(R.layout.ch8_3_2);
    }
    public void find(View v) {
        Intent intent = new Intent("android.intent.action.MAIN");
        intent.addCategory("sina.com.cn.GENG") ;//增加一个 category
        try {
            startActivity(intent);
        }
        catch(ActivityNotFoundException exp) {
            AlertDialog.Builder build = new AlertDialog.Builder(this);
            AlertDialog dialog = build.create();
            dialog.setTitle("can not find activity!");
            dialog.show();
        }
    }
}
```

(5) 启动 AVD，进入工程的根目录，用快捷方式编译工程、安装应用程序到 AVD(有关知识点参见 1.5 节)。对于本例子，用命令行进入 D:\2000\ch8_3_2，执行如下命令：

```
D:\2000>ch8_3_2>ant debug install
```

8.4　内置动作与自定义动作

1. 内置动作

Intent 类用一些静态的 String 常量表示一些动作，这些动作称作内置动作，如果希望寻找系统内置的一些特殊的 Activity 对象，就可以在 Intent 对象中使用这些内置动作，例如打开内置的浏览器、拨号程序等(参见 8.6 节)。

2. 自定义动作

自定义 action 的目的不仅是让自己的动作不同于内置动作，而且也便于用户程序准确寻找到用户自己编写的 Activity 对象。action 是一个字符串，例如"android.intent.action.COMPUTE"，其中的 COMPUTE 习惯上用大写。action 的包名可任意给定，例如，包名是 com.sun.moon 的 action："com.sun.moon.LISTENER"。

3. 示例

例子 8-4 需要创建两个工程，其中第一个工程中的 Activity 对象使用 Intent 对象启动第二个工程中的 Activity 对象，该对象能输出英文字母表。第二个工程将自己属于 "android.intent.category.DEFAULT"范畴，并能进行如下的动作"android.intent.action.OUTPUT_ENGLISH"，运行效果如图 8.5(a)，8.5(b)所示。

(a) Example8_4_1中的Activity　　(b) Example8_4_2中的Activity

图 8.5　运行效果

例子 8-4

1) 工程一

(1) 创建名字为 ch8_4_1 的工程，主要 Activity 子类的名字为 Example8_4_1，使用的包名为 ch8.four_1。用命令行进入 D:\2000，创建工程 D:\2000＞android create project -t 3 -n ch8_4_1 -p ./ch8_4_1 -a Example8_4_1 -k ch8.four_1。

(2) 将下列和视图相关的 XML 文件保存到工程的\res\layout 目录中。

ch8_4_1.xml

```xml
<?xml version = "1.0" encoding = "utf-8"?>
< LinearLayout xmlns:android = "http://schemas.android.com/apk/res/android"
    android:orientation = "vertical"
    android:layout_width = "match_parent"
    android:layout_height = "match_parent"
    android:background = "#87CEEB"       >
    < TextView
        android:id = "@ + id/text"
        android:layout_width = "wrap_content"
        android:layout_height = "wrap_content"
        android:background = "#FFFFFF"
        android:textColor = "#000000"
        android:textSize = "20sp"
        android:text = "启动 Example8_4_2 的 Activity 对象\n 查看英文字母表"/>
    < Button
        android:layout_width = "wrap_content"
        android:layout_height = "wrap_content"
        android:text = "启动 Activity 对象"
        android:onClick = "find"         />
</LinearLayout >
```

(3) 修改工程\src\ch8\four_1 目录下的 Example8_4_1.java 文件，修改后的内容如下：

Example8_4_1.java

```java
package ch8.four_1;
import android.content.*;
import android.app.*;
import android.os.Bundle;
import android.view.*;
import android.widget.*;
public class Example8_4_1 extends Activity {
    TextView text;
    public void onCreate(Bundle savedInstanceState) {
        super.onCreate(savedInstanceState);
        setContentView(R.layout.ch8_4_1);
    }
    public void find(View v) {
        Intent intent = new Intent("sohu.com.cn.OUTPUT_ENGLISH");
        try {
            startActivity(intent);
        }
        catch(ActivityNotFoundException exp) {
            AlertDialog.Builder build = new AlertDialog.Builder(this);
            AlertDialog dialog = build.create();
            dialog.setTitle("can not find activity!");
            dialog.show();
        }
    }
}
```

(4) 启动 AVD，进入工程的根目录，用快捷方式编译工程、安装应用程序到 AVD（有关知识点参见 1.5 节）。对于本例子，用命令行进入 D:\2000\ch8_4_1，执行如下命令：

D:\2000＞ch8_4_1＞ant debug install

2）工程二

(1) 创建名字为 ch8_4_2 的工程，主要 Activity 子类的名字为 Example8_4_2，使用的包名为 ch8.four_2。用命令行进入 D:\2000，创建工程 D:\2000＞android create project -t 3 -n ch8_4_2 -p ./ch8_4_2 -a Example8_4_2 -k ch8.four_2。

(2) 修改工程 ch8_4_2 中的 AndroidManifest.xml 配置文件（在工程的根目录中，本例就是 ch8_4_2 中），新增一个 OUTPUT_ENGLISH 动作，新增一个 DEFAULT 范畴，修改后的内容如下：

AndroidManifest.xml

```xml
<?xml version = "1.0" encoding = "utf-8"?>
<manifest xmlns:android = "http://schemas.android.com/apk/res/android"
    package = "ch8.four_2"
    android:versionCode = "1"
    android:versionName = "1.0">
  <application android:label = "@string/app_name" android:icon = "@drawable/ic_launcher">
    <activity android:name = "Example8_4_2"
              android:label = "@string/app_name">
```

```xml
            <intent-filter>
                <action android:name = "android.intent.action.MAIN" />
                <action android:name = "sohu.com.cn.OUTPUT_ENGLISH" />
                <category android:name = "android.intent.category.LAUNCHER" />
                <category android:name = "android.intent.category.DEFAULT" />
            </intent-filter>
        </activity>
    </application>
</manifest>
```

(3) 将下列和视图相关的 XML 文件保存到工程的\res\layout 目录中。

ch8_4_2.xml

```xml
<?xml version = "1.0" encoding = "utf-8"?>
<LinearLayout xmlns:android = "http://schemas.android.com/apk/res/android"
    android:orientation = "vertical"
    android:layout_width = "match_parent"
    android:layout_height = "match_parent"
    android:background = "#88CEAB"    >
    <TextView
        android:id = "@+id/text"
        android:layout_width = "match_parent"
        android:layout_height = "match_parent"
        android:background = "#AFFA00"
        android:textColor = "#FFFFFF"
        android:textSize = "22sp"/>
</LinearLayout>
```

(4) 修改工程\src\ch8\four_2 目录下的 Example8_4_2.java 文件,修改后的内容如下:

Example8_4_2.java

```java
package ch8.four_2;
import android.app.Activity;
import android.os.Bundle;
import android.widget.*;
public class Example8_4_2 extends Activity {
    public void onCreate(Bundle savedInstanceState) {
        super.onCreate(savedInstanceState);
        setContentView(R.layout.ch8_4_2);
        TextView text = (TextView)findViewById(R.id.text);
        for(char c = 'A';c <= 'Z';c++)
            text.append("\t" + c);
    }
}
```

(5) 启动 AVD,进入工程的根目录,用快捷方式编译工程、安装应用程序到 AVD(有关知识点参见 1.5 节)。对于本例子,用命令行进入 D:\2000\ch8_4_2,执行如下命令:

D:\2000>ch8_4_2>ant debug install

8.5　Intent 对象的附加数据

Intent 对象的附加数据不是寻找 Activity 对象的条件,是传递给所找到的 Activity 对象的数据,被找到的 Activity 对象可以根据需要决定是否接收这些数据。

1. 添加与获取附加数据

Intent 对象 intent 可以使用 putExtra 方法添加附加数据,并为这个附加数据指定一个 String 类型的检索关键字,例如:

```
intent.putExtra("PI",3.1415926);
intent.putExtra("HelloI",true);
```

被找到的 Activity 对象可以根据需要决定是否接收这些数据,如果希望使用 3.1415926 和 true 这两个值,首先得到启动当前 Activity 对象的 Intent 对象:

```
Intent intent = getIntent();
```

然后 intent 调用诸如 getXXX Extra（关键字）的方法,通过关键字 key 得到想要的数据,例如:

```
double pi = intent.getDoubleExtra ("PI",3.14); //如果获取不到,就用默认值 3.14
boolean boo = intent. getBooleanExtra ("Hello");
```

另外,也可以事先创建一个 Bundle 对象:

```
Bundle bundle = new Bundle();
```

bundle 对象调用诸如 putXXX(关键字 key,数据 data)的方法将数据 data 放入 bundle 对象中,并为该数据指定一个 String 类型的检索关键字,例如:

```
bundle.putDouble("PI",3.1415926);
bundle.putString("HelloI",true);
```

然后,Intent 对象使用 putExtras(Bundle extras)方法将 bundle 对象放入当前 Intent 对象中,例如:

```
intent. putExtras(bundle);
```

被找到的 Activity 对象可以根据需要决定是否接收这些数据,如果希望使用 3.1415926 和 true 这两个值,首先得到启动当前 Activity 对象的 Intent 对象:

```
Intent intent = getIntent();
```

然后 intent 调用 Bundle getExtras()的方法返回其中的 bundle 对象,例如:

```
Bundle bundle = intent. getExtras();
```

最后,bundle 对象调用诸如 getXXX(String key,…)的方法,通过关键字 key 得到想要的数据:

```
double pi = bundle.getDouble("PI",3.14); //如果获取不到,就用默认值 3.14
boolean boo = bundle. getBoolean ("Hello");
```

2. 示例

例子 8-5 需要创建两个工程,其中第一个工程中的 Activity 对象使用 Intent 对象启动第二个工程中的 Activity 对象。第二个工程中的 Activity 对象能计算矩形的面积,第一个工程中的 Activity 对象在启动第二个工程中的 Activity 对象时将矩形的宽和高作为附加数据传递给第二个工程中的 Activity 对象。第二个工程将自己属于"android.intent.category.DEFAULT"范畴,并只能进行如下的动作:"android.intent.action.COMPUTE_AREA",即第二个工程中没有 MAIN 动作,因此系统不提供应用程序的图标,该工程得到的 Activity 对象只能由其他 Activity 对象启动。运行效果如图 8.6(a),8.6(b)所示。

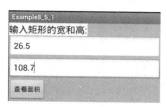
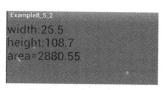

(a) 向Intent对象放入附加数据　　(b) 得到Intent对象的附加数据

图 8.6　运行效果

例子 8-5

1) 工程一

(1) 创建名字为 ch8_5_1 的工程,主要 Activity 子类的名字为 Example8_5_1,使用的包名为 ch8.five_1。用命令行进入 D:\2000,创建工程 D:\2000>android create project -t 3 -n ch8_5_1 -p ./ch8_5_1 -a Example8_5_1 -k ch8.five_1。

(2) 将下列和视图相关的 XML 文件保存到工程的\res\layout 目录中。

ch8_5_1.xml

```xml
<?xml version = "1.0" encoding = "utf-8"?>
<LinearLayout xmlns:android = "http://schemas.android.com/apk/res/android"
    android:orientation = "vertical"
    android:layout_width = "match_parent"
    android:layout_height = "match_parent"
    android:background = "#87CEEB"    >
    <TextView
        android:layout_width = "wrap_content"
        android:layout_height = "wrap_content"
        android:background = "#FFFFFF"
        android:textColor = "#000000"
        android:textSize = "20sp"
        android:text = "输入矩形的宽和高:"/>
    <EditText
        android:layout_width = "match_parent"
        android:layout_height = "wrap_content"
        android:inputType = "numberDecimal"
        android:id = "@+id/editWidth" />
    <EditText
        android:layout_width = "match_parent"
```

```
            android:layout_height = "wrap_content"
            android:inputType = "numberDecimal"
            android:id = "@ + id/editHeight" />
        <Button
            android:layout_width = "wrap_content"
            android:layout_height = "wrap_content"
            android:text = "查看面积"
            android:onClick = "find"          />
</LinearLayout>
```

(3) 修改工程\src\ch8\five_1 目录下的 Example8_5_1.java 文件，修改后的内容如下：

Example8_5_1.java

```
package ch8.five_1;
import android.content.*;
import android.app.*;
import android.os.Bundle;
import android.view.*;
import android.widget.*;import android.net.*;
public class Example8_5_1 extends Activity {
    EditText textWidth,textHeight;
    public void onCreate(Bundle savedInstanceState) {
        super.onCreate(savedInstanceState);
        setContentView(R.layout.ch8_5_1);
        textWidth = (EditText) findViewById(R.id.editWidth);
        textHeight = (EditText)findViewById(R.id.editHeight);
    }
    public void find(View v) {
        Intent intent = new Intent("android.intent.action.COMPUTE_AREA");
        double width = Double.parseDouble(textWidth.getText().toString());
        double height = Double.parseDouble(textHeight.getText().toString());
        try {
           intent.putExtra("width",width);
           intent.putExtra("height",height);
           startActivity(intent);
        }
        catch(ActivityNotFoundException exp) {
          AlertDialog.Builder build = new AlertDialog.Builder(this);
          AlertDialog dialog = build.create();
          dialog.setTitle("can not find activity!");
          dialog.show();
        }
    }
}
```

(4) 启动 AVD，进入工程的根目录，用快捷方式编译工程、安装应用程序到 AVD(有关知识点参见 1.5 节)。对于本例子，用命令行进入 D:\2000\ch8_5_1，执行如下命令：

D:\2000 > ch8_5_1 > ant debug install

2) 工程二

(1) 创建名字为 ch8_5_2 的工程，主要 Activity 子类的名字为 Example8_5_2，使用的包名为 ch8.four_2。用命令行进入 D:\2000，创建工程 D:\2000>android create project -t

3 -n ch8_5_2 -p . /ch8_5_2 -a Example8_5_2 -k ch8.five_2。

（2）修改工程 ch8_5_2 中的 AndroidManifest.xml 配置文件（在工程的根目录中，本例就是 ch8_5_2 中），去掉 MAIN 动作和 LAUNCHER 范畴，新增一个 COMPUTE_AREA 动作和一个 DEFAULT 范畴，修改后的内容如下：

AndroidManifest.xml

```xml
<?xml version = "1.0" encoding = "utf-8"?>
<manifest xmlns:android = "http://schemas.android.com/apk/res/android"
    package = "ch8.five_2"
    android:versionCode = "1"
    android:versionName = "1.0">
    <application android:label = "@string/app_name" android:icon = "@drawable/ic_launcher">
        <activity android:name = "Example8_5_2"
                android:label = "@string/app_name">
            <intent-filter>
                <action android:name = "android.intent.action.COMPUTE_AREA" />
                <category android:name = "android.intent.category.DEFAULT" />
            </intent-filter>
        </activity>
    </application>
</manifest>
```

（3）将下列和视图相关的 XML 文件保存到工程的 \res\layout 目录中。

ch8_5_2.xml

```xml
<?xml version = "1.0" encoding = "utf-8"?>
<LinearLayout xmlns:android = "http://schemas.android.com/apk/res/android"
    android:orientation = "vertical"
    android:layout_width = "match_parent"
    android:layout_height = "match_parent"
    android:background = "#88CEAB"    >
    <TextView
        android:id = "@+id/text"
        android:layout_width = "match_parent"
        android:layout_height = "match_parent"
        android:background = "#FF00FF"
        android:textColor = "#000000"
        android:textSize = "25sp"/>
</LinearLayout>
```

（4）修改工程 \src\ch8\five_2 目录下的 Example8_5_2.java 文件，修改后的内容如下：

Example8_5_2.java

```java
package ch8.five_2;
import android.content.*;
import android.app.*;
import android.os.Bundle;
import android.widget.*;
public class Example8_5_2 extends Activity {
    public void onCreate(Bundle savedInstanceState) {
        super.onCreate(savedInstanceState);
        setContentView(R.layout.ch8_5_2);
        TextView text = (TextView)findViewById(R.id.text);
```

```
            Intent intent = getIntent();
            double width = intent.getDoubleExtra("width",0); //如果获取不到,就用默认值 0
            double height = intent.getDoubleExtra("height",0);
            double area   = width * height;
            text.append("width:" + width + "\n");
            text.append("height:" + height + "\n");
            text.append("area = " + area + "\n");
        }
    }
```

（5）启动 AVD,进入工程的根目录,用快捷方式编译工程、安装应用程序到 AVD(有关知识点参见 1.5 节)。对于本例子,用命令行进入 D:\2000\ch8_5_2,执行如下命令:

```
D:\2000>ch8_5_2>ant debug install
```

8.6 启动拨号的 Activity 对象

1. ACTION_DIAL

Intent 类的静态常量 ACTION_DIAL 表示一个动作,其值等于"android. intent. action. DIAL",使用该动作可以启动 Android 系统内负责拨号的 Activity 对象(该对象的相关配置文件 AndroidManifest. xml 中的 action 的值一定是"android. intent. action. DIAL")。下列代码将启动拨号的 Activity 对象:

```
Intent intent = new Intent(Intent.ACTION_DIAL);
//或 Intent intent = new Intent("android.intent.action.DIAL");
startActivity(intent);
```

2. 示例

在下面的例子 8-6 中,Activity 对象中有一个 ListView 视图,该视图上列出了一些大学的联系电话,选择一个大学之后,可以启动拨号的 Activity 对象。运行效果如图 8.7(a)、8.7(b)所示。

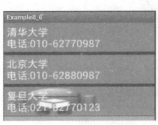

(a) 选择大学的Activity　　　　(b) 拨号的Activity

图 8.7 运行效果

例子 8-6

（1）创建名字为 ch8_6 的工程，主要 Activity 子类的名字为 Example8_6，使用的包名为 ch8.six。用命令行进入 D:\2000，创建工程 D:\2000＞android create project -t 3 -n ch8_6 -p ./ch8_6 -a Example8_6 -k ch8.six。

（2）增加值资源。将下列 array.xml 保存到值资源中，即保存到工程的\res\values 目录中（有关知识点参见 2.6 节）。

array.xml

```xml
<?xml version = "1.0" encoding = "utf-8"?>
<resources>
    <array name = "university_list">
        <item>清华大学 \n 电话:010-62770987</item>
        <item>北京大学 \n 电话:010-62880987</item>
        <item>复旦大学 \n 电话:021-82770123</item>
    </array>
</resources>
```

（3）将下列和视图相关的 XML 文件保存到工程的\res\layout 目录中。

ch8_6.xml

```xml
<?xml version = "1.0" encoding = "utf-8"?>
<LinearLayout xmlns:android = "http://schemas.android.com/apk/res/android"
    android:orientation = "vertical"
    android:layout_width = "match_parent"
    android:layout_height = "match_parent"
    android:background = "#87CEEB">
    <ListView
        android:layout_width = "wrap_content"
        android:layout_height = "wrap_content"
        android:divider = "#0000FF"
        android:dividerHeight = "6dp"
        android:background = "#22bbcc"
        android:listSelector = "@drawable/ic_launcher"
        android:id = "@+id/my_list"
        android:entries = "@array/university_list" />
</LinearLayout>
```

（4）修改工程\src\ch8\six 目录下的 Example8_6.java 文件，修改后的内容如下：

Example8_6.java

```java
package ch8.six;
import android.app.*;
import android.os.Bundle;
import android.widget.*;
import android.view.*;
import android.content.*;
import android.net.*;
public class Example8_6 extends Activity implements AdapterView.OnItemClickListener {
    ListView  listView;
    String nameAndPhone;
```

```java
public void onCreate(Bundle savedInstanceState) {
    super.onCreate(savedInstanceState);
    setContentView(R.layout.ch8_6);
    listView =  (ListView)findViewById(R.id.my_list);
    listView.setOnItemClickListener(this);
}
public void onItemClick(AdapterView parent,View view,int pos,long id) {
    nameAndPhone = listView.getItemAtPosition(pos).toString();
    Uri uri = Uri.parse("tel:" + nameAndPhone);
    Intent intent = new Intent(Intent.ACTION_DIAL,uri);
    try {
        startActivity(intent);
    }
    catch(ActivityNotFoundException exp) {
        AlertDialog.Builder build = new AlertDialog.Builder(this);
        AlertDialog dialog = build.create();
        dialog.setTitle("can not find activity!");
        dialog.show();
    }
}
```

(5) 启动 AVD，进入工程的根目录，用快捷方式编译工程、安装应用程序到 AVD(有关知识点参见 1.5 节)。对于本例子，用命令行进入 D:\2000\ch8_6，执行如下命令：

```
D:\2000 > ch8_6 > ant debug install
```

8.7 启动发送短信的 Activity 对象

1. ACTION_SENDTO

静态常量 ACTION_SENDTO 表示一个动作，其值等于"android.intent.action.SENDTO"，使用该动作可以启动 Android 系统内负责发送短信的 Activity 对象。下列代码将启动发送短信的 Activity 对象。

```java
//启动发短信的 Activity 对象
Uri uri = Uri.parse("smsto:");
Intent intent = new Intent("android.intent.action.SENDTO",uri);
//启动发短信的 Activity 对象,并给 10086 发送内容为"Hello"的短信:
Uri uri = Uri.parse("smsto:10086");
Intent intent = new Intent(Intent.ACTION_SENDTO,uri);
intent.putExtra("sms_body","Hello");
startActivity(intent);
```

2. 示例

在下面的例子 8-7 中，Activity 对象中有一个 ListView 视图，该视图上列出了几个电视栏目的联系电话，选择一个电视栏目后，可以启动发短信的 Activity 对象。运行效果如图 8.8(a),8.8(b)所示。

(a) 选择栏目的Activity对象　　　(b) 发短信的Activity对象

图 8.8　运行效果

例子 8-7

（1）创建名字为 ch8_7 的工程，主要 Activity 子类的名字为 Example8_7，使用的包名为 ch8.seven。用命令行进入 D:\2000，创建工程 D:\2000＞android create project -t 3 -n ch8_7 -p ./ch8_7 -a Example8_7 -k ch8.seven。

（2）增加值资源。将下列 array.xml 保存到值资源中，即保存到工程的\res\values 目录中（有关知识点参见 2.6 节）。

array.xml

```
<?xml version = "1.0" encoding = "utf-8"?>
<resources>
    <array name = "tv_list">
        <item>足球之夜 \n 电话:010 - 62770987</item>
        <item>实话实说 \n 电话:010 - 62880987</item>
        <item>非你莫属\n 电话:021 - 82770123</item>
    </array>
</resources>
```

（3）将下列和视图相关的 XML 文件保存到工程的\res\layout 目录中。

ch8_7.xml

```
<?xml version = "1.0" encoding = "utf-8"?>
<LinearLayout xmlns:android = "http://schemas.android.com/apk/res/android"
        android:orientation = "vertical"
        android:layout_width = "match_parent"
        android:layout_height = "match_parent"
        android:background = "#87CEEB">
    <ListView
            android:layout_width = "wrap_content"
            android:layout_height = "wrap_content"
            android:divider = "#0000FF"
            android:dividerHeight = "5dp"
            android:background = "#66bbcc"
            android:listSelector = "@drawable/ic_launcher"
            android:id = "@+id/my_list"
            android:entries = "@array/tv_list" />
</LinearLayout>
```

（4）修改工程\src\ch8\seven 目录下的 Example8_7.java 文件,修改后的内容如下:

Example8_7.java

```java
package ch8.seven;
import android.app.*;
import android.os.Bundle;
import android.widget.*;
import android.view.*;
import android.content.*;
import android.net.*;
public class Example8_7 extends Activity implements AdapterView.OnItemClickListener {
    ListView  listView;
    String nameAndPhone;
    public void onCreate(Bundle savedInstanceState) {
        super.onCreate(savedInstanceState);
        setContentView(R.layout.ch8_7);
        listView =  (ListView)findViewById(R.id.my_list);
        listView.setOnItemClickListener(this);
    }
    public void onItemClick(AdapterView parent,View view,int pos,long id) {
        nameAndPhone = listView.getItemAtPosition(pos).toString();
        nameAndPhone = nameAndPhone.replaceAll("[^0123456789 - ] + ","");
        Uri uri = Uri.parse("smsto:" + nameAndPhone);
        Intent intent = new Intent("android.intent.action.SENDTO",uri);
        intent.putExtra("sms_body","Hello");
        try {
            startActivity(intent);
        }
        catch(ActivityNotFoundException exp) {
          AlertDialog.Builder build = new AlertDialog.Builder(this);
          AlertDialog dialog = build.create();
          dialog.setTitle("can not find activity!");
          dialog.show();
        }
    }
}
```

（5）启动 AVD,进入工程的根目录,用快捷方式编译工程、安装应用程序到 AVD(有关知识点参见 1.5 节)。对于本例子,用命令行进入 D:\2000\ch8_7,执行如下命令:

D:\2000 > ch8_7 > ant debug install

8.8 启动播放视频的 Activity 对象

1. ACTION_VIEW

静态常量 ACTION_VIEW 表示一个动作,其值等于"android.intent.action.VIEW",使用 VIEW 动作,并结合一个视频内容的 Uri 数据(Android 手机支持的格式,有关内容参见 5.7 节),可以启动 Android 系统内负责播放视频的 Activity 对象。

下列代码将启动播放视频的 Activity 对象。

```
Uri uri =
Uri.parse("http://f3.3g.56.com/15/15/JGfMspPbHtzoqpzseFTPGUsKCEqMXFTW_smooth.3gp");
intent.setDataAndType(uri,"video/*");
startActivity(intent);
```

2. 示例

在下面的例子 8-8 中，Activity 对象中有一个 EditText 视图，在该视图内输入视频的地址，然后启动负责播放视频的 Activity 对象。运行效果如图 8.9(a)，图 8.9(b)所示。

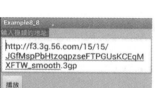 

(a) 输入视频地址的Activity对象　　(b) 播放视频的Activity对象

图 8.9　运行效果

例子 8-8

（1）创建名字为 ch8_8 的工程，主要 Activity 子类的名字为 Example8_8，使用的包名为 ch8.eight。用命令行进入 D:\2000，创建工程 D:\2000＞android create project -t 3 -n ch8_8 -p ./ch8_8 -a Example8_8 -k ch8.eight。

（2）将下列和视图相关的 XML 文件保存到工程的\res\layout 目录中。

ch8_8.xml

```xml
<?xml version = "1.0" encoding = "utf-8"?>
<LinearLayout xmlns:android = "http://schemas.android.com/apk/res/android"
    android:orientation = "vertical"
    android:layout_width = "match_parent"
    android:layout_height = "match_parent"
    android:background = "#87CEEB">
    <TextView
     android:background = "#555555"
     android:layout_width = "wrap_content"
     android:layout_height = "wrap_content"
     android:text = "输入视频的地址:" />
```

```xml
<EditText
    android:id = "@ + id/edit"
    android:layout_width = "match_parent"
    android:layout_height = "wrap_content"
    android:text =
"http://f3.3g.56.com/15/15/JGfMspPbHtzoqpzseFTPGUsKCEqMXFTW_smooth.3gp"/>
<Button
    android:layout_width = "wrap_content"
    android:layout_height = "wrap_content"
    android:text = "播放"
    android:onClick = "play"        />
</LinearLayout>
```

(3) 修改工程\src\ch8\eight 目录下的 Example8_8.java 文件，修改后的内容如下：

Example8_8.java

```java
package ch8.eight;
import android.app.*;
import android.os.Bundle;
import android.widget.*;
import android.view.*;
import android.content.*;
import android.net.*;
public class Example8_8 extends Activity   {
    EditText address;
    public void onCreate(Bundle savedInstanceState) {
        super.onCreate(savedInstanceState);
        setContentView(R.layout.ch8_8);
        address = (EditText)findViewById(R.id.edit);
    }
    public void play(View view) {
      Intent intent = new Intent(Intent.ACTION_VIEW);
      String s = address.getText().toString();
      Uri uri =   Uri.parse(s);
      intent.setDataAndType(uri,"video/*");
      try {
          startActivity(intent);
       }
       catch(ActivityNotFoundException exp) {
          AlertDialog.Builder build = new AlertDialog.Builder(this);
          AlertDialog dialog = build.create();
          dialog.setTitle("can not find activity!");
          dialog.show();
       }
    }
}
```

(4) 启动 AVD(要保证 PC 的 Internet 是连接状态)，进入工程的根目录，用快捷方式编译工程、安装应用程序到 AVD(有关知识点参见 1.5 节)。对于本例子，用命令行进入 D:\2000\ch8_8，执行如下命令：

```
D:\2000 > ch8_8 > ant debug install
```

8.9　启动使用 Google 地图的 Activity 对象

1. ACTION_VIEW

静态常量 ACTION_VIEW 表示一个动作，其值等于"android.intent.action.VIEW"使用 VIEW 动作，并结合一个 Google 地图内容的 Uri 数据（Google 地图的地址），可以启动 Android 系统内负责使用 Google 地图的 Activity 对象。

下列代码将启动使用 Google 地图的 Activity 对象：

```
Uri uri = Uri.parse("http://maps.google.com");
Intent intent = new Intent(Intent.ACTION_VIEW,uri);
startActivity(intent);
```

下列代码将启动使用 Google 地图的 Activity 对象，并给出从出发地：北纬 39.9°，东经 116.3°，到目的地：北纬 31.2°，东经 121.4°的路线详情。

```
Uri uri = Uri.parse("http://maps.google.com/maps?f = d&saddr = 39.9 116.3&daddr = 31.2 121.4");
Intent intent = new Intent(Intent.ACTION_VIEW,uri);
startActivity(intent);
```

2. 示例

在下面的例子 8-9 中，Activity 对象中用户单击名字是"直接进入 Google 地图"的按钮，将启动使用 Google 地图的 Activity 对象，如果输入出发地和目的地的北纬度和东经度后，单击名字是"查看 Google 地图给出的路线"的按钮，将启动使用 Google 地图的 Activity 对象，并给出具体的路线图。运行效果如图 8.10(a)，8.10(b)，8.10(c)所示。

(a) 准备使用Google地图

(b) Google地图

(c) 路线详情

图 8.10　运行效果

例子 8-9

（1）创建名字为 ch8_9 的工程，主要 Activity 子类的名字为 Example8_9，使用的包名为 ch8.nine。用命令行进入 D:\2000，创建工程 D:\2000＞android create project -t 3 -n

ch8_9 -p ./ch8_9 -a Example8_9 -k ch8.nine。
（2）将下列和视图相关的 XML 文件保存到工程的\res\layout 目录中。
ch8_9.xml

```xml
<?xml version="1.0" encoding="utf-8"?>
<LinearLayout xmlns:android="http://schemas.android.com/apk/res/android"
    android:orientation="vertical"
    android:layout_width="match_parent"
    android:layout_height="match_parent"
    android:background="#87CEEB">
    <TextView
     android:background="#FFFFFF"
     android:layout_width="match_parent"
     android:layout_height="wrap_content"
     android:textSize="15sp"
     android:text="使用 Google 地图" />
    <Button
        android:layout_width="match_parent"
        android:layout_height="wrap_content"
        android:text="直接进入 Google 地图"
        android:onClick="toGoogleMap"      />
    <TextView
     android:background="#FFFFFF"
     android:layout_width="match_parent"
     android:layout_height="wrap_content"
     android:textSize="15sp"
     android:text="输入出发地的北纬和东经度(用空格分隔):" />
    <EditText
     android:id="@+id/editStart"
     android:layout_width="match_parent"
     android:layout_height="wrap_content"
     android:inputType="numberDecimal"
     android:text="39.01 116.3"/>
    <TextView
     android:background="#FFFFFF"
     android:layout_width="match_parent"
     android:layout_height="wrap_content"
     android:textSize="15sp"
     android:text="输入目的地北纬和东经度(用空格分隔):" />
    <EditText
     android:id="@+id/editEnd"
     android:layout_width="match_parent"
     android:layout_height="wrap_content"
     android:inputType="numberDecimal"
     android:text="41.8 123.38"/>
    <Button
        android:layout_width="match_parent"
        android:layout_height="wrap_content"
        android:text="查看 Google 地图给出的路线"
        android:onClick="toGoogleMapLookRoad"      />
```

```
</LinearLayout>
```

(3) 修改工程\src\ch8\nine 目录下的 Example8_9.java 文件,修改后的内容如下:
Example8_9.java

```java
package ch8.nine;
import android.app.*;
import android.os.Bundle;
import android.widget.*;
import android.view.*;
import android.content.*;
import android.net.*;
public class Example8_9 extends Activity   {
    EditText startAddress,endAddress;
    public void onCreate(Bundle savedInstanceState) {
        super.onCreate(savedInstanceState);
        setContentView(R.layout.ch8_9);
        startAddress = (EditText)findViewById(R.id.editStart);
        endAddress = (EditText)findViewById(R.id.editEnd);
    }
    public void toGoogleMap(View view) {
        Uri uri = Uri.parse("http://maps.google.com");
        Intent intent = new Intent(Intent.ACTION_VIEW,uri);
        try {
            startActivity(intent);
        }
        catch(ActivityNotFoundException exp) {
            AlertDialog.Builder build = new AlertDialog.Builder(this);
            AlertDialog dialog = build.create();
            dialog.setTitle("can not find activity!");
            dialog.show();
        }
    }
    public void toGoogleMapLookRoad(View view) {
        String start = startAddress.getText().toString().trim();
        String end = endAddress.getText().toString().trim();
        Uri uri =
        Uri.parse("http://maps.google.com/maps?f = d&saddr = " + start + "&daddr = " + end);
        Intent intent = new Intent(Intent.ACTION_VIEW,uri);
        try {
            startActivity(intent);
        }
        catch(ActivityNotFoundException exp) {
            AlertDialog.Builder build = new AlertDialog.Builder(this);
            AlertDialog dialog = build.create();
            dialog.setTitle("can not find activity!");
            dialog.show();
        }
    }
}
```

（4）启动 AVD（要保证 PC 的 Internet 是连接状态），进入工程的根目录，用快捷方式编译工程、安装应用程序到 AVD（有关知识点参见 1.5 节）。对于本例子，用命令行进入 D:\2000\ch8_9，执行如下命令：

```
D:\2000>ch8_9>ant debug install
```

8.10 启动使用浏览器的 Activity 对象

1. ACTION_VIEW

静态常量 ACTION_VIEW 表示一个动作，其值等于"android.intent.action.VIEW"，使用 VIEW 动作，并结合一个 Uri 数据（Internet 的地址），可以启动 Android 系统内使用浏览器的 Activity 对象。

下列代码将启动使用浏览器的 Activity 对象，并默认访问 Google 的官方网址。

```
Uri uri = Uri.parse("http: ");
Intent intent  = new Intent(Intent.ACTION_VIEW,uri);
startActivity(intent);
```

下列代码将启动使用浏览器的 Activity 对象，并访问 http://www.tup.com.cn。

```
Uri uri = Uri.parse("http://www.tup.com.cn");
Intent intent  = new Intent(Intent.ACTION_VIEW,uri);
startActivity(intent);
```

2. 示例

下面的例子 8-10 中，Activity 对象中有一个 ListView 视图，该视图上列出了一些网址，选择一个网址之后，可以启动使用浏览器的 Activity 对象。运行效果如图 8.11(a)、8.11(b) 所示。

(a) 选择网址

(b) 访问人人网

图 8.11 运行效果

例子 8-10

（1）创建名字为 ch8_10 的工程，主要 Activity 子类的名字为 Example8_10，使用的包名为 ch8.ten。用命令行进入 D:\2000，创建工程 D:\2000＞android create project -t 3 -n ch8_10 -p ./ch8_10 -a Example8_10 -k ch8.ten。

（2）增加值资源。将下列 array.xml 保存到值资源中，即保存到工程的\res\values 目录中（有关知识点参见 2.6 节）。

array.xml

```xml
<?xml version = "1.0" encoding = "utf-8"?>
<resources>
    <array name = "net_list">
        <item>土豆网:http://www.tudou.com</item>
        <item>搜狐网:http://sohu.com.cn</item>
        <item>人人网:http://www.renren.com</item>
    </array>
</resources>
```

（3）将下列和视图相关的 XML 文件保存到工程的\res\layout 目录中。

ch8_10.xml

```xml
<?xml version = "1.0" encoding = "utf-8"?>
<LinearLayout xmlns:android = "http://schemas.android.com/apk/res/android"
        android:orientation = "vertical"
        android:layout_width = "match_parent"
        android:layout_height = "match_parent"
        android:background = "#87CEEB">
    <ListView
        android:id = "@+id/my_list"
        android:layout_width = "wrap_content"
        android:layout_height = "wrap_content"
        android:divider = "#0000FF"
        android:dividerHeight = "6dp"
        android:background = "#22bbcc"
        android:listSelector = "@drawable/ic_launcher"
        android:entries = "@array/net_list" />
</LinearLayout>
```

（4）修改工程\src\ch8\ten 目录下的 Example8_10.java 文件，修改后的内容如下：

Example8_10.java

```java
package ch8.ten;
import android.app.*;
import android.os.Bundle;
import android.widget.*;
import android.view.*;
import android.content.*;
import android.net.*;
public class Example8_10 extends Activity implements AdapterView.OnItemClickListener {
    ListView listView;
    String netAddress;
```

```
    public void onCreate(Bundle savedInstanceState) {
        super.onCreate(savedInstanceState);
        setContentView(R.layout.ch8_10);
        listView = (ListView)findViewById(R.id.my_list);
        listView.setOnItemClickListener(this);
    }
    public void onItemClick(AdapterView parent,View view,int pos,long id) {
        netAddress = listView.getItemAtPosition(pos).toString();
        int index = netAddress.indexOf("http://");
        netAddress = netAddress.substring(index);
        Uri uri = Uri.parse(netAddress);
        Intent intent = new Intent("android.intent.action.VIEW",uri);
        try {
            startActivity(intent);
        }
        catch(ActivityNotFoundException exp) {
            AlertDialog.Builder build = new AlertDialog.Builder(this);
            AlertDialog dialog = build.create();
            dialog.setTitle("can not find activity!");
            dialog.show();
        }
    }
}
```

(5)启动 AVD(要保证 PC 的 Internet 是连接状态),进入工程的根目录,用快捷方式编译工程、安装应用程序到 AVD(有关知识点参见 1.5 节)。对于本例子,用命令行进入 D:\2000\ch8_10,执行如下命令:

```
D:\2000>ch8_10>ant debug install
```

8.11 启动发送 E-mail 的 Activity 对象

1. 设置手机(或 AVD)中的 E-mail 账户

首先要保证手机(或 AVD)至少有一个 E-mail 账号。可以在手机(或 AVD)上选择 E-mail 程序,添加一个或多个 E-mail 账户,如图 8.12 所示。启动 E-mail 程序后,如果未曾设置过 E-mail 账户,将出现设置 E-mail 账户的界面,如图 8.13 所示。在图 8.13 所示界面中输入你曾注册过的有效邮箱(这里使用的是 xygeng0629@sina.com)和密码,单击 Next 按钮进入下一步,出现如图 8.14 所示的界面,在该界面选择接收 E-mail 的服务器使用的协议,几乎常用的邮件服务器都使用 POP3 协议接收邮件(比如 sina,126,163,sohu 等),因此在图中选择 POP3,然后逐条确定你的信息后(POP3 服务器使用端口 Port 是 110),将出现如图 8.15 所示的界面。

图 8.12 打开 E-mail 程序

在图 8.15 所示的界面选择 Next 按钮,系统将登录 POP3 服务器进行验证,如果验证被通过,那么接收邮件的有关信息就设置好了,然后将出现设置发送邮件相关数据的界面,如图 8.16 所示。

图 8.13 设置一个账户

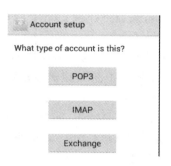

图 8.14 选择 POP3 协议

图 8.15 确认信息

图 8.16 确认信息

发送邮件使用的是 STMP 协议,默认端口是 25,在图 8.16 所示界面上逐条确认你的信息时,要把 STMP 服务器的端口 Port 设置为 25(如图 8.16 所示)。在图 8.16 所示界面上选择 Next 按钮,系统将登录 STMP 服务器进行验证,如果验证被通过,那么发送邮件的有关信息就设置好了。

验证被通过后,将出现让你设置的邮箱 xygeng0629@sina.com 相关信息的界面(这里我们省略了有关界面),你可以浏览所注册的邮箱中的邮件并使用该邮箱发送邮件。

2. ACTION_SEND

静态常量 ACTION_SEND 表示一个动作,其值等于"android.intent.action.SEND",使用 SEND 动作,可以启动 Android 系统内发送 E-mail 或 Message 的 Activity 对象。

下列代码将启动使用 E-mail 或 Message 的 Activity 对象:

```
Intent intent = new Intent(Intent.ACTION_SEND);
intent.setType("text/plain");
intent.putExtra(Intent.EXTRA_SUBJECT,"Subject");   //该代码不是必须的
intent.putExtra(Intent.EXTRA_TEXT,"Hello");        //该代码不是必须的
startActivity(intent);
```

代码运行后,将提示用户在发送 E-mail 或 Message 的 Activity 对象中选择其一,选择发送 E-mail 的 Activity 对象的效果如图 8.17 所示。

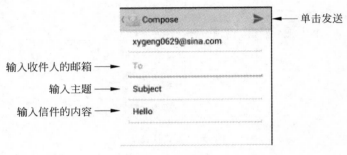

图 8.17　发送 E-mail

3. 示例

下面的例子 8-11 中,Activity 对象中有一个 Button 视图,单击 Button 视图之后,可以选择启动发送 E-mail 的 Activity 对象。运行效果如图 8.18(a),8.18(b)所示。

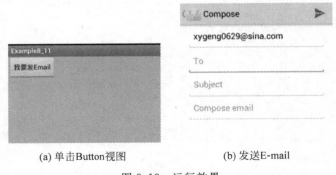

(a) 单击Button视图　　　　(b) 发送E-mail

图 8.18　运行效果

例子 8-11

(1) 创建名字为 ch8_11 的工程,主要 Activity 子类的名字为 Example8_11,使用的包

名为 ch8.eleven。用命令行进入 D:\2000，创建工程 D:\2000>android create project -t 3 -n ch8_11 -p ./ch8_11 -a Example8_11 -k ch8.eleven。

（2）将下列和视图相关的 XML 文件保存到工程的\res\layout 目录中。

ch8_11.xml

```xml
<?xml version = "1.0" encoding = "utf-8"?>
<LinearLayout xmlns:android = "http://schemas.android.com/apk/res/android"
    android:orientation = "vertical"
    android:layout_width = "match_parent"
    android:layout_height = "match_parent"
    android:background = "#87CEEB"   >
    <Button
        android:layout_width = "wrap_content"
        android:layout_height = "wrap_content"
        android:text = "我要发 Email"
        android:onClick = "email"    />
</LinearLayout>
```

（3）修改工程\src\ch8\eleven 目录下的 Example8_11.java 文件，修改后的内容如下：

Example8_11.java

```java
package ch8.eleven;
import android.content.*;
import android.app.*;
import android.os.Bundle;
import android.widget.*;
import android.view.*;
public class Example8_11 extends Activity {
    public void onCreate(Bundle savedInstanceState) {
        super.onCreate(savedInstanceState);
        setContentView(R.layout.ch8_11);
    }
    public void email(View v) {
        Intent intent = new Intent(Intent.ACTION_SEND);
        intent.setType("text/plain");
        try {
            startActivity(intent);
        }
        catch(ActivityNotFoundException exp) {
            AlertDialog.Builder build = new AlertDialog.Builder(this);
            AlertDialog dialog = build.create();
            dialog.setTitle("can not find activity!");
            dialog.show();
        }
    }
}
```

（4）启动 AVD（要保证 PC 的 Internet 是连接状态），进入工程的根目录，用快捷方式编译工程、安装应用程序到 AVD（有关知识点参见 1.5 节）。对于本例子，用命令行进入 D:\2000\ch8_11，执行如下命令：

D:\2000 > ch8_11 > ant debug install

8.12 具有多个 Activity 对象的程序

一个应用程序可以包含若干个 Activity 对象（参见 2.1 节），那么一个应用程序中的一个 Activity 对象可能需要启动当前应用程序中的其他的 Activity 对象。这些 Activity 对象是应用程序中最重要的部分。一个 Android 应用程序中必须有一个 Activity 对象被设置为主要的 Activity 对象，Android 运行环境（AVD 模拟器或 Android 手机）通过加载主要的 Activity 对象开始运行一个 Android 应用程序（在创建项目时给出的 Activity 对象是默认的主要的 Activity 对象）。Android 应用程序中的一个 Activity 对象可以请求加载它包含的其他的 Activity 对象，Android 应用程序正是通过其中的这些 Activity 对象来体现自身的功能。本章将讲解在应用程序中让某个 Activity 对象使用 Intent 对象来启动其他的 Activity 对象。

1. Intent 的一个特殊构造方法

Intent(Context packageContext,Class<?> cls)构造方法创建的 Intent 对象体现的意图被习惯地称为显式意图（Explicit Intents），即非常准确地给出了要启动的 Activity 对象。该构造方法的参数 packageContext 是当前应用程序所在的上下文，参数 cls 是打算启动的当前应用程序中 Activity 对象的类的名字（该类负责创建要启动的 Activity 对象），比如，当前应用程序中有名字为 Hello 的 Activity 的子类，该类负责创建某个 Activity 对象，那么可如下创建一个 Intent 对象：

Intent intent = new Intent(this,Hello.class);

注：也可以使用隐式的 Intent 对象启动程序中的其他的 Activity 对象，但需要在配置文件中明确给出 Activity 对象的 action 和 category，有关内容参见 8.3 节和 8.4 节。

2. AndroidManifest.xml 配置文件与应用程序中的 Activity 对象

一个应用程序可以包含多个 Activity 对象，用户必须编辑和该应用程序相关的 AndroidManifest.xml 配置文件，在该文件中标明当前应用程序可以有哪些 Activity 对象。

AndroidManifest.xml 文件中的标记<application …>…</application>中的子标记<activty>…</activity>说明应用程序有几个 Activity 对象，并且哪个是主要的 Activity 对象。如果某个 < activity > … </activity > 包含有 < intent-filter > 子标记，并使用 <action…/>空标记的 name 属性说明自己是 MAIN,使用<category…/>空标记的 name 属性说明自己是 LAUNCHER,那么,Android 运行环境将首先加载这个 Activity 对象,即这个 Activity 对象是应用程序中的主要的 Activity 对象。

3. 示例

在下面的例子 8-12 中，应用程序中一共有两个的 Activity 对象。主要的 Activity 对象由 Example8_12 类负责创建，另一个 Activity 对象由 Beijing 类创建。主要的 Activity 对象可以启动 Beijing 创建的 Activity 对象，Beijing 创建的 Activity 对象也可以启动主要的 Activity 对象。运行效果如图 8.19(a),8.19(b)所示。

(a) 主要的Activity　　　　　　(b) Beijing的Activity

图 8.19　运行效果

例子 8-12

(1) 创建名字为 ch8_12 的工程，主要 Activity 子类的名字为 Example8_12，使用的包名为 ch8.twelve。用命令行进入 D:\2000，创建工程 D:\2000>android create project -t 3 -n ch8_12 -p ./ch8_12 -a Example8_12 -k ch8.twelve。

(2) 修改配置文件。修改工程根目录下的 AndroidManifest.xml 配置文件，必须在配置文件中用<activty>…</activity>标明应用程序中包含的两个 Activity 对象。修改后的内容如下：

AndroidManifest.xml

```xml
<?xml version = "1.0" encoding = "utf-8"?>
<manifest xmlns:android = "http://schemas.android.com/apk/res/android"
    package = "ch8.twelve"
    android:versionCode = "1"
    android:versionName = "1.0">
    <application android:label = "@string/app_name" android:icon = "@drawable/ic_launcher">
        <activity android:name = "Example8_12"
                android:label = "@string/app_name">  <!--主要的 Activity 对象 -->
            <intent-filter>
                <action android:name = "android.intent.action.MAIN" />
                <category android:name = "android.intent.category.LAUNCHER" />
            </intent-filter>
        </activity>
        <activity android:name = "Beijing" android:label = "北京"> <!--新添加的 Activity 对象 -->
        </activity>
    </application>
</manifest>
```

(3) 将下列和视图相关的 XML 文件 ch8_12.xml、beijing.xml 保存到工程的\res\layout 目录中。ch8_12.xml 是主要 Activity 对象使用的视图文件，beijing.xml 是 Beijing 类创建的 Activity 对象使用的视图文件。

ch8_12.xml

```xml
<?xml version = "1.0" encoding = "utf-8"?>
<LinearLayout xmlns:android = "http://schemas.android.com/apk/res/android"
    android:orientation = "vertical"
    android:layout_width = "match_parent"
    android:layout_height = "match_parent"
    android:background = "#87CEEB">
    <TextView
        android:background = "#B3C1BF"
        android:textColor = "#0000FF"
```

```
            android:textSize = "22sp"
            android:layout_width = "match_parent"
            android:layout_height = "wrap_content"
            android:text = "我是主要的Activity" />
    <Button
            android:layout_width = "match_parent"
            android:layout_height = "wrap_content"
            android:text = "去北京看看"
            android:onClick = "goBeijing"/>
</LinearLayout>
```

beijing.xml

```
<?xml version = "1.0" encoding = "utf-8"?>
<LinearLayout xmlns:android = "http://schemas.android.com/apk/res/android"
        android:orientation = "vertical"
        android:layout_width = "match_parent"
        android:layout_height = "match_parent"
        android:background = "#87CEEB">
    <TextView
            android:background = "#B3C1BF"
            android:textColor = "#FF0000"
            android:layout_width = "match_parent"
            android:layout_height = "wrap_content"
            android:id = "@+id/text"
            android:textSize = "22sp"
            android:text = "我是北京"/>
    <Button
            android:layout_width = "match_parent"
            android:layout_height = "wrap_content"
            android:text = "去主要的Activity看看"
            android:onClick = "goMain"/>
</LinearLayout>
```

(4) 修改工程\src\ch8\twevle目录下的Example8_12.java文件,并将下列Bejing.java保存到工程\src\ch8\twevle目录下。修改后Example8_12.java和Beijing.java文件的内容如下:

Example8_12.java

```
package ch8.twelve;
import android.content.*;
import android.app.*;
import android.os.Bundle;
import android.widget.*;
import android.view.*;
public class Example8_12 extends Activity {
    public void onCreate(Bundle savedInstanceState) {
        super.onCreate(savedInstanceState);
        setContentView(R.layout.ch8_12);
    }
    public void goBeijing(View v) {
```

```
            Intent intent = new Intent(this,Beijing.class);
            try {
                startActivity(intent);
            }
            catch(ActivityNotFoundException exp) {
              AlertDialog.Builder build = new AlertDialog.Builder(this);
              AlertDialog dialog = build.create();
              dialog.setTitle("can not find activity!");
              dialog.show();
            }
        }
    }
```

Beijing.java

```
package ch8.twelve;
import android.content.*;
import android.app.*;
import android.os.Bundle;
import android.widget.*;
import android.view.*;
public class Beijing extends Activity {
    public void onCreate(Bundle savedInstanceState) {
        super.onCreate(savedInstanceState);
        setContentView(R.layout.beijing);
    }
    public void goMain(View v) {
        Intent intent = new Intent(this,Example8_12.class);
        try {
             startActivity(intent);
        }
        catch(ActivityNotFoundException exp) {
          AlertDialog.Builder build = new AlertDialog.Builder(this);
          AlertDialog dialog = build.create();
          dialog.setTitle("can not find activity!");
          dialog.show();
        }
    }
}
```

（5）启动 AVD，进入工程的根目录，用快捷方式编译工程、安装应用程序到 AVD（有关知识点参见 1.5 节）。对于本例子，用命令行进入 D:\2000\ch8_12，执行如下命令：

D:\2000>ch8_12>ant debug install

8.13 让 Activity 对象返回数据

在 8.5 节中，我们介绍了怎样向 startActivity(Intent intent)方法启动 Activity 对象传递附加数据。程序有时候可能还需要被启动 Activity 对象返回某些数据给启动它的

Activity 对象。本节介绍怎样让 Activity 对象返回数据给启动它的 Activity 对象。

1. startActivityForResult 与 onActivityResult 方法

如果当前 Activity 对象希望启动某个其他的 Activity 对象,并希望被启动的 Activity 对象返回某些数据,那么它需要使用 public void startActivityForResult(Intent intent, int requestCode)方法来启动其他的 Activity 对象。其中的参数 requestCode 应当是非负整数,否则该方法的作用等同于 startActivity(Intent intent)。如果参数 requestCode 取值是非负整数,表示当前 Activity 对象等待被启动的 Activity 对象返回数据给它。

负责创建当前 Activity 对象的 Activity 对象的子类必须重写 Activity 类的 void onActivityResult(int requestCode, int resultCode, Intent data)方法,通过该方法得到当前 Activity 对象启动的 Activity 对象返回的数据。当前对象在使用 public void startActivityForResult(Intent intent, int requestCode)方法启动其他的 Activity 对象时,如果参数 requestCode 取值是非负整数,那么当前对象就开始等待调用 onActivityResult()方法,一旦当前对象启动 Activity 对象结束生命周期,当前对象就会调用 onActivityResult(int requestCode, int resultCode, Intent data)方法处理返回的数据。

2. 返回数据和得到返回数据的机制

当前 Activity 对象使用 startActivityForResult()方法启动另一个 Activity 对象后,就会等待调用 onActivityResult()方法,因此系统不会杀死当前的 Activity 对象。被启动的 Activity 对象必须将自己要返回的数据作为附加数据存放到一个没有任何意图的 Intent 对象中,例如,存放到 Intent 对象 backIntent 中:

```
Intent backIntent = new Intent();
backIntent.putExtra("number",3.1415926);
```

然后为 backIntent 指定一个 int 型的结果码(resultCode),并调用 final void setResult(int resultCode, Intent data)方法返回这个 Intent 对象 backIntent,例如:

```
int resultCode = 100;
setResult(resultCode,backIntent);
```

如果需要返回多个 Intent 对象,需要为这些不同的 Intent 对象指定不同的结果码。

最后,返回数据的 Activity 对象需要调用 finish 方法结束自己的生命周期,以便等待返回数据的 Activity 对象调用 onActivityResult()方法处理返回的数据。

等待返回数据的 Activity 对象调用 onActivityResult(int requestCode, int resultCode, Intent data)方法处理返回的数据,其中的参数 requestCode 的值是当初请求启动 Activity 对象时使用的值,resultCode 是被启动的 Activity 对象返回的 Intent 对象对应的结果码。onActivityResult 方法处理返回的数据时可能需要根据参数 resultCode 的值,来决定使用怎样的诸如 getXXX(String key)的方法得到 Intent 对象中的数据。

需要特别注意的是,负责返回数据的 Activity 对象不可以使用有意图的 Intent 对象来启动曾启动过它的 Activity 对象,否则会导致启动它的 Activity 对象重新开始自己的生命周期,即导致重新执行 onCreate()等方法。

3. 示例

例子 8-13 中,让应用程序本身包含两个 Activity 对象。主要的 Activity 对象启动其中

Computer 类创建的 Activity 对象，并将一个 double 值传递给所启动的 Activity 对象。Computer 类创建的 Activity 对象计算 double 值的平方和平方根，Computer 类创建的 Activity 对象可以选择将计算出的 double 值的平方和平方根返回给主要的 Activity 对象，也可以选择返回一句问候语给主要的 Activity 对象。运行效果如图 8.20(a)，8.20(b) 和 8.20(c) 所示。

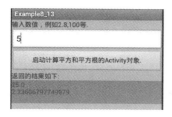
(a) 得到返回的平方和平方根

(b) 得到返回的问候语

(c) 返回平方、平方根或问候语

图 8.20　运行效果

例子 8-13

(1) 创建名字为 ch8_13 的工程，主要 Activity 子类的名字为 Example8_13，使用的包名为 ch8.thirteen。用命令行进入 D:\2000，创建工程 D:\2000>android create project -t 3 -n ch8_13 -p ./ch8_13 -a Example8_13 -k ch8.thirteen。

(2) 修改配置文件。修改工程根目录下的 AndroidManifest.xml 配置文件，在配置文件中用 <activty>…</activity> 标明应用程序中包含的两个 Activity 对象。修改后的内容如下：

AndroidManifest.xml

```
<?xml version = "1.0" encoding = "utf-8"?>
<manifest xmlns:android = "http://schemas.android.com/apk/res/android"
      package = "ch8.thirteen"
      android:versionCode = "1"
      android:versionName = "1.0">
  <application android:label = "@string/app_name" android:icon = "@drawable/ic_launcher">
      <activity android:name = "Example8_13"
            android:label = "@string/app_name">
          <intent-filter>
              <action android:name = "android.intent.action.MAIN" />
              <category android:name = "android.intent.category.LAUNCHER" />
          </intent-filter>
      </activity>
       <activity android:name = "Computer"
            android:label = "计算平方和平方根"><!--新添加的 Activity 对象 -->
      </activity>
  </application>
</manifest>
```

(3) 将下列和视图相关的 XML 文件 ch8_13.xml、computer.xml 保存到工程的 \res\layout 目录中。ch8_13.xml 是主要 Activity 对象使用的视图文件，computer.xml 是

Computer 类创建的 Activity 对象使用的视图文件。

ch8_13.xml

```xml
<?xml version = "1.0" encoding = "utf-8"?>
<LinearLayout xmlns:android = "http://schemas.android.com/apk/res/android"
    android:orientation = "vertical"
    android:layout_width = "match_parent"
    android:layout_height = "match_parent"
    android:background = "#87CEEB">
    <TextView
        android:background = "#B3C1BF"
        android:textColor = "#0000FF"
        android:layout_width = "match_parent"
        android:layout_height = "wrap_content"
        android:text = "输入数值,例如 2.8,100 等."/>
    <EditText
        android:layout_width = "match_parent"
        android:layout_height = "wrap_content"
        android:inputType = "numberDecimal|textMultiLine"
        android:text = "1.0"
        android:id = "@+id/edit"/>
    <Button
        android:layout_width = "match_parent"
        android:layout_height = "wrap_content"
        android:text = "启动计算平方和平方根的 Activity 对象."
        android:onClick = "goComputer"/>
    <TextView
        android:background = "#B3C1BF"
        android:textColor = "#0000FF"
        android:layout_width = "match_parent"
        android:layout_height = "wrap_content"
        android:text = "返回的结果如下:"/>
    <TextView
        android:background = "#00AA99"
        android:textColor = "#FF0000"
        android:layout_width = "match_parent"
        android:layout_height = "wrap_content"
        android:id = "@+id/textResult"/>
</LinearLayout>
```

computer.xml

```xml
<?xml version = "1.0" encoding = "utf-8"?>
<LinearLayout xmlns:android = "http://schemas.android.com/apk/res/android"
    android:orientation = "vertical"
    android:layout_width = "match_parent"
    android:layout_height = "match_parent"
    android:background = "#87CEEB">
    <TextView
        android:background = "#B3C1BF"
        android:textColor = "#FF0000"
```

```xml
            android:layout_width = "match_parent"
            android:layout_height = "wrap_content"
            android:id = "@ + id/text"
            android:textSize = "22sp"
            android:text = "我能计算平方和平方根."/>
    <Button
            android:layout_width = "match_parent"
            android:layout_height = "wrap_content"
            android:text = "返回平方和平方根"
            android:onClick = "backData"/>
    <Button
            android:layout_width = "match_parent"
            android:layout_height = "wrap_content"
            android:text = "返回问候语"
            android:onClick = "backHello"/>
</LinearLayout>
```

（4）修改工程\src\ch8\thirteen 目录下的 Example8_13.java 文件，并将下列 Computer.java 保存到工程\src\ch8\thirteen 目录下。修改后的 Example8_13.java 和 Computer.java 文件的内容如下：

Example8_13.java

```java
package ch8.thirteen;
import android.content. * ;
import android.app. * ;
import android.os.Bundle;
import android.widget. * ;
import android.view. * ;
public class Example8_13 extends Activity {
    EditText edit;
    TextView showData;
    public void onCreate(Bundle savedInstanceState) {
        super.onCreate(savedInstanceState);
        setContentView(R.layout.ch8_13);
        edit = (EditText)findViewById(R.id.edit);
        showData = (TextView)findViewById(R.id.textResult);
    }
    public void goComputer(View v) {
        String s = edit.getText().toString();
        if(s == null)
            s = "1.0";
        double n = Double.parseDouble(s);
        Intent intent = new Intent(this,Computer.class);
        intent.putExtra("number",n);
        try {
            startActivityForResult(intent,0);
        }
        catch(ActivityNotFoundException exp) {
            AlertDialog.Builder build = new AlertDialog.Builder(this);
            AlertDialog dialog = build.create();
```

```java
            dialog.setTitle("can not find activity!");
            dialog.show();
        }
    }
    public void onActivityResult(int requestCode,int resultCode,Intent data){
        if(resultCode == 1) {
            double pingfang = data.getDoubleExtra ("pingfang",0);
            double pingfangGen = data.getDoubleExtra ("pingfangGen",0);
            showData.setText("");
            showData.append(pingfang + "\n");
            showData.append(pingfangGen + "\n");
        }
        if(resultCode == 2) {
            String hello = data.getStringExtra ("Hello");
            showData.setText("");
            showData.append(hello + "\n");
        }
    }
}
```

Computer.java

```java
package ch8.thirteen;
import android.content.*;
import android.app.*;
import android.os.Bundle;
import android.widget.*;
import android.view.*;
public class Computer extends Activity {
    double n;
    public void onCreate(Bundle savedInstanceState) {
        super.onCreate(savedInstanceState);
        setContentView(R.layout.computer);
        TextView text = (TextView)findViewById(R.id.text);
        Intent intent = getIntent();
        n = intent.getDoubleExtra("number",0); //如果获取不到,就用默认值0
    }
    public void backData(View v) {
      Intent intent = new Intent();
      intent.putExtra("pingfang",n * n);
      intent.putExtra("pingfangGen",Math.sqrt(n));
      int resultCode = 1;
      setResult(resultCode,intent);
      finish();
    }
    public void backHello(View v) {
      Intent intent = new Intent();
      intent.putExtra("Hello","How are you\nI am glad to meet you!");
      int resultCode = 2;
      setResult(resultCode,intent);
      finish();
```

            }
    }

(5) 启动 AVD,进入工程的根目录,用快捷方式编译工程、安装应用程序到 AVD(有关知识点参见 1.5 节)。对于本例子,用命令行进入 D:\2000\ch8_13,执行如下命令:

```
D:\2000>ch8_13>ant debug install
```

8.14 启动使用照相机的 Activity 对象

1. ACTION_IMAGE_CAPTURE

android.provider.MediaStore 类的静态常量 ACTION_IMAGE_CAPTURE 表示一个动作,其值等于"android.media.action.IMAGE_CAPTURE",使用 ACTION_IMAGE_CAPTURE 动作,可以启动 Android 系统中使用照相机的 Activity 对象。

下列代码将启动使用照相机的 Activity 对象:

```
Intent intent = new Intent(MediaStore.ACTION_IMAGE_CAPTURE);
startActivity(intent);
```

2. 得到照相机拍照的图像

Android 系统中使用照相机的 Activity 对象将拍照的图像封装成 Bitmap 类型的对象(位图图像),并能返回一个 Intent 对象给启动它的 Activity 对象,该 Intent 对象中含有 Bitmap 类型的对象,即含有照相机拍照得到的图像。

因此,当前 Activity 对象如果希望得到照相机拍照后得到的图像,就需要使用 public void startActivityForResult(Intent intent, int requestCode)方法启动使用照相机的 Activity 对象,并在 onActivityResult(int requestCode, int resultCode, Intent intent)方法中使用如下代码获得返回的 Bitmap 类型的对象:

```
Bundle extras = intent.getExtras();
Bitmap bitmap = (Bitmap) extras.get("data");
```

使用照相机的 Activity 对象被启动后,如果拍照之后,单击照相机上的确认按钮(√),使用照相机的 Activity 对象将返回一个 Intent 对象给启动它的 Activity 对象,并结束自己的生命周期。启动使用照相机的 Activity 对象将执行 onActivityResult 方法,此时方法中的 resultCode 参数的值是 Activity.RESULT_OK,如果拍照之后,单击照相机上的取消按钮(×),resultCode 参数的值是 Activity.RESULT_CANCELED。

注:照相机拍照的图像默认被保存到 Android 手机的 SD 卡中,可以使用 Android 手机中的 Gallery(相册)程序查看相机拍照的图像或删除相机拍照的图像。

3. 示例

例子 8-14 中主要的 Activity 对象可以启动 Android 系统中使用照相机的 Activity 对象(AVD 提供了模拟拍照),主要的 Activity 对象将照相机拍照得到的图像显示在一个 ImageButton 视图上,如果用户拍照了多个图像,就会产生多个 ImageButton 视图,因此主要的 Activity 对象使用 HorizontalScrollView 视图放置多个 ImageButton 视图,以便用户能使用水平滚动方式观看拍照所得到的图像。运行效果如图 8.21(a),8.21(b)所示。

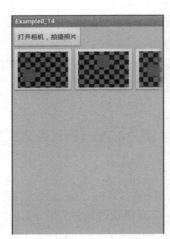 

(a) 打开相机并得到照片　　　　(b) 拍照，并返回照片

图 8.21　运行效果

例子 8-14

(1) 创建名字为 ch8_14 的工程，主要 Activity 子类的名字为 Example8_14，使用的包名为 ch8. fourteen，用命令行进入 D:\2000，创建工程 D:\2000＞android create project -t 3 -n ch8_14 -p ./ch8_14 -a Example8_14 -k ch8. fourteen。

(2) 将下列和视图相关的 XML 文件保存到工程的\res\layout 目录中。

ch8_14. xml

```
<?xml version = "1.0" encoding = "utf-8"?>
<LinearLayout xmlns:android = "http://schemas.android.com/apk/res/android"
    android:orientation = "vertical"
    android:layout_width = "match_parent"
    android:layout_height = "match_parent"
android:background = "#87CEEB">
   <Button  android:layout_width = "wrap_content"
        android:layout_height = "wrap_content"
        android:text = "打开相机,拍摄照片"
        android:onClick = "open"           />
   <HorizontalScrollView   android:layout_width = "match_parent"
        android:layout_height = "match_parent">
      <LinearLayout
          android:id = "@+id/layout"
          android:orientation = "horizontal"
          android:layout_width = "match_parent"
          android:layout_height = "match_parent">
      </LinearLayout>
   </HorizontalScrollView>
</LinearLayout>
```

(3) 修改工程\src\ch8\fourteen 目录下的 Example8_14. java 文件，修改后的内容如下：

Example8_14.java

```java
package ch8.fourteen;
import android.app.*;
import android.os.Bundle;
import android.widget.*;
import android.view.*;
import android.content.*;
import android.graphics.Bitmap;
import android.provider.MediaStore;
public class Example8_14 extends Activity {
    LinearLayout   layout;
    public void onCreate(Bundle savedInstanceState) {
        super.onCreate(savedInstanceState);
        setContentView(R.layout.ch8_14);
        layout =   (LinearLayout)findViewById(R.id.layout);
    }
    public void open(View v) {
        Intent intent = new Intent(MediaStore.ACTION_IMAGE_CAPTURE);
        startActivityForResult(intent,0);
    }
    public void onActivityResult(int requestCode,int resultCode,Intent intent){
        if(resultCode == Activity.RESULT_OK) {
            Bundle extras = intent.getExtras();
            Bitmap bitmap = (Bitmap)extras.get("data");
            ImageButton imageButton = new ImageButton(this);
            imageButton.setImageBitmap(bitmap);
            layout.addView(imageButton);
        }
    }
}
```

（4）启动 AVD，进入工程的根目录，用快捷方式编译工程、安装应用程序到 AVD(有关知识点参见 1.5 节)。对于本例子，用命令行进入 D:\2000\ch8_14，执行如下命令：

D:\2000 > ch8_14 > ant debug install

习 题 8

1. "使用 Intent 对象最多能找到一个 Activity 对象"是正确的说法吗？

2. 编写一个程序，用户单击一个按钮视图，程序使用 Intent 对象寻找能进行"android.intent.action.PICK"动作，并且在"android.intent.category.DEFAULT"范畴内的 Activity 对象。

3. 编写一个程序，借助 Intent 对象启动负责拨号的 Activity 对象，并给你的同学打个电话。

4. 编写程序，借助 Intent 对象启动使用 Google 地图的 Activity 对象，并显示北京市的地图。

5. 编写程序，借助 Intent 对象启动使用浏览器的 Activity 对象，并让浏览器显示自己学校的主页。

第 9 章 常用后台对象

主要内容：
- Activity 对象与 Service 对象、BroadcastReceiver 对象；
- Service 对象及生命周期；
- 使用多个 Service 对象；
- IntentService 类；
- AsyncTask 类；
- 广播及接收；
- PendingIntent 类。

一个应用程序不仅可以包含若干个 Activity 对象（参见 2.1 节），也可以包含若干个 Service 对象（习惯上称为一个服务）和 BroadcastReceiver 对象（习惯上称为一个接收者）。Activity 对象负责提供和用户交互的视图（View），Service 对象不提供视图（View），目的是帮助 Activity 对象完成一些后台的任务。BroadcastReceiver 对象负责接收程序的广播信息。当 Activity 对象需要处理一些和界面无关的工作时，可以启动一个 Service 对象，让它帮助完成相应的工作（人们习惯地称 Service 对象是一个没有视图的 Activity 对象）。当程序需要接收广播信息时，就需要包含相应的 BroadcastReceiver 对象。

本章的一个重要内容之一是讲解怎样在应用程序中使用当前应用程序中的 Service 对象，这样的 Service 对象也称为本地 Service 对象。相对当前应用程序而言，其他应用程序中的 Service 对象，被称作远程 Service 对象。在许多实际问题中，只需使用本地 Service 对象即可，因此本章主要讲解怎样使用本地 Service 对象。应用程序可以使用 Intent 对象帮助当前应用程序启动所需要的本地 Service 对象，Intent 对象不仅是 Activity 对象进行交往、互通信息的"桥梁"，也是 Activity 对象与 Service 对象互通信息的"桥梁"。

9.1 Activity 对象与 Service 对象、BroadcastReceiver 对象

应用程序中的 Activity 对象由 Activity 类的子类负责创建，Service 对象由 Service 类的子类负责创建（参见稍后的 9.2 节和 9.3 节），BroadcastReceiver 对象由 BroadcastReceiver 类的子类负责创建。Android 系统加载应用程序之后，将启动一个线程，称作主线程（习惯称作 UI 线程），负责管理和运行其中的 Activity、Service 和 BroadcastReceiver 对象，即应用程序中的 Activity、Service 和 BroadcastReceiver 对象都运行在一个主线程中，Android 系统没有为每个 Activity、Service 或 BroadcastReceiver 对象提供单独的线程，简单地说，Activity，

Service 或 BroadcastReceiver 对象都运行在主线程中,只不过这些对象都有自己的生命周期(Activity 对象的生命周期参见 2.4 节,Service 对象的生命周期参见稍后的 9.2 节,BroadcastReceiver 对象的生命周期参见稍后的 9.6 节),Activity、Service 和 BroadcastReceiver 对象通过 Intent 对象实现通信,如图 9.1 所示。

图 9.1　应用程序中的 Activity,Service 和 BroadcastReceiver 对象

9.2　Service 对象及生命周期

Service 类的继承关系如下:

```
android.content.Context
    ┕ android.content.ContextWrapper
        ┕ android.app.Service
```

1. 在配置文件中给出<service>标记

和 Activity 对象类似,应用程序中的 Service 对象由 Service 类的子类负责创建,因此用户需要编写 Service 类的子类,并重写 Service 类的一些重要方法。

Service 对象一定与应用程序的配置文件 AndroidManifest.xml 中的一个<service>标记相对应,否则,应用程序无法加载使用 Service 对象。假设创建 Service 对象的类是 NumberService(Service 类的一个子类),那么配置文件应当包含如下的<service>标记:

< service android:name = "NumberService" android:enabled = "true" />

其中,参数 android:name 的值必须是创建 Service 对象的类的名字(这里的名字是 NumberService),如下所示:

AndroidManifest.xml

```
<?xml version = "1.0" encoding = "utf-8"?>
< manifest xmlns:android = http://schemas.android.com/apk/res/android
    ……
    < application android:label = "@string/app_name" android:icon = "@drawable/ic_launcher">
        < activity ……>
            ……
        </activity>
        < service android:name = "NumberService" android:enabled = "true" /> <!-- service 标记 -->
    </application>
</manifest>
```

2. Service 对象的生命周期

以下介绍 Service 类的子类需要重写的重要方法,以及这些方法和 Service 对象的生命周期的关系。

1) 新建状态与 void onCreate()方法

public void onCreate()方法是 Service 类的一个重要方法,子类必须重写这个方法。该方法负责完成 Service 对象的初始化工作。当程序请求启动 Service 对象时,如果系统未曾创建要启动的 Service 对象,那么系统将用配置文件中指定的、负责创建 Service 对象的类创建这个 Service 对象,并立刻让这个 Service 对象调用 onCreate()方法完成必要的初始化工作。如果系统已经创建了要启动的 Service 对象,并执行过 onCreate()方法,即程序请求启动的 Service 对象已经启动过了,而且该 Service 对象没有死亡,那么该 Service 对象不会再调用 onCreate()方法。简单地说,Service 对象在其生命周期内只调用一次 onCreate()方法。

2) 工作状态与 int onStartCommand(Intent intent,int flags,int startId)方法

Service 对象启动后,即执行完 onCreate()方法后,将会立刻执行 int onStartCommand(Intent intent,int flags,int startId)方法,用户需要把分配给 Service 对象的任务写在该方法内,Service 对象执行该方法时,称 Service 对象处于工作状态。需要特别注意的是,如果 Service 对象已经启动,并且没有死亡,如果程序再次请求启动该 Service 对象,那么 Service 对象不会执行 onCreate()方法,但却会再次执行 onStartCommand 方法。简单地说,Service 对象在其生命周期内可能多次调用 onStartCommand 方法。该方法中的参数 intent 是启动 Service 对象的 Intent 对象,startId 是该方法被执行的次数。

3) 死亡状态与 void onDestroy()方法

一旦应用程序请求停止 Service 对象,那么 Service 对象立刻调用执行 void onDestroy()方法,然后进入死亡状态,即释放当前 Service 对象占用的内存。因此,如果希望 Service 对象死亡之前处理好善后工作,就需要在 onDestroy()方法中编写相应的代码。

注:系统可能根据当前应用程序占用内存的情况或各个 Activity 以及 Service 的重要程度让当前 Service 对象进入死亡状态,而不是等待程序主动让 Service 对象进入死亡状态。

4) onBind(Intent intent)方法

另外,子类还必须要重写 public IBinder onBind(Intent intent)方法,如果 Service 对象不准备做远程对象,重写该方法时返回 null 即可。

Service 对象的生命周期及涉及的方法如图 9.2 所示。

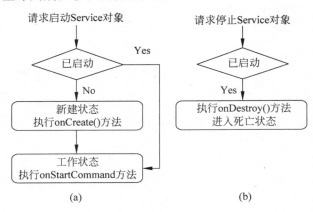

图 9.2 Service 对象的生命周期

3. 启动与停止 Service 对象

应用程序中的 Activity 对象或 Service 对象可以使用 startService(Intent intent)方法启动一个其他的 Service 对象。例如，假设 NumberService 类是负责创建 Service 对象的类，那么当前 Activity 对象可用如下代码启动一个 NumberService 类创建的 Service 对象：

```
Intent intent = new Intent(this,NumberService.class);
startService(intent) ;
```

执行 startService(intent)后，如果系统未曾创建要启动的 Service 对象，那么系统将创建这个 Service 对象，并立刻让这个 Service 对象调用 onCreate()方法，如果 startService(Intent service)方法请求启动的 Service 对象已经启动过了，而且该 Service 对象没有死亡，那么该 Service 对象不会再调用 onCreate()方法，而是直接调用 onStartCommand 方法。

应用程序中的 Activity 对象或 Service 对象可以使用 stopService(Intent intent)方法停止曾启动的 Service 对象，即执行 stopService(Intent intent)方法会导致 Service 对象执行 void onDestroy()方法并进入死亡状态。

Service 对象本身也可以在自己的生命周期内随时调用 final void stopSelf()方法让自己进入死亡状态，即执行 stopSelf()方法会导致 Service 对象执行 void onDestroy()方法并进入死亡状态。

4. 向 Service 对象传递附加数据

Activity 对象或 Service 对象使用 startService(Intent intent)方法启动一个其他的 Service 对象时，可以通过向 Intent 对象添加附加值来向启动的 Service 对象传递值(有关知识点参见 8.5 节)。需要特别注意的是，Activity 对象或 Service 对象不能获得被启动的 Service 对象存放到 Intent 对象中的附加值，即传值是单向传值。

5. 示例

在下面的例子 9-1 中，程序中的 Activity 对象向启动的 Service 对象传递一个数字 5，Service 对象计算 5 的平方，并将 5 的平方显示在 Toast 漂浮条中(Toast 参见 5.5 节)。运行效果如图 9.3 所示。

图 9.3 Activity 对象反复启动 Service 对象 3 次

例子 9-1

(1) 创建名字为 ch9_1 的工程，主要 Activity 子类的名字为 Example9_1，使用的包名为 ch9.one。用命令行进入 D:\2000，创建工程 D:\2000＞android create project -t 3 -n ch9_1 -p ./ch9_1 -a Example9_1 -k ch9.one。

(2) 将下列和视图相关的 XML 文件保存到工程的\res\layout 目录中。

ch9_1.xml

```
<?xml version = "1.0" encoding = "utf-8"?>
<LinearLayout xmlns:android = "http://schemas.android.com/apk/res/android"
    android:orientation = "vertical"
    android:layout_width = "fill_parent"
    android:layout_height = "fill_parent"
    android:background = "#87CEEB">
    <TextView
```

```xml
            android:id = "@ + id/text"
            android:layout_width = "fill_parent"
            android:layout_height = "wrap_content"
            android:background = "#AA00EB"
            android:textSize = "20sp"
            android:text = "单击启动服务按钮,查看 5 的平方"/>
    <Button
            android:layout_width = "wrap_content"
            android:layout_height = "wrap_content"
            android:text = "启动服务查看结果"
            android:onClick = "startService"   />
    <Button
            android:layout_width = "wrap_content"
            android:layout_height = "wrap_content"
            android:text = "停止服务"
            android:onClick = "stopService"   />
</LinearLayout>
```

（3）应用程序中的 Activity 对象需要启动 Service 对象,将下列 NumberService.java 保存到工程\src\ch9\one 目录下,NumberService.java 内容如下：

NumberService.java

```java
package ch9.one;
import android.app.*;
import android.os.*;
import android.content.*;
import android.widget.*;
import android.view.*;
public class NumberService extends Service {
    int n;
    public void onCreate() {
        super.onCreate();
    }
    public int onStartCommand(Intent intent, int flags, int startId) {
        super.onStartCommand(intent, flags, startId);
        n = intent.getIntExtra("number", 0);
        int result = n * n;
        Toast toast =
        Toast.makeText (this, "n * n = " + result + " \ntimes = " + startId, Toast.LENGTH_LONG);
        toast.setGravity(Gravity.TOP, 60, 160);
        toast.show();
        return START_STICKY;
    }
    public void onDestroy() {
        Toast toast = Toast.makeText (this, "Service is die ", Toast.LENGTH_LONG);
        toast.show();
    }
    public IBinder onBind(Intent intent) {
        return null;
    }
}
```

(4) 修改工程\src\ch9\one 目录下的 Example9_1.java 文件,修改后的内容如下:

Example9_1.java

```java
package ch9.one;
import android.os.*;
import android.content.*;
import android.app.*;
import android.widget.*;
import android.view.*;
public class Example9_1 extends Activity {
    Intent intent;
    public void onCreate(Bundle savedInstanceState) {
        super.onCreate(savedInstanceState);
        setContentView(R.layout.ch9_1);
        intent = new Intent(this,NumberService.class);
        intent.putExtra("number",5);
    }
    public void startService(View v) {
        startService(intent) ;
    }
    public void stopService(View v) {
        stopService(intent) ;
    }
}
```

(5) 修改配置文件 AndroidManifest.xml(在工程的根目录下,即 ch9_1 目录下),在配置文件中增加一个<service>标记,修改后的 AndroidManifest.xml 内容如下:

AndroidManifest.xml.java

```xml
<?xml version = "1.0" encoding = "utf-8"?>
<manifest xmlns:android = "http://schemas.android.com/apk/res/android"
    package = "ch9.one"
    android:versionCode = "1"
    android:versionName = "1.0">
    <application android:label = "@string/app_name" android:icon = "@drawable/ic_launcher">
        <activity android:name = "Example9_1"
            android:label = "@string/app_name">
            <intent-filter>
                <action android:name = "android.intent.action.MAIN" />
                <category android:name = "android.intent.category.LAUNCHER" />
            </intent-filter>
        </activity>
        <service android:name = "NumberService" android:enabled = "true" /> <!-- service 标记 -->
    </application>
</manifest>
```

(6) 启动 AVD,进入工程的根目录,用快捷方式编译工程、安装应用程序到 AVD(有关知识点参见 1.5 节)。对于本例子,用命令行进入 D:\2000\ch9_1,执行如下命令:

```
D:\2000 > ch9_1 > ant debug install
```

9.3 使用多个 Service 对象

1. 配置多个 Service 对象

应用程序可能需要多个 Service 对象完成后台的工作，比如下载视频、保存文件等，那么就需要在配置文件 AndroidManifest.xml 中给出多个＜service＞标记，每个标记对应一个 Service 对象。不仅 Activity 对象可以启动一个 Service 对象，而且一个 Service 对象也可以启动另外一个 Service 对象，这些 Service 对象可以协同工作完成某些任务。

2. 示例

在下面的例子 9-2 中，程序中有两个 Service 对象，分别由 ServiceOne 和 ServiceTwo 负责创建，其中 ServiceOne 创建的 Service 对象可以计算不大于 10 的数的平方。ServiceTwo 类创建的 Service 对象可以计算不大于 100 的数的平方。

例子 9-2 中的 Activity 对象首先启动 ServiceOne 创建的 Service 对象，并请求它计算一个正整数的平方，如果 ServiceOne 创建的 Service 对象能完成计算任务，就用 Toast 显示结果，否则 ServiceOne 创建的 Service 对象负责启动 ServiceTwo 类创建的 Service 对象，并让它去完成计算任务，同时让自己进入死亡状态。如果 ServiceTwo 创建的 Service 对象能完成计算任务，就用 Toast 显示结果，否则 ServiceTwo 就用 Toast 显示"refuse compute"，同时让自己进入死亡状态。运行效果如图 9.4 所示。

图 9.4 启动多个 Service 对象

例子 9-2

（1）创建名字为 ch9_2 的工程，主要 Activity 子类的名字为 Example9_2，使用的包名为 ch9.two。用命令行进入 D:\2000，创建工程 D:\2000＞android create project -t 3 -n ch9_2 -p ./ch9_2 -a Example9_2 -k ch9.two。

（2）将下列和视图相关的 XML 文件保存到工程的\res\layout 目录中。

ch9_2.xml

```
<?xml version = "1.0" encoding = "utf-8"?>
<LinearLayout xmlns:android = "http://schemas.android.com/apk/res/android"
    android:orientation = "vertical"
    android:layout_width = "fill_parent"
    android:layout_height = "fill_parent"
    android:background = "#87CEEB">
    <TextView
        android:layout_width = "fill_parent"
        android:layout_height = "wrap_content"
        android:background = "#AC00E0"
        android:textSize = "20sp"
        android:text = "输入一个正整数"/>
    <EditText
        android:id = "@+id/edit"
        android:layout_width = "match_parent"
        android:layout_height = "wrap_content"
```

```
            android:text = "6"
            android:inputType = "number" />
    < Button
            android:layout_width = "wrap_content"
            android:layout_height = "wrap_content"
            android:text = "启动服务查看数的平方"
            android:onClick = "startService"   />
</LinearLayout >
```

(3) 应用程序中需要两个 Service 对象,将下列 ServiceOne.java 和 ServiceTwo.java 保存到工程\src\ch9\two 目录下。

ServiceOne.java

```
package ch9.two;
import android.app.*;
import android.os.*;
import android.content.*;
import android.widget.*;
import android.view.*;
public class ServiceOne extends Service {
    int result = 1 ;
    public void onCreate() {
        super.onCreate();
    }
    public int onStartCommand(Intent intent,int flags,int startId) {
        result = 1;
        int n = intent.getIntExtra("number",0);
        if(n > 10){
            Intent nextService = new Intent(this,ServiceTwo.class);
            nextService.putExtra("number",n);
            startService(nextService);
            stopSelf();
            return START_STICKY;
        }
        else {
            result = n * n;
            Toast toast =
            Toast.makeText (this,"ServiceOne give result:\n" + result,Toast.LENGTH_LONG);
            toast.setGravity(Gravity.TOP,90,160);
            toast.show();
            return START_STICKY;
        }
    }
    public void onDestroy() {
        Toast toast = Toast.makeText (this,"ServiceOne is die ",Toast.LENGTH_LONG);
        toast.setGravity(Gravity.TOP,100,260);
        toast.show();
    }
    public IBinder onBind(Intent intent) {
        return null;
    }
```

}

ServiceTwo.java

```java
package ch9.two;
import android.app.*;
import android.os.*;
import android.content.*;
import android.widget.*;
import android.view.*;
public class ServiceTwo extends Service {
    long result = 1 ;
    public void onCreate() {
        super.onCreate();
    }
    public int onStartCommand(Intent intent, int flags, int startId) {
        result = 1;
        int n = intent.getIntExtra("number",0);
        if(n > 100){
            Toast toast =
            Toast.makeText (this,"ServiceTwo refuse compute",Toast.LENGTH_LONG);
            toast.setGravity(Gravity.TOP,60,260);
            stopSelf();
            return START_STICKY;
        }
        else {
            result = n * n;
            Toast toast =
            Toast.makeText (this,"ServiceTwo give result:\n" + result,Toast.LENGTH_LONG);
            toast.setGravity(Gravity.TOP,90,160);
            toast.show();
            return START_STICKY;
        }
    }
    public void onDestroy() {
        Toast toast = Toast.makeText (this,"ServiceThree is die ",Toast.LENGTH_LONG);
        toast.setGravity(Gravity.TOP,100,260);
        toast.show();
    }
    public IBinder onBind(Intent intent) {
        return null;
    }
}
```

(4) 修改工程\src\ch9\two 目录下的 Example9_2.java 文件,修改后的内容如下:

Example9_2.java

```java
package ch9.two;
import android.os.*;
import android.content.*;
import android.app.*;
import android.widget.*;
```

```java
import android.view.*;
public class Example9_2 extends Activity {
    Intent intent;
    EditText edit;
    public void onCreate(Bundle savedInstanceState) {
        super.onCreate(savedInstanceState);
        setContentView(R.layout.ch9_2);
        edit = (EditText)findViewById(R.id.edit);
        intent = new Intent(this,ServiceOne.class);
    }
    public void startService(View v) {
        String s = edit.getText().toString();
        intent.putExtra("number",Integer.parseInt(s));
        startService(intent) ;
    }
}
```

（5）修改配置文件 AndroidManifest.xml（在工程的根目录下，即 ch9_2 目录下），在配置文件中增加两个＜service＞标记，修改后的 AndroidManifest.xml 内容如下：

AndroidManifest.xml.java

```xml
<?xml version = "1.0" encoding = "utf-8"?>
<manifest xmlns:android = "http://schemas.android.com/apk/res/android"
    package = "ch9.two"
    android:versionCode = "1"
    android:versionName = "1.0">
    <application android:label = "@string/app_name" android:icon = "@drawable/ic_launcher">
        <activity android:name = "Example9_2"
            android:label = "@string/app_name">
            <intent-filter>
                <action android:name = "android.intent.action.MAIN" />
                <category android:name = "android.intent.category.LAUNCHER" />
            </intent-filter>
        </activity>
        <service android:name = "ServiceOne" android:enabled = "true" />
        <service android:name = "ServiceTwo" android:enabled = "true" />
    </application>
</manifest>
```

（6）启动 AVD，进入工程的根目录，用快捷方式编译工程、安装应用程序到 AVD（有关知识点参见 1.5 节）。对于本例子，用命令行进入 D:\2000\ch9_2，执行如下命令：

D:\2000 > ch9_2 > ant debug install

9.4 IntentService 类

1. 问题的提出

我们已经知道，应用程序中的 Activity 对象以及 Service 对象都运行在同一个线程中（参见 9.1 节），即都运行在主线程（UI 线程）中，也就是说，这些对象的方法都运行在主线程中。那么就可能涉及这样的问题，当 Activity 对象启动一个 Service 对象后，Service 对象在工作状态时会执行 onStartCommand() 方法，如果该方法需要较长的时间（比如超过 5 秒）才能返回，那么主线程中的 Activity 对象中的视图就得不到及时刷新的机会，用户会感觉程序似乎被挂起，可能导致用户关闭程序。因此应当避免在 onStartCommand() 方法执行非常耗时的操作。

2. IntentService 类的工作状态

解决我们前面提出的问题的办法之一是使用 Android 提供的 IntentService 类，该类是 Service 类的子类。IntentService 类有一个非常重要的方法：void onHandleIntent(Intent intent)。

IntentService 类的子类创建的 Service 对象进入工作状态时，系统会单独启动一个工作线程（主线程之外再启动的线程，习惯地被称作一个 work thread），并在该工作线程中运行 onHandleIntent(Intent intent)。因此，用户在编写 IntentService 类的子类时，不要重写 onStartCommand() 方法（该方法运行在主线程中），也就是说，将分配给 Service 对象的任务写在 onHandleIntent(Intent intent) 方法中。这样一来，系统会协调主线程和工作线程，使得二者都有机会使用 CPU 资源，用户就可以把比较耗时的操作放在 onHandleIntent(Intent intent) 方法中。另外，和 Service 子类创建的 Service 对象不同的是，IntentService 类的子类创建的 Service 对象完成了工作之后，即 onHandleIntent(Intent intent) 返回之后，系统将让该 Service 对象进入死亡状态。

另外需要特别注意的是，IntentService 类只有一个带参数的构造方法 IntentService(String name)，因此子类必须要有显示的构造方法，并使用 super 调用 IntentService 类的 IntentService(String name) 构造方法。

3. 示例

我们知道，函数递归是比较耗时的操作，尤其是 2 次递归，比如，用函数递归计算 Fibinacii 数列的第 n 项所需要的时间大约是 $2^n \times unitTime$，其中 unitTime 是计算完成一次操作所用的时间。例子 9-3 中的 Activity 对象启动 IntentService 类的子类 ComputerService 创建的 Service 对象，并请求它计算 Fibinacii 数列的第 n 项。运行效果如图 9.5 所示。

图 9.5 启用 IntentService 创建的 Service 对象

例子 9-3

（1）创建名字为 ch9_3 的工程，主要 Activity 子类的名字为 Example9_3，使用的包名为 ch9.three。用命令行进入 D:\2000，创建工程 D:\2000＞android create project -t 3 -n ch9_3 -p ./ch9_3 -a Example9_3 -k ch9.three。

（2）将下列和视图相关的 XML 文件保存到工程的 \res\layout 目录中。

ch9_3.xml

```
<?xml version = "1.0" encoding = "utf-8"?>
<LinearLayout xmlns:android = "http://schemas.android.com/apk/res/android"
    android:orientation = "vertical"
    android:layout_width = "fill_parent"
```

```
    android:layout_height = "fill_parent"
    android:background = "#87CEEB">
    <TextView
        android:layout_width = "fill_parent"
        android:layout_height = "wrap_content"
        android:background = "#AC00E0"
        android:textSize = "20sp"
        android:text = "计算出 Fibinacii 数列的第 n 项\n输入 n 的值:"/>
    <EditText
        android:id = "@+id/edit"
        android:layout_width = "match_parent"
        android:layout_height = "wrap_content"
        android:text = "6"
        android:inputType = "number" />
    <Button
        android:layout_width = "wrap_content"
        android:layout_height = "wrap_content"
        android:text = "启动服务查看第 n 项的值"
        android:onClick = "startService"   />
</LinearLayout>
```

（3）应用程序中需要一个 Service 对象，该 Service 对象由 IntentService 类的子类 ComputerService 负责创建。将下列 ComputerService.java 保存到工程\src\ch9\three 目录下。

ComputerService.java

```
package ch9.three;
import android.app.*;
import android.os.*;
import android.content.*;
import android.widget.*;
import android.view.*;
public class ComputerService extends IntentService {
    long result = 1 ;
    public ComputerService() {
        super("ok");
    }
    public void onCreate() {
        super.onCreate();
    }
    public void onHandleIntent(Intent intent) {
        int n = intent.getIntExtra("number",1);
        Toast wait =
        Toast.makeText (this,"waiting...",Toast.LENGTH_LONG);
        wait.setGravity(Gravity.TOP,90,260);
        wait.show();
        result = fibinacci(n);
        Toast toast =
        Toast.makeText (this,"result:\n" + result,Toast.LENGTH_LONG);
        toast.setGravity(Gravity.TOP,90,160);
```

```
            toast.show();
        }
        public void onDestroy() {
            Toast toast = Toast.makeText (this,"Service is die ",Toast.LENGTH_LONG);
            toast.setGravity(Gravity.TOP,100,260);
            toast.show();
        }
        public IBinder onBind(Intent intent) {
            return null;
        }
        long fibinacci(int n) {
            long result = 1;
            if(n == 1||n == 2)
                result = 1;
            else
                result = fibinacci(n - 1) + fibinacci(n - 2);
            return result;
        }
    }
```

(4) 修改工程\src\ch9\three 目录下的 Example9_3.java 文件,修改后的内容如下:

Example9_3.java

```
package ch9.three;
import android.os.*;
import android.content.*;
import android.app.*;
import android.widget.*;
import android.view.*;
public class Example9_3 extends Activity {
    Intent intent;
    EditText edit;
    public void onCreate(Bundle savedInstanceState) {
        super.onCreate(savedInstanceState);
        setContentView(R.layout.ch9_3);
        edit = (EditText)findViewById(R.id.edit);
        intent = new Intent(this,ComputerService.class);
    }
    public void startService(View v) {
        String s = edit.getText().toString();
        intent.putExtra("number",Integer.parseInt(s));
        startService(intent) ;
    }
}
```

(5) 修改配置文件 AndroidManifest.xml(在工程的根目录下,即 ch9_3 目录下),在配置文件中增加一个<service>标记,修改后的 AndroidManifest.xml 内容如下:

AndroidManifest.xml.java

```
<?xml version = "1.0" encoding = "utf - 8"?>
< manifest xmlns:android = "http://schemas.android.com/apk/res/android"
```

```
            package = "ch9.three"
            android:versionCode = "1"
            android:versionName = "1.0">
    < application android:label = "@string/app_name"
                  android:icon = "@drawable/ic_launcher">
        < activity android:name = "Example9_3"
                   android:label = "@string/app_name">
            < intent-filter >
                < action android:name = "android.intent.action.MAIN" />
                < category android:name = "android.intent.category.LAUNCHER" />
            </intent-filter >
        </activity >
        < service android:name = "ComputerService" android:enabled = "true" />
    </application >
</manifest >
```

（6）启动 AVD，进入工程的根目录，用快捷方式编译工程、安装应用程序到 AVD（有关知识点参见 1.5 节）。对于本例子，用命令行进入 D:\2000\ch9_3，执行如下命令：

```
D:\2000> ch9_3> ant debug install
```

9.5 AsyncTask 类

1. AsyncTask 类的特点

AsyncTask 类的对象就像它的名字一样，负责完成一些异步任务。我们已经知道应用程序中的 Activity 对象以及 Service 对象都运行在同一个线程中（参见 9.1 节），即都运行在主线程（UI 线程）中，也就是说，这些对象的方法都运行在主线程中。Service 对象主要是进行后台的一些计算或上传下载工作，无法操作 Activity 对象中的视图。程序可能需要在一个单独的工作线程中处理数据，同时又需要更新 Activity 对象中视图，那么 Service 对象就无法完成这样的工作。

需要注意的是，Android 系统的视图不是线程安全的（Android UI toolkit is not thread-safe），这意味着，如果想在主线程之外的其他工作线程中操作视图，就需要进行比较复杂的编码，需要使用 android.os 包中的 HandlerThread、Looper 和 Message 共同协调工作，而且编码也有一定的难度（讲解 HandlerThread、Looper 和 Message 类不在本书范畴内）。为此，Android 提供了 AsyncTask 类。AsyncTask 类中的方法分成两部分，其中一部分运行在主线程中，另一部分运行在一个工作线程中（用户不必关心该工作线程是如何使用 HandlerThread、Looper 和 Message 来协调工作的）。这样一来，用户可以把比较耗时的代码放到在工作线程中运行的方法中，将需要操作视图的代码放到在主线程中运行的方法中（如后面的图 9.6 所示）。

2. AsyncTask 类的重要方法

AsyncTask 类是泛型类，AsyncTask<Params, Progress, Result>，继承关系如下：

```
java.lang.Object
    └ android.os.AsyncTask< Params, Progress, Result >
```

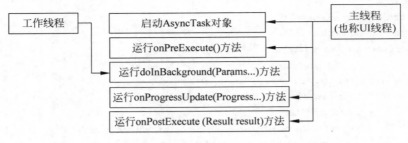

图 9.6 运行机制

其中 Params,Progress 和 Result 是泛型的名字,表示我们在构造 AsyncTask 类的对象时,可以指定 Params,Progress 和 Result 代表的具体类型(可以是任何 Object 类的子类,但不能是基本型,如 int 型,double 型等)。如果不需指定具体类型,可以让 Params,Progress 或 Result 是 Void 型。

1) 启动 AsyncTask 对象

必须在主线程中创建 AsyncTask 对象,并由主线程启动该对象,例如在 Activity 对象中创建一个 AsyncTask 对象,并让该对象执行 execute(Params... params)方法。简单地说,主线程启动一个 AsyncTask 对象,就相当于开始了一个异步任务(运行在主线中和工作线程中的方法不需要互相等待)。

2) AsyncTask 对象运行机制

AsyncTask 对象被启动后,将按下列机制自动调用如下的方法。

• void onPreExecute()

AsyncTask 对象首先调用 onPreExecute()方法,该方法将运行在主线程中(如图 9.6 所示)。因此,用户在编写 AsyncTask 类的子类时,如果需要对某些视图进行必要的初始化操作,就需要重写该方法,否则直接继承该方法即可。

• Result doInBackground(Params... params)

该方法运行在系统提供的一个工作线程中(如图 9.6 所示)。AsyncTask 对象调用 onPreExecute()方法后,将立刻调用 doInBackground(Params...)方法。用户在编写 AsyncTask 类的子类中,如果需要做一些耗时的操作,就需要重写该方法,否则直接继承该方法即可。doInBackground(Params...)方法的参数得到的值是启动 AsyncTask 对象的 execute(Params... params)方法中的参数 params 的值。Result doInBackground(Params... params)必须返回 Result 类型,如果泛型 Result 指定的类型是 View 类型,那么方法必须返回一个 View 子类创建的对象。

• void onProgressUpdate(Progress...)

onProgressUpdate(Progress...)方法将运行在主线程中(如图 9.6 所示)。AsyncTask 对象在调用 doInBackground(Params...)方法的过程中,可以需要随时调用 final void publishProgress(Progress... values)方法,publishProgress 方法是父类的 final 方法,不允许子类重写。调用 publishProgress 方法的作用就是让 AsyncTask 对象调用 onProgressUpdate(Progress...)方法,并将 values 传递给 onProgressUpdate(Progress...)方法的参数。需要注意的是,onProgressUpdate(Progress...)方法在主线程中运行时,doInBackground(Params...)

方法会继续在工作线程中运行(不是同步的)，如果恰好 doInBackground(Params…)方法执行完毕，那么 onProgressUpdate(Progress…)方法会立刻停止执行。用户在编写 AsyncTask 类的子类中，如果需要使用 doInBackground 方法执行过程中的数据随时更新视图，就需要在 doInBackground 方法中随时调用 publishProgress 方法，并在重写的 onProgressUpdate(Progress…)方法中编写更新视图的代码。如果不需要及时更新视图，就不需要在执行 doInBackground 方法中调用 publishProgress 方法，也没有必要重写 onProgressUpdate(Progress…)方法。

- void onPostExecute(Result result)

该方法将运行在主线程中(如图 9.6 所示)。AsyncTask 对象执行完 doInBackground(Params…)方法后，即运行在工作线程中的 doInBackground(Params…)方法结束后，AsyncTask 对象将立刻调用 onPostExecute(Result result)方法，其中的参数就是 doInBackground(Params…)方法返回的 Result。因此，用户在编写 AsyncTask 类的子类时，如果需要在 doInBackground(Params…)方法结束后，对视图进行更新操作，就要重写该方法，否则直接继承该方法即可。

3. 示例

例子 9-4 中有一个进度条 bar，我们使用进度条 bar 显示程序用递归计算 Fibinacii 数列前 35 项的进度情况。由于递归计算 Fibinacii 数列的第 n 项的时间随着 n 的增大将需要更长的时间，因此需要将计算工作放到一个工作线程中，在本例中，放到一个 AsyncTask 对象的 doInBackground(Params…)方法中，每当计算出一项，就需要让进度条前进一个单位，因此需要及时调用 onProgressUpdate(Progress…)方法，其理由是 onProgressUpdate(Progress…)方法运行在主线程中，可以更新进度条 bar。当 doInBackground(Params…)执行完毕后，AsyncTask 对象将调用 onPostExecute(Result result)方法，该方法运行在主线程中，因此可以更新视图，本例子中，我们在 onPostExecute(Result result)方法中将 TextView 视图中的文本更新为"work is finished"。运行效果如图 9.7(a)，9.7(b)所示。

(a) 后台计算并更新前台进度条

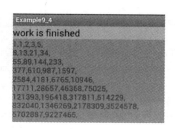
(b) 后台计算结束，更新前台视图

图 9.7 运行效果

例子 9-4

(1) 创建名字为 ch9_4 的工程，主要 Activity 子类的名字为 Example9_4，使用的包名为 ch9.four。用命令行进入 D:\2000，创建工程 D:\2000＞android create project -t 3 -n ch9_4 -p ./ch9_4 -a Example9_4 -k ch9.four。

(2) 将下列和视图相关的 XML 文件保存到工程的 \res\layout 目录中。

ch9_4.xml

```xml
<LinearLayout xmlns:android="http://schemas.android.com/apk/res/android"
    android:orientation="vertical"
    android:layout_width="match_parent"
    android:layout_height="match_parent">
    <TextView
        android:id="@+id/textOne"
        android:layout_width="match_parent"
        android:layout_height="wrap_content"
        android:background="#87CEEB"
        android:textSize="20sp"
        android:textColor="#000000"
        android:text="输出 Fibinacii 数列.\n 以下是进度情况:" />
    <TextView
        android:id="@+id/textTwo"
        android:layout_width="match_parent"
        android:layout_height="match_parent"
        android:background="#999999"
        android:textSize="16sp"
        android:textColor="#0000FF" />
</LinearLayout>
```

(3) 修改工程 \src\ch9\four 目录下的 Example9_4.java 文件，修改后的内容如下：

Example9_4.java

```java
package ch9.four;
import android.app.*;
import android.os.*;
import android.widget.*;
import android.view.*;
import java.util.*;
public class Example9_4 extends Activity {
    TextView textOne,textTwo;
    ProgressDialog dialog;
    MyComputerTask task;
    int N=35;
    public void onCreate(Bundle savedInstanceState) {
        super.onCreate(savedInstanceState);
        setContentView(R.layout.ch9_4);
        textOne=(TextView)findViewById(R.id.textOne);
        textTwo=(TextView)findViewById(R.id.textTwo);
        dialog=new ProgressDialog(this,ProgressDialog.THEME_HOLO_LIGHT);
        dialog.setProgressStyle(ProgressDialog.STYLE_HORIZONTAL);
        dialog.setIndeterminate(false);
        dialog.setCancelable(false);
        dialog.setMax(N);
        dialog.setTitle("showing Progress of giving "+N+" items");
        dialog.setIcon(R.drawable.ic_launcher);
        task=new MyComputerTask();
```

```
            task.textOne = textOne;
            task.textTwo = textTwo;
            task.dialog = dialog;
            task.N = N;
            task.execute(null);
        }
    }
    class MyComputerTask extends AsyncTask<Object,Object,Object> {
        TextView textOne,textTwo;
        ProgressDialog dialog;
        ArrayList<Long> list = new ArrayList<Long>();
        long item = 1;
        int i = 0,N;
        public void onPreExecute() {
            dialog.show();
        }
        public Object doInBackground(Object... params) {
            for(i = 1;i <= N;i++) {
                item = fibinacci(i);
                list.add(item);
                publishProgress(params) ;
            }
            return null;
        }
        public void onProgressUpdate(Object... progress) {
            dialog.setProgress(i);
        }
        public void onPostExecute(Object result) {
            textOne.setText("work is finished");
            for(int i = 0;i < list.size();i++) {
                textTwo.append(list.get(i) + ",") ;
                if(i % 4 == 0&&i!= 0)
                    textTwo.append("\n") ;
            }
            dialog.setProgress(N);
            dialog.hide();
        }
        long fibinacci(int n) {
            long result = 1;
            if(n == 1||n == 2)
                result = 1;
            else
                result = fibinacci(n - 1) + fibinacci(n - 2);
            return result;
        }
    }
```

(4) 启动 AVD,进入工程的根目录,用快捷方式编译工程、安装应用程序到 AVD(有关知识点参见 1.5 节)。对于本例子,用命令行进入 D:\2000\ch9_4,执行如下命令:

```
D:\2000>ch9_4>ant debug install
```

9.6 广播及接收

广播和接收是指某个应用程序可以让自己的 Activity 对象调用广播的方法进行广播，即广播信息，那么当前应用程序或其他应用程序中的接收者(receiver)就可以接收到有关的信息。接收者是一个特殊的对象(BroadcastReiceiver 类的实例)，使用特殊的方法接收信息。

1. 广播

1) 广播的方法

android.content.Context 类提供了一个常用的广播方法：void sendBroadcast(Intent intent)，该方法向 intent 找到的接收者广播信息(信息就是 intent 中的附加数据)。需要特别注意的是，sendBroadcast(Intent intent)方法中的 Intent 对象和 startActivity(Intent intent)中的不同，sendBroadcast(Intent intent)方法中的 Intent 对象体现的意图是寻找接收者(是应用程序中 BroadcastReceiver 类的实例，而不是 Activity 对象)。

2) 广播中的 Intent 对象

广播方法 sendBroadcast(Intent intent)中的 Intent 对象负责寻找接收者，而且接收者能获得 sendBroadcast(Intent intent)中 Intent 对象的附加信息(即获得广播的信息)。

与 Intent 对象寻找 Activity 对象类似，sendBroadcast(Intent intent)方法中的 Intent 对象经常使用下列条件寻找接收者(有关细节参见 8.2 节，8.3 节和 8.4 节)。

- action(动作)

该内容给出的信息是一个字符串，该字符串被习惯地称作 Intent 对象中的 action，是指 Intent 对象用 action 要求的所寻找的接收者(receiver)应当能进行的动作。action 的包名可任意给定，例如，包名是 com.sun.moon 的动作"com.sun.moon.LISTENER"。

- category(范畴)

该内容给出的信息是一个字符串，被称作 Intent 对象中的一个 category，是指 Intent 对象用 category 要求的所寻找的接收者(receiver)应在的范畴。和寻找 Activity 对象不同的是，sendBroadcast(Intent intent)方法在寻找接收者时，不会给参数 intent 增加额外的范畴(有关细节参见 8.2 节)。category 的包名可任意给定，例如，包名是 sohu.com.cn 的范畴 "sohu.com.cn.BIRD"。

- 指定应用程序

如果希望在某个特定应用程序中寻找接收者，Intent 对象 intent 可以事先使用 setPackage(String packageName)方法给出应用程序的包名。

2. 接收者

接收者只有满足 sendBroadcast(Intent intent)方法中 intent 给出的条件，才能接收到信息，接收者是应用程序中 BroadcastReceiver 类的实例。

1) BroadcastReceiver 类

用户需要编写 BroadcastReceiver 类的子类，以便系统使用这样的子类创建接收者。BroadcastReceiver 类的继承关系如下：

```
java.lang.Object
    └ android.content.BroadcastReceiver
```

BroadcastReceiver 有一个重要的方法如下：

void onReceive(Context context,Intent intent);

当接收者接收到广播时，即 Activity 对象或 Service 对象调用 sendBroadcast(Intent intent)方法找到接收者时，接收者立刻执行 onReceive(Context context,Intent intent)方法，该方法的参数 intent 就是广播方法中使用的 Intent 对象，参数 context 是接收者所在的上下文对象。用户在编写 BroadcastReceiver 类的子类时必须要重写 onReceive(Context context,Intent intent)方法，并通过参数 intent 接收广播的信息，比如，让 intent 调用诸如 getXXXExa()方法获得广播方法 sendBroadcast(Intent intent)放到 intent 中的附加数据（有关附加数据的知识参见 8.5 节）。

2) 在配置文件中注册接收者

接收者也是由主线程负责管理的对象（参见 9.1 节），可以在项目的配置文件中增加＜receiver＞标记来注册一个接收者，比如，注册一个接收者的＜receiver＞标记如下（完整的配置文件内容见稍后的例子中的.xml 文件）：

```
<receiver android:name = "ReceiverOne">
    <intent-filter>
        <!-- 能进行的 action -->
        <action android:name = " geng.xiang.yi.LISTENER" />
    </intent-filter>
</receiver>
```

上述＜receiver＞标记注册的接收者由 ReceiverOne 类（BroadcastReceiver 类的一个子类）负责创建。＜receiver＞标记需要使用＜intent-filter＞子标记给出自己能被 Intent 对象找到的条件（有关细节参见 8.2 节）。

3) 接收者的生命周期

接收者也是运行于主线中的一个对象，当广播方法找到接收者后，系统用配置文件中的＜receiver＞标记中指定的类创建一个接收者，并让该接收者执行 onReceive(Context context,Intent intent)方法，该方法执行完毕后，主线程将立刻释放当前接收者，即让当前接收者进入死亡状态。

3. 示例

例子 9-5 中有两个应用程序，第一个应用程序中的 Activity 对象使用广播方法寻找第二个程序中的接收者。广播的信息是"It is raining"，第二个应用程序中有一个接收者负责接收信息。接收者用 Toast 显示接收到的信息。首先要运行第一个应用程序，然后进行广播。运行效果如图 9.8 所示。

图 9.8 广播与接收者

例子 9-5

1) 工程 1

(1) 创建名字为 ch9_5_1 的工程，主要的 Activity 子类的名字为 Example9_5_1,使用的包名为 ch9.five_1。用命令行进入 D:\2000,创建工程 D:\2000>android create project -t 3 -n ch9_5_1 -p ./ch9_5_1 -a Example9_5_1 -k ch9.five_1。

(2) 将下列和视图相关的 XML 文件 ch9_5_1.xml 保存到工程的\res\layout 目录中。

ch9_5_1.xml

```xml
<?xml version = "1.0" encoding = "utf-8"?>
<LinearLayout xmlns:android = "http://schemas.android.com/apk/res/android"
    android:orientation = "vertical"
    android:layout_width = "fill_parent"
    android:layout_height = "fill_parent"
    android:background = "#87CEEB">
    <Button
        android:layout_width = "wrap_content"
        android:layout_height = "wrap_content"
        android:text = "开始广播"
        android:onClick = "startBroadcast" />
</LinearLayout>
```

(3) 修改工程\src\ch9\five_1 目录下的 Example9_5_1.java 文件，修改后的内容如下：

Example9_5_1.java

```java
package ch9.five_1;
import android.app.*;
import android.os.Bundle;
import android.view.*;
import android.content.*;
public class Example9_5_1 extends Activity {
    public void onCreate(Bundle savedInstanceState) {
        super.onCreate(savedInstanceState);
        setContentView(R.layout.ch9_5_1);
    }
    public void startBroadcast(View v) {
        Intent intent = new Intent("geng.xiang.yi.LISTENER");  //接收者能进行 LISTENER 动作
        intent.putExtra("mess","It is raining");
        sendBroadcast(intent);
    }
}
```

(4) 启动 AVD，进入工程的根目录，用快捷方式编译工程、安装应用程序到 AVD(有关知识点参见 1.5 节)。对于本例子，用命令行进入 D:\2000\ch9_5_1，执行如下命令：

```
D:\2000 > ch9_5_1 > ant debug install
```

2) 工程 2

(1) 创建名字为 ch9_5_2 的工程，主要的 Activity 子类的名字为 Example9_5_2，使用的包名为 ch9.five_2。用命令行进入 D:\2000，创建工程 D:\2000> android create project -t 3 -n ch9_5_2 -p ./ch9_5_2 -a Example9_5_2 -k ch9.five_2。

(2) 将负责创建接收者的 BoyReceiver.java 保存到工程\src\ch9\five_2 目录下。BoyReceiver.java 内容如下：

BoyReceiver.java

```java
package ch9.five_2;
```

```java
import android.widget.*;
import android.view.*;
import android.content.*;
public class BoyReceiver extends BroadcastReceiver {
    public void onReceive(Context context, Intent intent) {
        String str = intent.getStringExtra("mess");
        Toast toast = Toast.makeText (context,str,Toast.LENGTH_LONG);
        toast.setGravity(Gravity.TOP,30,100);
        toast.show();
    }
}
```

(3) 修改工程的配置文件 AndroidManifest.xml(在工程的根目录下,本例就是 ch9_5_2 目录下),注册一个能进行 LISTENER 动作的接收者,即增加一个＜receiver＞标记,修改后的内容如下：

AndroidManifest.xml.java

```xml
<?xml version = "1.0" encoding = "utf-8"?>
<manifest xmlns:android = "http://schemas.android.com/apk/res/android"
    package = "ch9.five_2"
    android:versionCode = "1"
    android:versionName = "1.0">
    <application android:label = "@string/app_name" android:icon = "@drawable/ic_launcher">
        <activity android:name = "Example9_5_2"
                android:label = "@string/app_name">
            <intent-filter>
                <action android:name = "android.intent.action.MAIN" />
                <category android:name = "android.intent.category.LAUNCHER" />
            </intent-filter>
        </activity>
        <receiver android:name = "BoyReceiver">
            <intent-filter>
                <action android:name = "geng.xiang.yi.LISTENER" />
            </intent-filter>
        </receiver>
    </application>
</manifest>
```

(4) 启动 AVD,进入工程的根目录,用快捷方式编译工程、安装应用程序到 AVD(有关知识点参见 1.5 节)。对于本例子,用命令行进入 D:\2000\ch9_5_2,执行如下命令：

D:\2000＞ch9_5_2＞ant debug install

9.7 PendingIntent 类

1. PendingIntent 对象的作用

Service 和 BroadcastReceiver 都是后台对象。有时这些后台对象可能希望将自己得到或处理的数据显示在前台,即显示在某个 Activity 对象中,这就需要启动一个 Activity 对象,但是 Service 和 BroadcastReceiver 后台对象所在的上下文(Context)不能使用 startActivity(Intent intent)启动一个 Activity 对象(会发生运行异常)。

Android 提供了一个 android.app.PendingIntent 类,如果在 Service 和 BroadcastReceiver 对象所在的上下文中想启动其他应用程序或本应用程序中的 Activity 对象,就需要将负责寻找 Activity 对象的 Intent 对象事先封装到 PendingIntent 对象中,人们习惯地称 PendingIntent 对象是"准备式"Intent 对象。"准备式"Intent 对象调用 send()方法将委托其封装的 Intent 对象去寻找 Activity 对象,其作用相当于 startActivity(Intent intent)方法。

PendingIntent 类调用静态(static)方法 PendingIntent getActivity(Context context, int requestCode, Intent intent, int flags)可以返回一个"准备式"Intent 对象,其中参数 context 可以取当前 Service 对象和 BroadcastReceiver 对象所在的上下文(Context),参数 requestCode 取 0 即可,参数 flags 取值 Intent.FLAG_ACTIVITY_NEW_TASK。例如,假设当前上下文是 context,准备启动的 Activity 对象是 Example9_6_2.class,代码如下:

```
Intent newIntent = new Intent(context,Example9_6_2.class);  //寻找 Example9_6_2.class 类创建
                                                            //的 activity
PendingIntent pendingIntent =
PendingIntent.getActivity(context,0,newIntent,Intent.FLAG_ACTIVITY_NEW_TASK);
```

注:Service 类是 Context 类的子类,而 BroadcastReceiver 类的 onReceive(Context context,Intent intent)方法中提供了上下文对象。

2. 示例

例子 9-6 中有两个应用程序,第一个应用程序中的 Activity 对象有两个按钮视图,单击一个按钮,第一个应用程序将使用其中 Service 对象计算 Fibinacii 数列,计算出数列的某项后,比如第 31 项后,使用"准备式"Intent 对象重新启动当前应用程序中的 Activity 对象,并让该 Activity 对象显示数列的第 31 项。

单击第一个应用程序中的另一个按钮,第一个应用程序使用广播方法寻找第二个程序中的接收者。广播的信息是"It is raining",第二个应用程序中的接收者收到信息后,用"准备式"Intent 对象启动第二个应用程序中的 Activity 对象,并让该 Activity 对象显示接收者收到的信息。运行效果如图 9.9(a),9.9(b)所示。

(a) 启动服务后的效果　　　　　(b) 开始广播后的效果

图 9.9　运行效果

例子 9-6

1) 工程 1

(1) 创建名字为 ch9_6_1 的工程,主要的 Activity 子类的名字为 Example9_6_1,使用的包名为 ch9.six_1。用命令行进入 D:\2000,创建工程 D:\2000>android create project -t 3 -n ch9_6_1 -p ./ch9_6_1 -a Example9_6_1 -k ch9.six_1。

(2) 将下列和视图相关的 XML 文件 ch9_6_1.xml 保存到工程的 \res\layout 目录中。

ch9_6_1.xml

```xml
<?xml version = "1.0" encoding = "utf-8"?>
<LinearLayout xmlns:android = "http://schemas.android.com/apk/res/android"
    android:orientation = "vertical"
    android:layout_width = "fill_parent"
    android:layout_height = "fill_parent"
    android:background = "#87CEEB">
    <Button
        android:layout_width = "wrap_content"
        android:layout_height = "wrap_content"
        android:text = "启动服务"
        android:onClick = "startServer" />
    <Button
        android:layout_width = "wrap_content"
        android:layout_height = "wrap_content"
        android:text = "开始广播"
        android:onClick = "startBroadcast" />
    <TextView
        android:id = "@+id/text"
        android:layout_width = "fill_parent"
        android:layout_height = "wrap_content"
        android:background = "#A900EB"
        android:textSize = "20sp" />
</LinearLayout>
```

(3) 第一个应用程序中需要一个 Service 对象，该 Service 对象由 IntentService 类的子类 ComputerService 负责创建。将下列 ComputerService.java 保存到工程 \src\ch9\six_1 目录下。

ComputerService.java

```java
package ch9.six_1;
import android.app.*;
import android.os.*;
import android.content.*;
import android.widget.*;
import android.view.*;
public class ComputerService extends IntentService {
    long result = 1;
    public ComputerService() {
        super("ok");
    }
    public void onCreate() {
        super.onCreate();
    }
    public void onHandleIntent(Intent intent) {
        int n = intent.getIntExtra("number",1);
        result = fibinacci(n);
        Intent newIntent = new Intent(this,Example9_6_1.class);
```

```
            newIntent.putExtra("result",result);
            PendingIntent pendingIntent =
            PendingIntent.getActivity(this,0,newIntent,Intent.FLAG_ACTIVITY_NEW_TASK);
            try {
                pendingIntent.send();
            }
            catch(PendingIntent.CanceledException exp){}
        }
        public void onDestroy() {
            Toast toast = Toast.makeText (this,"Service is die ",Toast.LENGTH_LONG);
            toast.setGravity(Gravity.TOP,100,260);
            toast.show();
        }
        public IBinder onBind(Intent intent) {
            return null;
        }
        long fibinacci(int n) {
            long result = 1;
            if(n == 1||n == 2)
                result = 1;
            else
                result = fibinacci(n - 1) + fibinacci(n - 2);
            return result;
        }
    }
```

（4）修改工程\src\ch9\six_1 目录下的 Example9_6_1.java 文件。Example9_6_1 创建的 Activity 对象需要显示 Service 对象计算数列的结果，修改后的内容如下：

Example9_6_1.java

```
package ch9.six_1;
import android.app.*;
import android.os.*;
import android.content.*;
import android.widget.*;
import android.view.*;
public class Example9_6_1 extends Activity {
    TextView text;
    int N = 31;
    public void onCreate(Bundle savedInstanceState) {
        super.onCreate(savedInstanceState);
        setContentView(R.layout.ch9_6_1);
        text = (TextView)findViewById(R.id.text);
        Intent intent = getIntent();
        long result = intent.getLongExtra("result",0);
        if(result!= 0)
            text.setText("Result: " + result) ;
    }
    public void startServer(View v) {
        Intent intent = new Intent(this,ComputerService.class);
        intent.putExtra("number",N);
        startService(intent) ;
```

```
        text.setText("Waiting ...result") ;
    }
    public void startBroadcast(View v) {
        Intent intent = new Intent("tom.cat.TOM");      //寻找能进行 TOM 动作的广播接收者
        intent.setPackage("ch9.six_2");
        intent.putExtra("mess","It is raining");
        sendBroadcast(intent);
    }
}
```

(5) 修改配置文件 AndroidManifest.xml(在工程的根目录下,即 ch9_6_1 目录下),在配置文件中增加一个＜service＞标记,修改后的 AndroidManifest.xml 内容如下:

AndroidManifest.xml.java

```
<?xml version = "1.0" encoding = "utf-8"?>
<manifest xmlns:android = "http://schemas.android.com/apk/res/android"
    package = "ch9.six_1"
    android:versionCode = "1"
    android:versionName = "1.0">
    <application android:label = "@string/app_name" android:icon = "@drawable/ic_launcher">
        <activity android:name = "Example9_6_1"
            android:label = "@string/app_name">
            <intent-filter>
                <action android:name = "android.intent.action.MAIN" />
                <category android:name = "android.intent.category.LAUNCHER" />
            </intent-filter>
        </activity>
        <service android:name = "ComputerService" android:enabled = "true" /> <!-- service 标记 -->
    </application>
</manifest>
```

(6) 启动 AVD,进入工程的根目录,用快捷方式编译工程、安装应用程序到 AVD(有关知识点参见 1.5 节)。对于本例子,用命令行进入 D:\2000\ch9_6_1,执行如下命令:

```
D:\2000 > ch9_6_1 > ant debug install
```

2) 工程 2

(1) 创建名字为 ch9_6_2 的工程,主要的 Activity 子类的名字为 Example9_6_2,使用的包名为 ch9.six_2。用命令行进入 D:\2000,创建工程 D:\2000＞ android create project -t 3 -n ch9_6_2 -p ./ch9_6_2 -a Example9_6_2 -k ch9.six_2。

(2) 将下列和视图相关的 XML 文件 ch9_6_2.xml 保存到工程的 \res\layout 目录中。

ch9_6_2.xml

```
<?xml version = "1.0" encoding = "utf-8"?>
<LinearLayout xmlns:android = "http://schemas.android.com/apk/res/android"
    android:orientation = "vertical"
    android:layout_width = "fill_parent"
    android:layout_height = "fill_parent"
```

```
            android:background = " #87CAEB">
    <TextView
        android:id = "@ + id/text"
        android:layout_width = "fill_parent"
        android:layout_height = "wrap_content"
        android:background = " #AC00E0"
        android:textSize = "20sp"
        android:text = "得到的广播信息:\n"/>
</LinearLayout>
```

(3) 将负责创建接收者的 GirlReceiver.java 保存到工程\src\ch9\six_2 目录下。GirlReceiver.java 内容如下:

GirlReceiver.java

```
package ch9.six_2;
import android.content.*;
import android.app.*;
public class GirlReceiver extends BroadcastReceiver {
    public void onReceive(Context context,Intent intent) {
        String str = intent.getStringExtra("mess");
        Intent newIntent = new Intent(context,Example9_6_2.class);
        newIntent.putExtra("mess",str);
        PendingIntent pendingIntent =
        PendingIntent.getActivity(context,0,newIntent,Intent.FLAG_ACTIVITY_NEW_TASK);
        try {
            pendingIntent.send();
        }
        catch(PendingIntent.CanceledException exp){}
    }
}
```

(4) 修改工程\src\ch9\six_2 目录下的 Example9_6_2.java 文件。Example9_6_2 创建的 Activity 对象需要显示接收者接收到的广播信息,修改后的内容如下:

Example9_6_2.java

```
package ch9.six_2;
import android.app.*;
import android.os.*;
import android.content.*;
import android.widget.*;
import android.view.*;
public class Example9_6_2 extends Activity {
    TextView text;
    public void onCreate(Bundle savedInstanceState) {
        super.onCreate(savedInstanceState);
        setContentView(R.layout.ch9_6_2);
        text = (TextView)findViewById(R.id.text);
        Intent intent = getIntent();
        String s = intent.getStringExtra("mess");
        if(s!= null) {
            text.append("\n" + s) ;
```

            }
        }
}
```

(5) 修改工程的配置文件 AndroidManifest.xml（在工程的根目录下,本例就是 ch9_6_2 目录下）,注册一个能进行 TOM 动作的接收者,即增加一个＜receiver＞标记,修改后的内容如下：

**AndroidManifest.xml.java**

```
<?xml version = "1.0" encoding = "utf-8"?>
<manifest xmlns:android = "http://schemas.android.com/apk/res/android"
        package = "ch9.six_2"
        android:versionCode = "1"
        android:versionName = "1.0">
    <application android:label = "@string/app_name" android:icon = "@drawable/ic_launcher">
        <activity android:name = "Example9_6_2"
                android:label = "@string/app_name">
            <intent-filter>
                <action android:name = "android.intent.action.MAIN" />
                <category android:name = "android.intent.category.LAUNCHER" />
            </intent-filter>
        </activity>
        <receiver android:name = "GirlReceiver">
            <intent-filter>
                <action android:name = "tom.cat.TOM" />
            </intent-filter>
        </receiver>
    </application>
</manifest>
```

(6) 启动 AVD,进入工程的根目录,用快捷方式编译工程、安装应用程序到 AVD（有关知识点参见 1.5 节）。对于本例子,用命令行进入 D:\2000\ch9_6_2,执行如下命令：

D:\2000＞ch9_6_2＞ant debug install

习 题 9

1. Service 对象能单独运行在一个工作线程中吗？
2. 在 Service 对象的生命周期内,Service 对象会多次调用 onCreate()方法吗？会多次调用 onStartCommand 方法吗？
3. 一个 Service 对象能启动另外一个 Service 对象吗？
4. IntentService 类创建的 Service 对象有什么特点？
5. AsyncTask 类创建的对象需要在哪个线程中启动运行？
6. 编写程序,程序中的 Activity 对象启动 IntentService 类的子类 ComputerService 创建的 Service 对象,并请求它计算出半径是 100 的圆的面积,这个 Service 对象使用 PendingIntent 对象将计算出的结果显示在 Activity 对象的 TextView 视图中。
```

第 10 章　使用 SD 卡

主要内容：
- 设置 SD 卡的大小；
- 上传文件到 SD 卡；
- 查看 SD 卡中的文件；
- 显示 SD 卡中的图像；
- 播放 SD 卡中的视频或 MP3。

手机的内置存储空间毕竟有限，因此经常需要将一些比较大的文件，如图片、音频和视频等文件，放置在 SD 卡中。

10.1　设置 SD 卡的大小

开发 Android 程序经常需要使用 AVD 模拟 SD 卡，程序开发人员在创建一个 AVD（虚拟设备，即手机模拟器）时，简单地指定该 AVD 中的 SD 卡的大小即可（如图 10.1 所示或参见 1.3 节）。在设置 SD 卡大小时，系统要求的最小值是 9MB，目前 Android 支持 9MB～128GB 的 SD 卡。

图 10.1　设置 SD 卡的大小

需要注意的是，当我们建立一个 AVD 后，PC 会在系统的磁盘上给出 AVD 占用的空间，包括模拟 SD 卡占用的空间（PC 需保证有足够的磁盘空间），对于 Win 7 可以在 C:\Users\asus\.android\avd 目录下看到和 AVD 相关的配置文件。对于 Windows XP 可以在 C:\Documents and Setting\Adnimistrator\.android\avd 目录下看到和 AVD 相关的配置文件。如果删除 \.android\avd 目录或其下和 AVD 相关的文件，就需要重新创建 AVD。

10.2 上传文件到 SD 卡

Android 操作系统是基于 Linux 的操作系统,它将 SD 卡挂接在系统的/sdcard 目录中。我们无法在 AVD 中操作 SD 卡,因为 AVD 没有提供类似 PC 的资源管理器来管理系统中的文件,比如复制、剪贴、删除文件等。

可以使用 adb(Android Debug Bridge)指令上传文件到 SD 卡,我们曾在 1.4 节介绍过 adb 命令。adb 命令在 Android SDK 安装目录的 platform-tools 文件夹中,要确保 Android SDK 安装目录\platform-tools 是环境变量 path 的一个值(参见 1.2 节),如果读者没有做到这一点,就需要用命令行进入 abd 命令所在的目录来执行 abd 命令,例如,本教材中 adb 所在的目录是 D:\android-sdk-windows\platform-tools。

sdcard 目录中有许多子目录(如 Pictures,Movies,Music 等),可以使用 adb 命令将 PC 上的文件上传到 sdcard 根目录或其子目录中。需要注意的是,在上传文件之前要确保 AVD 已经启动,文件名中不要有中文。

上传到 sdcard 的根目录中的语法格式如下:

adb push PC 文件的路径 /sdcard/

上传到 sdcard 的子目录中的语法格式如下:

adb push PC 文件的路径 /sdcard/子目录

例如,使用 adb 命令将 D:\pic 目录中 koala.jpg 上传到 SD 卡的根目录:

adb push D:\pic\koala.jpg /sdcard/

上传到 SD 卡的 Pictures 子目录:

adb push D:\pic\koala.jpg /sdcard/Pictures

图 10.2 Eclipse 环境上传文件到 SD 卡

如果读者使用了 Eclipse,那么上传文件到 SD 卡就更加方便。启动 Eclipse,打开 DDMS 视图,选择 File Explorer,出现图 10.2 所示的界面,在该界面上选中 sdcard 文件夹,将要上传的文件复制到 sdcard 文件夹即可。

10.3 查看 SD 卡中的内容

Android 操作系统是基于 Linux 的操作系统,如果读者熟悉 Linux,那么就可以方便地访问操作 Android 系统中的文件,比如建立新的文件夹,删除拷贝文件等。我们简要介绍几个简单的命令,以便查看 sdcard 中的文件。

首先需要执行如下命令 adb shell 进入 Linux 的 shell,如图 10.3 所示。退出 shell 执行 exit 即可。

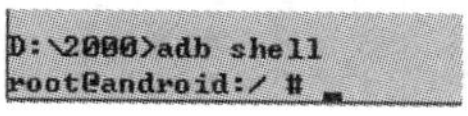

图 10.3 启动 shell

可以执行 Linux 的 ls 命令列出 Android 系统中的全部目录,如图 10.4 所示,比如 data\app 目录中存放着我们使用 ant debug install 命令安装到 AVD 中的全部应用程序。

在图 10.4 中最下方(图 10.4 只截取了一部分目录),我们可以看到系统中 SD 卡对应的子目录 sdcard。使用"cd 目录"命令进入到 sdcard:

cd sdcard

此时如果执行 ls 命令可以列出 sdcard 目录中的全部子目录和文件,如图 10.5 所示。

图 10.4　系统中的文件和目录　　　　图 10.5　sdcard 目录中的文件及子目录

10.4　显示 SD 卡中的图像

1. 使用 SD 卡的 Pictures 目录

图像、音频和视频等文件都占用较大的磁盘空间,因此程序可以将这些文件存放到 SD 卡中的相应目录中,比如 Pictures,Movies 或 Music 等目录中。如果希望制作一个基于 SD 卡的相册,那么就应该事先将图像保存到 SD 卡中的 Pictures 目录中。然后程序使用相应的视图显示 Pictures 目录中的图像。

2. 示例

例子 10-1 中将几幅图像上传到 SD 卡,程序显示 SD 卡中的图像。运行效果如图 10.6 所示。

例子 10-1

(1) 启动 AVD,将 D:\pic 中的名字为 pic1.jpg,pic2.jpg,pic3.jpg 和 pic4.jpg 的图像文件上传到 AVD 的 SD 卡的 Pictures 子目录中,即/sdcard/Pictures 中(上传花费的时间依赖文件的大小,需耐心等待)。在命令行分别执行如下命令:

```
adb push D:\pic\pic1.jpg /sdcard/Pictures
adb push D:\pic\pic2.jpg /sdcard/Pictures
adb push D:\pic\pic3.jpg /sdcard/Pictures
adb push D:\pic\pic4.jpg /sdcard/Pictures
```

(2) 创建名字为 ch10_1 的工程,主要 Activity 子类的名字为 Example10_1,使用的包名为 ch10.one。用命令行进入

图 10.6　显示 SD 卡中的图像

D:\2000,创建工程 D:\2000＞android create project -t 3 -n ch10_1 -p ./ch10_1 -a Example10_1 -k ch10.one。

(3) 将下列和视图相关的 XML 文件保存到工程的\res\layout 目录中。

ch10_1.xml

```xml
<?xml version = "1.0" encoding = "utf-8"?>
<LinearLayout xmlns:android = "http://schemas.android.com/apk/res/android"
    android:orientation = "vertical"
    android:layout_width = "match_parent"
    android:layout_height = "match_parent"
    android:background = "#87CEEB">
    <HorizontalScrollView android:layout_width = "match_parent"
        android:layout_height = "match_parent">
        <LinearLayout
            android:id = "@+id/layout"
            android:orientation = "horizontal"
            android:layout_width = "match_parent"
            android:layout_height = "match_parent">
        </LinearLayout>
    </HorizontalScrollView>
</LinearLayout>
```

(4) 修改工程\src\ch10\one 目录下的 Example10_1.java 文件,修改后的内容如下:

Example10_1.java

```java
package ch10.one;
import android.app.*;
import android.os.Bundle;
import android.widget.*;
import android.graphics.drawable.Drawable;
import java.io.*;
public class Example10_1 extends Activity {
    LinearLayout layout;
    public void onCreate(Bundle savedInstanceState) {
        super.onCreate(savedInstanceState);
        setContentView(R.layout.ch10_1);
        layout = (LinearLayout)findViewById(R.id.layout);
        setPic("pic1.jpg");
        setPic("pic2.jpg");
        setPic("pic3.jpg");
        setPic("pic4.jg");
    }
    void setPic(String imageName) {
      ImageButton imageButton = new ImageButton(this);
      File f = new File("/sdcard/Pictures/" + imageName);
      Drawable drawable = Drawable.createFromPath(f.getAbsolutePath());
      imageButton.setImageDrawable(drawable);
      layout.addView(imageButton);
    }
}
```

(5) 启动 AVD,进入工程的根目录,用快捷方式编译工程、安装应用程序到 AVD(有关知识点参见 1.5 节)。对于本例子,用命令行进入 D:\2000\ch10_1,执行如下命令:

D:\2000 > ch10_1 > ant debug install

10.5　播放 SD 卡中的视频或 MP3

1. 使用 SD 卡的 Movies 目录

视频和 MP3 文件都占用较大的磁盘空间,因此程序可以将这些文件存放到 SD 卡中的 Movies 目录中。如果希望制作一个基于 SD 卡的 MP3 播放器或视频播放,那么就应该事先将 MP3 或视频文件(Android 支持的视频格式文件,参见 5.7 节)保存到 SD 卡中的 Movies 目录中。然后程序使用相应的视图播放 Movies 目录中的视频/MP3。

2. 示例

在下面的例子 10-2 中,我们将一个视频和一个 MP3 文件上传到 SD 卡中,然后使用程序播放 SD 卡中的视频或 MP3。运行效果如图 10.7(a),10.7(b)所示。

(a) 用列表显示视频名称　　　　　　(b) 播放所选择的视频

图 10.7　运行效果

例子 10-2

(1) 启动 AVD,将 D:\00 中的名字为 ok.3gp 和 pingan.mp3 文件上传到 AVD 的 SD 卡的 Movies 子目录中,即/sdcard/Movies 中(上传花费的时间依赖文件的大小,需耐心等待)。在命令行分别执行如下命令:

D:\2000 > adb push D:\00\ok.3gp /sdcard/Movies
D:\2000 > adb push D:\00\pingan.mp3 /sdcard/Movies

(2) 创建名字为 ch10_2 的工程,主要 Activity 子类的名字为 Example10_2,使用的包名为 ch10.two。用命令行进入 D:\2000,创建工程 D:\2000>android create project -t 3 -n ch10_2 -p ./ch10_2 -a Example10_2 -k ch10.two。

(3) 将下列和视图相关的 XML 文件保存到工程的\res\layout 目录中。

ch10_2.xml

```xml
<?xml version = "1.0" encoding = "utf-8"?>
<LinearLayout xmlns:android = "http://schemas.android.com/apk/res/android"
    android:orientation = "vertical"
    android:layout_width = "match_parent"
    android:layout_height = "match_parent"
    android:background = "#87CEEB">
    <ListView
        android:layout_width = "wrap_content"
        android:layout_height = "wrap_content"
        android:divider = "#0000FF"
        android:dividerHeight = "6dp"
        android:background = "#777777"
        android:id = "@+id/my_list" />
</LinearLayout>
```

（4）修改工程\src\ch10\two 目录下的 Example10_2.java 文件，修改后的内容如下：

Example10_2.java

```java
package ch10.two;
import android.app.*;
import android.os.Bundle;
import android.widget.*;
import android.view.*;
import android.content.*;
import android.net.*;
import java.util.*;
import java.io.*;
public class Example10_2 extends Activity implements AdapterView.OnItemClickListener {
    ListView listView;
    MyAdapter adapter;          //适配器
    ArrayList<String> listItem;
    public void onCreate(Bundle savedInstanceState) {
        super.onCreate(savedInstanceState);
        setContentView(R.layout.ch10_2);
        listView = (ListView)findViewById(R.id.my_list);
        listView.setOnItemClickListener(this);
        listItem = new ArrayList<String>();
        File dir = new File("/sdcard/Movies");
        File file[] = dir.listFiles();
        for(int i = 0;i < file.length;i++)
            listItem.add(file[i].getName());
        adapter = new MyAdapter();
        adapter.setContext(this);
        adapter.setArrayList(listItem);
        listView.setAdapter(adapter);
    }
    public void onItemClick(AdapterView parent,View view,int pos,long id) {
        String selectedItem = parent.getItemAtPosition(pos).toString();
        File f = new File("/sdcard/Movies/" + selectedItem);
```

```
            Intent intent = new Intent(Intent.ACTION_VIEW);
            Uri uri = Uri.parse(f.getAbsolutePath());
            intent.setDataAndType(uri,"video/*");
            startActivity(intent);
        }
    }
    class MyAdapter extends BaseAdapter {
        ArrayList<String> list;
        Context context;
        public void setContext(Context context) {
            this.context = context;
        }
        public void setArrayList(ArrayList<String> list) {
            this.list = list;
        }
        public int getCount() {
            return list.size();
        }
        public String getItem (int position) {
            String item = list.get(position);
            return item;
        }
        public long getItemId (int position) {
            return 0;
        }
        public View getView (int position,View convertView,ViewGroup parent) {
            TextView text = new TextView(context);
            text.setTextSize(1,30);
            text.setText(getItem(position));
            return text;
        }
    }
```

（5）启动 AVD，进入工程的根目录，用快捷方式编译工程、安装应用程序到 AVD（有关知识点参见 1.5 节）。对于本例子，用命令行进入 D:\2000\ch10_2，执行如下命令：

```
D:\2000 > ch10_2 > ant debug install
```

习 题 10

1. 目前 Android 支持多大容量的 SD 卡？
2. 使用 adb 命令上传到 sdcard 的根目录中的语法格式是怎样的？
3. 将一个 Java 源文件 E.java 上传到 AVD 的 SD 卡 sdcard 的根目录中，然后编写一个程序，用 TextView 视图显示 sdcard 的根目录中的 E.java 的内容。

第 11 章　文件的读写

主要内容：
- 使用输入/输出流在数据区读写文件；
- 使用 SharedPreferences 在数据区读写文件；
- 在 SD 卡中读写文件；
- 读取 assets(资产)中的文件；
- 读取\res\raw(原始资源)中的文件；
- 读取\res\xml 中的 XML 文件；
- 基于文件的电话簿；
- 基于 XML 数据库的英-汉字典。

应用程序可能需要将数据以文件的形式存储到磁盘空间或从文件读取自己需要的数据，这就涉及文件的读写操作，本章将介绍 Android 系统提供的文件读写方法，包括在系统提供的数据区读写文件、在 SD 卡中读写文件以及读取特殊资源中的文件等。

在学习本章的过程中，建议读者启动 AVD 后，时常在命令行窗口使用 adb 命令启动 Linux 的 shell：adb shell，然后使用下列几个常用的 Linux 命令查看 Android 系统目录中的文件或删除用户程序写的文件，以便于读者调试程序的效果。常见的几个命令如下所示。

ls：列出当前目录下的所有文件和子目录。
cd 子目录名：进入当前目录的子目录(注意 cd 和子目录之间有一个空格)。
cd ..：从当前目录退回到上一层目录(注意 cd 和..之间有一个空格)。
rm 文件名：删除文件(注意 rm 和文件名之间有一个空格)。

11.1　使用输入/输出流在数据区读写文件

程序在系统提供的数据内进行文件读写有两个方式，第一种方式就是使用输入、输出流，第二种方式是使用 SharedPreferences 类的实例，本节介绍第一种方式，下一节介绍第二种方式。

1. 写文件

Android 系统允许应用程序使用输入、输出流在系统提供的\data\data\应用程序包名\files 目录中进行写文件操作。

1) 使用 Context 类的方法打开输出流

在系统提供的\data\data\应用程序包名\files 目录中进行写文件操作时，可以使用

android.content 包中的 Context 类提供的 FileOutputStream openFileOutput(String name,int mode)方法打开指向文件的文件输出流(不必给出文件的目录路径)。

例如,假设应用程序的包名为 tom.jiafei,openFileOutput(String name,int mode)方法的参数 name 给出的文件名为 myDog.txt,那么该文件所在的目录结构是/data/data/tom.jiafei/files/,即 myDog.txt 文件的路径是/data/data/tom.jiafei/files/myDog.txt。

openFileOutput(String name,int mode)方法中的参数取值意义如下。

- 参数 name 是文件的名称。
- 参数 mode 可取 Context 类的下列静态常量,意义如下。
  ◆ 参数 mode 取 Context.MODE_PRIVATE:文件是私有式,即所写文件被定性为私有式文件。只有本应用程序可以读取私有式文件。
  ◆ 参数 mode 取 Context.MODE_WORLD_READABLE:文件是开放式的,其他应用程序也可以读取(比如,通常使用 IO 流读文件)。
  ◆ 参数 mode 取 Context.MODE_WORLD_WRITEABLE:文件是开放式的,其他应用程序也可以对其进行写操作(比如,通常使用 IO 流写文件)。
  当参数取上述常量值时,如果所写文件不存在,就建立该文件,若文件已存在就更新该文件(使文件的长度变为 0 字节),然后向文件写入新内容。
  ◆ 参数 mode 取 Context.MODE_APPEND:如果所写文件不存在,就建立该文件,若文件已存在,不更新该文件(保持文件的长度不变),继续向文件尾加新内容。

参数也可以取上述常量的和值,例如:

MODE_PRIVATE+MODE_APPEND
MODE_WORLD_READABLE+MODE_APPEND
MODE_WORLD_READABLE+MODE_WORLD_WRITEABLE
MODE_APPEND+MODE_WORLD_READABLE+MODE_WORLD_WRITEABLE

2) 使用 java.io 中的 API 打开输出流

在系统提供的/data/data/应用程序包名/files 目录中进行写文件操作时,也可以使用 java.io 包提供 API 打开指向文件的文件输出流(需要给出文件的目录路径),而且所写文件一定被定性为是私有式的文件,不能是开放式的文件。

例如,假设当前应用程序(包名为 ch11.one),准备向文件 tom.txt 写入数据,并刷新 tom.txt,那么代码如下:

```
File file = new File("/data/data/ch11.one/files","tom.txt");
FileOutputStream out = new FileOutputStream(file,false);
```

假设当前应用程序(包名为 ch11.one),准备向文件 tom.txt 尾加数据,不刷新 tom.txt,那么代码如下:

```
File file = new File("/data/data/ch11.one/files","tom.txt");
FileOutputStream out = new FileOutputStream(file,true);
```

2. 读文件

当准备读取系统/data/data/应用程序包名/files 目录中的文件时,可以使用 android.content 包中的 Context 类提供的 FileInputStream openFileInput(String name)方法打开指

向文件的文件输入流（不必给出文件的目录路径）。例如，假设读取应用程序 ch11.one（应用程序的包名）曾存放在/data/data/ch11.one/files 目录中的 zhang.txt 文件，那么打开指向文件的输入流的代码如下：

```
FileInputStream in = openFileInput("zhang.txt");
```

也可以使用 Java.io 包中提供的 API，比如文件输入流，读取/data/data/应用程序包名/files 目录中的文件。读取时，需给出文件所在的准确位置（文件的路径），例如，假设读取应用程序 tom.jiafei（应用程序的包名）存放在/data/data/tom.jiafei/files 目录中的 bird.txt 文件，代码如下：

```
File file = new File("/data/data/tom.jiafei/files","bird.txt");
FileInputStream = new FileInputStream(file);
```

注：若某个应用程序存放的文件是私有式文件，那么其他应用程序无法读取到文件的内容。

3．列出或删除文件

Context 类提供了列出用户程序在/data/data/程序包名/files 目录下的私有式文件的 public String[] fileList()方法，也提供了删除/data/data/程序包名/files 目录下的私有式文件的 public boolean deleteFile(String name)方法。当前应用程序可以使用 java.io 中的 API 对文件进行操作，比如读写或删除等操作。对于开放式文件，其他应用程序可以使用 java.io 中的 API 操作当前应用程序中的开放式文件，比如进行读写，甚至删除等操作。

4．示例

例子 11-1 中，用户单击"写私有式文件"按钮，程序向私有式文件尾加字符序列"hello"，用户单击"读取私有式文件"按钮，程序显示私有式文件中的内容。运行效果如图 11.1 所示。

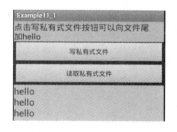

图 11.1 读写私有式文件

例子 11-1

（1）创建名字为 ch11_1 的工程，主要 Activity 子类的名字为 Example11_1，使用的包名为 ch11.one。用命令行进入 D:\2000，创建工程 D:\2000＞android create project -t 3 -n ch11_1 -p ./ch11_1 -a Example11_1 -k ch11.one。

（2）将下列和视图相关的 XML 文件保存到工程的\res\layout 目录中。

ch11_1.xml

```
<?xml version = "1.0" encoding = "utf-8"?>
< LinearLayout xmlns:android = "http://schemas.android.com/apk/res/android"
    android:orientation = "vertical"
    android:layout_width = "match_parent"
    android:layout_height = "match_parent">
    < TextView
        android:background = "#87CEEB"
        android:textColor = "#000000"
        android:layout_width = "match_parent"
        android:layout_height = "wrap_content"
```

```xml
        android:textSize = "18sp"
        android:text = "单击写私有式文件按钮可以向文件尾加 hello" />
<Button
    android:layout_width = "match_parent"
    android:layout_height = "wrap_content"
    android:text = "写私有式文件"
    android:onClick = "writePrivateFile" />
<Button
    android:layout_width = "match_parent"
    android:layout_height = "wrap_content"
    android:text = "读取私有式文件"
    android:onClick = "readPrivateFile" />
<TextView
    android:id = "@ + id/showFile"
    android:background = "#AAEE00"
    android:textColor = "#000000"
    android:layout_width = "match_parent"
    android:layout_height = "wrap_content"
    android:textSize = "20sp" />
</LinearLayout>
```

(3) 修改工程\src\ch11\one 目录下的 Example11_1.java 文件，修改后的内容如下：

Example11_1.java

```java
package ch11.one;
import android.app.Activity;
import android.os.Bundle;
import android.widget.*;
import android.view.*;
import android.content.Context;
import java.io.*;
public class Example11_1 extends Activity {
    TextView showFile;
    public void onCreate(Bundle savedInstanceState) {
        super.onCreate(savedInstanceState);
        setContentView(R.layout.ch11_1);
        showFile = (TextView)findViewById(R.id.showFile);
    }
    public void writePrivateFile(View view) {
        try {
            FileOutputStream out =
            openFileOutput("zhang.txt",Context.MODE_PRIVATE + MODE_APPEND);
            String str = "hello";
            byte b [] = str.getBytes();
            out.write(b);
            out.write('\n');
                out.close();
        }
        catch(FileNotFoundException exp){}
        catch(IOException exp){}
    }
```

```java
public void readPrivateFile(View view) {
    showFile.setText(null);
    try {
        FileInputStream in = openFileInput("zhang.txt");
        byte b [] = new byte[8];
        int n = -1;
        while((n = in.read(b))!= -1) {
            String s = new String(b,0,n);
            showFile.append(s);
        }
        in.close();
    }
    catch(FileNotFoundException exp){}
    catch(IOException exp){}
}
```

(4) 启动 AVD，进入工程的根目录，用快捷方式编译工程、安装应用程序到 AVD(有关知识点参见 1.5 节)。对于本例子，用命令行进入 D:\2000\ch11_1，执行如下命令：

D:\2000 > ch11_1 > ant debug install

11.2 使用 SharedPreferences 对象在数据区读写文件

程序在系统提供的数据区进行文件读写的第二种方式是使用 SharedPreferences 类的实例。在使用 SharedPreferences 类读写文件时，用户程序不需要显示地使用输入、输出流，用户只需操作 SharedPreferences 类的对象，即让 SharedPreferences 类的对象调用相应的方法就可以实现对文件的读写操作。

SharedPreferences 类的对象操作文件的最大特点是将"键/值"数据写入到文件，并通过"键"读取"键/值"数据中的"值"(类似操作一个散列表)。

1. 写文件

1) 得到 SharedPreferences 对象

Android 系统允许应用程序使用 SharedPreferences 类的对象在系统提供的/data/data/应用程序包名/shared_prefs 目录中进行写文件操作。首先需要使用 android.content 包中的 Context 类提供的 SharedPreferences getSharedPreferences(String name,int mode) 方法，返回一个 SharedPreferences 类的对象，参数 name 不必给出文件的目录路径，也不要给出文件的扩展名，文件的扩展名被系统强制为.xml。

例如，假设应用程序的包名为 tom.jiafei，getSharedPreferences(String name,int mode) 方法的参数 name 给出的文件名为 myDog.txt，那么该文件所在的目录结构是/data/data/tom.jiafei/shared_prefs/，即 myDog.txt 文件的路径是/data/data/tom.jiafei/shared_prefs/myDog.txt。

getSharedPreferences(String name,int mode) 方法中的参数取值意义如下。

- 参数 name 是文件的名称。
- 参数 mode 可取 Context 类的下列静态常量，意义如下。

- 参数 mode 取 Context.MODE_PRIVATE：文件是私有式，即所写文件被定性为私有式文件。只有本应用程序可以读取私有式文件。
- 参数 mode 取 Context.MODE_WORLD_READABLE：文件是开放式的，其他应用程序也可以读取（比如，通常使用 IO 流读取文件）。
- 参数 mode 取 Context.MODE_WORLD_WRITEABLE：文件是开放式的，其他应用程序也可以对其进行写操作（比如，通常使用 IO 流写文件）。

当参数取上述常量值时，如果所写文件不存在，就建立该文件，若文件已存在也不更新该文件（保持已有文件的长度不变）。

参数也可以取上述常量的和值，例如：

MODE_WORLD_READABLE＋MODE_WORLD_WRITEABLE

2）得到 SharedPreferences.Editor 对象

使用 SharedPreferences 读写文件的特点是将"键/值"数据写入到文件，并通过"键"读取"键/值"数据中的"值"。在写"键/值"数据到文件之前，要首先将"键/值"数据存放到 SharedPreferences.Editor 对象中，然后再由 SharedPreferences.Editor 对象将其中的"键/值"数据提交给文件，即写入到文件。

- SharedPreferences.Editor 对象

SharedPreferences 对象需要返回一个 SharedPreferences.Editor 对象，SharedPreferences 对象调用方法 SharedPreferences.Editor edit()返回一个 SharedPreferences.Editor 对象。

- 准备数据

所谓准备数据，就是 SharedPreferences.Editor 对象调用诸如 putXXX(键,值)方法将"键/值"放到 SharedPreferences.Editor 对象中（不会放到文件中）。常用的 putXXX 方法如下：

```
putBoolean(String key,boolean value)
putFloat(String key,float value)
putInt(String key,int value)
putLong(String key,long value)
putString(String key,String value)
remove(String key)
```

- 提交数据

所谓提交数据，就是 SharedPreferences.Editor 对象调用 commit()方法，当调用 commit()方法时，就会将 SharedPreferences.Editor 对象准备好的数据，即 SharedPreferences.Editor 对象中的"键/值"数据写入到文件。

2. 读文件

当准备读取系统/data/data/应用程序包名/shared_prefs 目录中的文件时，需要使用 android.content 包中的 Context 类提供的 SharedPreferences getSharedPreferences(String name,int mode)方法返回一个 SharedPreferences 类的对象，参数 name 不必给出文件的目录路径，也不必给出文件的扩展名。例如，假设读取应用程序 ch11.two（应用程序的包名）曾使用 SharedPreferences 对象在/data/data/tom.jiafei/shared_prefs 目录中存放了 phone.xml 文件，那么返回 SharedPreferences 对象的代码如下：

```
SharedPreferences people = getSharedPreferences("phone",Context.MODE_PRIVATE);
```

那么 getSharedPreferences(String name,int mode)返回的 SharedPreferences 对象调用诸如 getXXX(键,默认值)方法读取文件中"键/值"数据中的"值",例如,假设在写文件时,曾使用 putInt("ok",123456)向文件写入"ok/123456"数据("键"是"ok",值是"123456"),如果执行下列代码:

```
int m = people.getInt("ok",0);
```

那么,m 的值就是 123456,如果文件中没有"ok/123456"数据,m 就是指定的默认值 0。

3. 遍历文件中的数据

我们知道 SharedPreferences 类的对象写入到文件中的数据都是"键/值"数据,如果希望遍历"键/值"数据中的"值",通过使用 getXXX(键,默认值)方法读取文件中"键/值"数据中的"值"不是个好办法,因为程序需要知道每个"值"的"键",才可以使用 getXXX(键,默认值)方法得到"键/值"数据中的"值"。

遍历文件中"键/值"数据中的"值"的步骤如下。

1)得到 HashMap(散列映射)对象

首先让 getSharedPreferences(String name,int mode)返回的 SharedPreferences 对象调用 getAll()方法返回一个 HashMap(散列映射)对象(HashMap 在 java.util 包中),例如:

```
HashMap map = (HashMap)people.getAll();
```

得到的 HashMap 对象 map 刚好存放着文件中的全部"键/值"数据(此时程序仍可以让 map 对象调用相应的方法得到某个"键"对应的"值")。

2)得到全部的"值"

HashMap 对象调用 values()方法,将返回其中的全部"键/值"数据中的"值",并将这些值存放在 Collection 对象中,代码如下:

```
Collection coll = map.values();
```

那么 Collection 对象 coll 存放着全部"键/值"数据中的"值"。

3)使用迭代器遍历"值"

让存放着全部"键/值"数据中的"值"的 Collection 对象,比如 coll,返回一个迭代器,代码如下:

```
Iterator iterator=coll.iterator();
```

然后迭代器可以依次地调用 next()方法返回 Collection 对象中存放的全部"值",假设 Collection 对象中的"值"都是 String 类型,那么代码如下:

```
while(iterator.hasNext()) {
    String s = (String)iterator.next();
        //其他操作
}
```

4. 示例

如果用户用手机学习英文单词,可以考虑使用例子 11-2 提供的简单的手机程序。在例子11-2中,用户可以把"单词"和"单词的例句"作为"键/值"数据保存到文件中,用户也可以查询(支持模糊查询)文件中的单词。这样一来,不断地累积单词,手机中的程序的文件就存储的单词越多,保存的单词越多,查询学习的功能也就强大。运行效果如图11.2(a),11.2(b)所示。

(a) 输入单词界面　　　　　(b) 模糊查询单词

图 11.2　运行效果

例子 11-2

例子11-2给出的应用程序中有两个Activity对象,一个是主要的Activity对象(由Example11_2类创建),负责录入单词,另一个Activity对象(由QueryWord类创建)负责查询单词。

(1) 创建名字为ch11_2的工程,主要Activity子类的名字为Example11_2,使用的包名为ch11.two。用命令行进入D:\2000,创建工程D:\2000>android create project -t 3 -n ch11_2 -p ./ch11_2 -a Example11_2 -k ch11.two。

(2) 将下列主要Activity对象以及QueryWord类创建的Activity对象使用的视图文件ch11_2.xml和find.xml保存到工程的\res\layout目录中。

ch11_2.xml

```xml
<?xml version = "1.0" encoding = "utf-8"?>
<LinearLayout xmlns:android = "http://schemas.android.com/apk/res/android"
    android:orientation = "vertical"
    android:layout_width = "match_parent"
    android:layout_height = "match_parent"
    android:background = "#87CEEB">
    <TextView
        android:background = "#000000"
        android:layout_width = "wrap_content"
        android:layout_height = "wrap_content"
        android:text = "输入单词:" />
    <EditText
        android:id = "@+id/english"
        android:layout_width = "match_parent"
        android:layout_height = "wrap_content" />
    <TextView
        android:background = "#000000"
        android:layout_width = "wrap_content"
        android:layout_height = "wrap_content"
        android:text = "输入解释:" />
```

```xml
<EditText
    android:id = "@+id/explain"
    android:layout_width = "match_parent"
    android:layout_height = "wrap_content" />
<Button
    android:layout_width = "match_parent"
    android:layout_height = "wrap_content"
    android:text = "保存"
    android:onClick = "writeFile"   />
<Button
    android:layout_width = "match_parent"
    android:layout_height = "wrap_content"
    android:text = "去查询单词"
    android:onClick = "queryWord"   />
</LinearLayout>
```

find.xml

```xml
<?xml version = "1.0" encoding = "utf-8"?>
<LinearLayout xmlns:android = "http://schemas.android.com/apk/res/android"
    android:orientation = "vertical"
    android:layout_width = "match_parent"
    android:layout_height = "match_parent"
    android:background = "#87CEEB">
    <TextView
        android:background = "#55955a"
            android:textColor = "#0000FF"
        android:layout_width = "wrap_content"
        android:layout_height = "wrap_content"
        android:text = "输入要查找的单词:" />
    <EditText
        android:id = "@+id/input"
        android:layout_width = "match_parent"
        android:layout_height = "wrap_content" />
    <LinearLayout android:layout_width = "match_parent"
            android:layout_height = "wrap_content">
        <Button
            android:layout_width = "wrap_content"
            android:layout_height = "wrap_content"
            android:text = "精确查询"
            android:onClick = "findWord" />
        <Button
            android:layout_width = "wrap_content"
            android:layout_height = "wrap_content"
            android:text = "模糊查询"
            android:onClick = "findStartWord" />
        <Button
            android:layout_width = "wrap_content"
            android:layout_height = "wrap_content"
            android:text = "输入单词界面"
            android:onClick = "goToMainActivity" />
```

```xml
        </LinearLayout>
        <ScrollView android:id = "@ + id/scrollView"
            android:layout_width = "match_parent"
            android:layout_height = "wrap_content"
            android:scrollbarStyle = "outsideOverlay">
            <TextView
                android:id = "@ + id/show"
                android:textColor = "#806400"
                android:layout_width = "match_parent"
                android:layout_height = "wrap_content"
                android:textSize = "20sp" />
        </ScrollView>
    </LinearLayout>
```

(3) 将负责创建查询单词的 Activity 对象的 QueryWord.java 保存到工程的 src\ch11\two 目录下，并修改工程\src\ch11\two 目录下的 Example11_2.java 文件。QueryWord.java 和修改后的 Example11_2.java 内容如下：

QueryWord.java

```java
package ch11.two;
import android.app.*;
import android.os.*;
import android.widget.*;
import android.view.*;
import android.content.*;
import java.util.*;
public class QueryWord extends Activity {
    EditText inputWord;
    TextView showExplain;
    public void onCreate(Bundle savedInstanceState) {
        super.onCreate(savedInstanceState);
        setContentView(R.layout.find);
        inputWord = (EditText)findViewById(R.id.input);
        showExplain = (TextView)findViewById(R.id.show);
    }
    public void goToMainActivity(View v) {
        Intent intent = new Intent(this,Example11_2.class);
        startActivity(intent);
    }
    public void findWord(View v) {
        String word = inputWord.getText().toString();
        SharedPreferences in = getSharedPreferences("word",Context.MODE_PRIVATE);
        String str = "can not find \"" + word + "\"";
        String explain = in.getString(word,str);
        showExplain.setText(explain);
    }
    public void findStartWord(View v) {
        String word = inputWord.getText().toString();
        SharedPreferences in = getSharedPreferences("word",Context.MODE_PRIVATE);
        String str = "can not find \"" + word + "\"";
```

```java
        HashMap map = (HashMap)in.getAll();
        Set keySet = map.keySet();
        Iterator iterator = keySet.iterator();
        boolean isFind = false;
        while(iterator.hasNext()) {
            String key = (String)iterator.next();
            if(key.contains(word)) {
              isFind = true;
              String explain = (String)map.get(key);
              showExplain.append(explain + "\n");
            }
        }
        if(isFind == false)
            showExplain.setText("can not find\n");
    }
}
```

Example11_2.java

```java
package ch11.two;
import android.app.*;
import android.os.Bundle;
import android.widget.*;
import android.view.*;
import android.content.*;
public class Example11_2 extends Activity {
    EditText englishEdit,explainEdit;
    public void onCreate(Bundle savedInstanceState) {
        super.onCreate(savedInstanceState);
        setContentView(R.layout.ch11_2);
        englishEdit = (EditText)findViewById(R.id.english);
        explainEdit = (EditText)findViewById(R.id.explain);
    }
    public void writeFile(View view) {
        String english = englishEdit.getText().toString();
        String explain = explainEdit.getText().toString();
        String mess = english + ":" + explain;
        SharedPreferences out = getSharedPreferences("word",Context.MODE_PRIVATE) ;
        SharedPreferences.Editor editFile = out.edit();
        if(english.length()> 0&&explain.length()> 0){
            editFile.putString(english,mess);
            editFile.commit();
            getAlertDialogWithButton("just save word to file!").show();
        }
        else {
            getAlertDialogWithButton("input word and explain please!").show();
        }
    }
    public void queryWord(View view) {
        Intent intent = new Intent(this,QueryWord.class);
        startActivity(intent);
    }
    AlertDialog getAlertDialogWithButton(String mess) {
        AlertDialog.Builder builder = new AlertDialog.Builder(this);
```

```
            builder.setTitle(mess);
            builder.setMessage(mess);
            AlertDialog dialog = builder.create();
            return dialog;
        }
    }
```

（4）程序包含了两个 Activity 对象，需要修改工程根目录下的配置文件 AndroidManifest. xml，加入一个＜activity＞标记，该标记对应着 QueryWord 类创建的 Activity 对象，修改后的配置文件的内容如下：

AndroidManifest. xml

```
<?xml version = "1.0" encoding = "utf - 8"?>
< manifest
        <!-- 此处省略了原有内容,以下是新增的 activity 标记 -->
        < activity android:name = "QueryWord"
               android:label = "查询单词">
        </activity>
    <!-- 此处省略了原有内容 -->
</manifest>;
```

（5）启动 AVD，进入工程的根目录，用快捷方式编译工程、安装应用程序到 AVD(有关知识点参见 1.5 节)。对于本例子，用命令行进入 D:\2000\ch11_2，执行如下命令：

D:\2000 > ch11_2 > ant debug install

11.3　在 SD 卡中读写文件

相对手机的内置数据存储器，SD 卡是手机的外挂存储器，程序在 SD 卡中读写文件更加灵活，而且 SD 卡的容量相对较大，可以读写比较大的文件。

1. 使用 java. io

在 SD 卡中读写文件的要求非常宽松，程序可以把 SD 卡看做一个普通存储磁盘，可以使用 java. io 包的 API 在 SD 卡上实现文件的读写操作。可以在 SD 卡中进行建立目录、删除目录、读写文件、删除文件等操作。有关 SD 卡的知识参见第 10 章。

与 Android 系统的/data/data/目录中文件相比(参见 11.1 节)，SD 卡中的文件没有定性私有式的权力，因此安全性下降，因为任何程序都可以操作 SD 卡中的文件。

2. 示例

例子 11-3 中，用户单击"将账单写入到文件"按钮，程序在 SD 卡中新建一个子目录 myfile，将一个账单 The television cost 1876 dollar . The milk cost 99 dollar. . 写入目录 myfile 下的 cost. txt 文件中。用户单击"读取账单中的价格信息"按钮，程序解析出 cost. txt 中的消费金额，并计算出账单的消费总额。运行效果如图 11.3 所示。

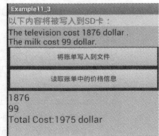

图 11.3　在 SD 卡中读写文件

例子 11-3

（1）创建名字为 ch11_3 的工程，主要 Activity 子类的名字为 Example11_3，使用的包名为 ch11.three。用命令行进入 D:\2000，创建工程 D:\2000＞android create project -t 3 -n ch11_3 -p ./ch11_3 -a Example11_3 -k ch11.three。

（2）将下列和视图相关的 XML 文件保存到工程的 \res\layout 目录中。

ch11_3.xml

```xml
<?xml version = "1.0" encoding = "utf-8"?>
<LinearLayout xmlns:android = "http://schemas.android.com/apk/res/android"
    android:orientation = "vertical"
    android:layout_width = "match_parent"
    android:layout_height = "match_parent">
    <TextView
        android:background = "#dddddd"
        android:textColor = "#FF0A00"
        android:layout_width = "match_parent"
        android:layout_height = "wrap_content"
        android:textSize = "18sp"
        android:text = "以下内容将被写入到 SD 卡："/>
    <TextView
        android:id = "@+id/contentFile"
        android:background = "#87CEEB"
        android:textColor = "#000000"
        android:layout_width = "match_parent"
        android:layout_height = "wrap_content"
        android:textSize = "18sp"
        android:text = "The television cost 1876 dollar.\nThe milk cost 99 dollar."/>
    <Button
        android:layout_width = "match_parent"
        android:layout_height = "wrap_content"
        android:text = "将账单写入到文件"
        android:onClick = "writeFileToSD"/>
    <Button
        android:layout_width = "match_parent"
        android:layout_height = "wrap_content"
        android:text = "读取账单中的价格信息"
        android:onClick = "readFileFromSD"/>
    <TextView
        android:id = "@+id/showFile"
        android:background = "#AAEE00"
        android:textColor = "#000000"
        android:layout_width = "match_parent"
        android:layout_height = "wrap_content"
        android:textSize = "20sp"/>
</LinearLayout>
```

（3）修改工程 \src\ch11\three 目录下的 Example11_3.java 文件，修改后的内容如下：

Example11_3.java

```
package ch11.three;
```

```java
import android.app.Activity;
import android.os.Bundle;
import android.widget.*;
import android.view.*;
import java.io.*;
import java.util.*;
public class Example11_3 extends Activity {
    TextView contentFile;
    TextView showFile;
    File dir;
    public void onCreate(Bundle savedInstanceState) {
        super.onCreate(savedInstanceState);
        setContentView(R.layout.ch11_3);
        contentFile = (TextView)findViewById(R.id.contentFile);
        showFile = (TextView)findViewById(R.id.showFile);
        dir = new File("/sdcard/myfile");
        dir.mkdir();
    }
    public void writeFileToSD(View view) {
        String str = contentFile.getText().toString();
        File file = new File(dir,"cost.txt");
        try {
            FileOutputStream out = new FileOutputStream(file);
            byte b [] = str.getBytes();
            out.write(b);
            out.close();
        }
        catch(IOException exp){
            showFile.setText("" + exp);
        }
    }
    public void readFileFromSD(View view) {
        showFile.setText(null);
        File file = new File(dir,"cost.txt");
        Scanner sc = null;
        int sum = 0;
        try { sc = new Scanner(file);
            while(sc.hasNext()){
                try{
                    int price = sc.nextInt();
                    sum = sum + price;
                    showFile.append(price + "\n");
                }
                catch(InputMismatchException exp){
                    String t = sc.next();
                }
            }
            showFile.append("Total Cost:" + sum + " dollar\n");
        }
        catch(Exception exp){
            showFile.setText("" + exp);
```

            }
        }
}

(4) 启动 AVD,进入工程的根目录,用快捷方式编译工程、安装应用程序到 AVD(有关知识点参见 1.5 节)。对于本例子,用命令行进入 D:\2000\ch11_3,执行如下命令:

D:\2000 > ch11_3 > ant debug install

11.4 读取 assets(资产)中的文件

当我们创建工程后,可以在工程的根目录下建立一个名字为 assets 的子目录,该子目录是和已有的资源目录 res 是同层次的目录,但所不同的是,编译器(Debug)不会为 assets 目录中的文件在系统的 R.java 文件中生成一个资源 ID(参见 2.8 节)。程序只可以使用输入流读取 assets 目录中的文件。

1. 只能读取

当编译器(Debug)发现程序读取 assets 目录或其子目录中的文件时,就会将程序读取的文件打包在应用程序中(apk 文件中),因此程序实际读取的是打包在应用程序中的文件,基于这样的机制,程序只能读取 assets 目录中的文件,不可以进行写文件操作(系统不允许动态改变 apk 文件的大小)。需要注意的是,保存在 assets 目录中的文件,如果是文本文件,并包含有中文字符,保存该文本文件时需要将编码选择为"UTF-8"编码。

2. 获得输入流

假设 assets 目录中有名字为 test.txt 的文件,获得指向 test.txt 文件的输入流的步骤如下。

1) 得到资源对象

android.content 包中的 Context 类以及 android.view 包中的 View 类提供了获取资源的 public Resources getResources()方法(参见 2.8 节),该方法返回一个 Resources 类的实例(Resource 类在 android.content.res 包中)。例如,Activity 类是 Context 类的子类,返回资源对象的代码如下:

Resources resource = getResources();

2) 得到 AssetManager 对象

资源对象 resource 调用 AssetManager getAssets()方法可以返回一个 AssetManager 类(在 android.content.res 包中)的实例,习惯称该实例为一个资产管理者,例如:

AssetManager assetManager = resource.getAssets();

3) 得到输入流

资产管理者 assetManager 调用 InputStream open(String fileName)方法可以返回指向文件的输入流,例如:

InputStream in = assetManageropen("test.txt");

如果用户在 assets 目录中又建立了子目录,比如 myMoon 子目录,并将 test.txt 存放到

myMoon子目录中,那么上述3)中获得输入流的代码中需给出子目录的名字,即给出文件的相对路径,代码如下:

```
InputStream in = assetManager.open("myMoon/test.txt");
```

3. 示例

例子11-4中,用户单击"读文件"按钮,程序将读取资产中名字为direction.txt文件。运行效果如图11.4所示。

例子11-4

(1)创建名字为ch11_4的工程,主要Activity子类的名字为Example11_4,使用的包名为ch11.four。用命令行进入D:\2000,创建工程D:\2000＞android create project -t 3 -n ch11_4 -p ./ch11_4 -a Example11_4 -k ch11.four。

图11.4　读取资产中的文件

(2)在工程的根目录下建立名字为assets的子目录,对于例子11-4,在ch11_4下建立名字为assets的子目录。将名字为direction.txt的文件保存在\assets目录中。本例子中direction.txt文件的内容是关于程序版权的信息。如果direction.txt中包含有中文字符,保存direction.txt时需要将编码选择为"UTF-8"编码。

(3)将下列和视图相关的XML文件保存到工程的\res\layout目录中。

ch11_4.xml

```xml
<?xml version = "1.0" encoding = "utf-8"?>
<LinearLayout xmlns:android = "http://schemas.android.com/apk/res/android"
    android:orientation = "vertical"
    android:layout_width = "match_parent"
    android:layout_height = "match_parent">
    <Button
        android:layout_width = "match_parent"
        android:layout_height = "wrap_content"
        android:text = "读取资产下的文件"
        android:onClick = "readFile" />
    <TextView
        android:background = "#dddddd"
        android:textColor = "#FF0A00"
        android:layout_width = "match_parent"
        android:layout_height = "wrap_content"
        android:textSize = "18sp"
        android:text = "文件的内容如下:" />
    <TextView
        android:id = "@+id/contentFile"
        android:background = "#87CEEB"
        android:textColor = "#000000"
        android:layout_width = "match_parent"
        android:layout_height = "wrap_content"
        android:textSize = "18sp" />
</LinearLayout>
```

(4)修改工程\src\ch11\four目录下的Example11_4.java文件,修改后的内容如下:

Example11_4.java

```java
package ch11.four;
import android.app.Activity;
import android.os.Bundle;
import android.widget.*;
import android.view.*;
import java.io.*;
public class Example11_4 extends Activity {
    TextView contentFile;
    public void onCreate(Bundle savedInstanceState) {
        super.onCreate(savedInstanceState);
        setContentView(R.layout.ch11_4);
        contentFile = (TextView)findViewById(R.id.contentFile);
    }
    public void readFile(View view) {
        contentFile.setText(null);
        try {
            InputStream in = getResources().getAssets().open("direction.txt");
            int n = -1;
            byte a[] = new byte[1024];
            while((n = in.read(a))!= -1)
                contentFile.append(new String(a,0,n));
        }
        catch(IOException exp){
            contentFile.setText("" + exp);
        }
    }
}
```

（5）启动 AVD，进入工程的根目录，用快捷方式编译工程、安装应用程序到 AVD（有关知识点参见 1.5 节）。对于本例子，用命令行进入 D:\2000\ch11_4，执行如下命令：

D:\2000 > ch11_4 > ant debug install

11.5 读取\res\raw（原始资源）中的文件

当我们创建工程后，工程的根目录下就有一个名字为 res（资源）的目录，即资源目录，和 assets（资产）目录不同的是，用户不能随意在 res 目录建立任意名字的子目录，只能建立系统允许的子目录，编译器（Debug）会为 res 的子目录中的文件在系统的 R.java 文件中生成一个资源 ID（参见 2.8 节）。系统允许在 res 下建立名字是 raw 的子目录，该子目录下可以存放视频、音频以及文本文件（在 5.7 节曾使用过 raw 子目录）。

1. 只能读取

当编译器（Debug）发现程序引用了\res\raw 目录中的文件时，就会将程序读取的文件打包在应用程序中（apk 文件中），因此程序实际读取的是打包在应用程序中的文件，基于这样的机制，程序只能读取\res\raw 目录中的文件，不可以进行写文件操作（系统不允许动态改变 apk 文件的大小）。

2. 获得输入流

假设\res\raw 目录中有 test.txt 文件,获得指向 test.txt 的输入流的步骤如下。

1) 得到资源对象

android.content 包中的 Context 类以及 android.view 包中的 View 类提供了获取资源的 public Resources getResources()方法(参见 2.8 节),该方法返回一个 Resources 类的实例(Resource 类在 android.content.res 包中)。例如,Activity 类是 Context 类的子类,返回资源对象的代码如下:

```
Resources resource = getResources();
```

2) 得到输入流

Resources 对象调用 InputStream openRawResource(int id)方法打开指向文件的输入流。编译器(Debug)会为\res\raw 目录中的文件在系统的 R.java 文件中生成一个资源 ID,假设保存在\res\raw 下的文件是 ok.txt,那么可以用 R 类得到资源 ID:R.raw.ok。下面是打开指向 ok.txt 的输入流的代码:

```
InputStream in = resource.openRawResource(R.raw.ok);
```

3. 示例

例子 11-5 中,我们将一幅图像文件(flower.jpg)和一个文本文件保存在/res/raw 目录中,用户单击"显示图像"按钮,程序将读取打包在程序文件中的图像文件,并显示图像,用户单击"显示文本"按钮,程序将读取打包在程序文件中的文本文件,并显示文本文件的内容。运行效果如图 11.5 所示。

图 11.5 读取原始资源

例子 11-5

(1) 创建名字为 ch11_5 的工程,主要 Activity 子类的名字为 Example11_5,使用的包名为 ch11.five。用命令行进入 D:\2000,创建工程 D:\2000>android create project -t 3 -n ch11_5 -p ./ch11_5 -a Example11_5 -k ch11.five。

(2) 将名字为 flower.jpg 和 ok.txt 的文件保存在工程的\res\raw 目录中。

(3) 将下列和视图相关的 XML 文件保存到工程的\res\layout 目录中。

ch11_5.xml

```
<?xml version = "1.0" encoding = "utf-8"?>
<LinearLayout xmlns:android = "http://schemas.android.com/apk/res/android"
    android:orientation = "vertical"
    android:layout_width = "match_parent"
    android:layout_height = "match_parent">
    <Button
        android:layout_width = "match_parent"
        android:layout_height = "wrap_content"
        android:text = "显示图像"
        android:onClick = "showImage" />
    <Button
        android:layout_width = "match_parent"
```

```xml
        android:layout_height = "wrap_content"
        android:text = "显示文本"
        android:onClick = "showText" />
    <TextView
        android:id = "@+id/text"
        android:background = "#dddddd"
        android:textColor = "#FF0A00"
        android:layout_width = "match_parent"
        android:layout_height = "wrap_content"
        android:textSize = "18sp" />
    <ImageButton
        android:id = "@+id/image"
        android:visibility = "visible"
        android:scaleType = "centerInside"
        android:layout_width = "match_parent"
        android:layout_height = "wrap_content"/>
</LinearLayout>
```

（4）修改工程\src\ch11\five 目录下的 Example11_5.java 文件，修改后的内容如下：

Example11_5.java

```java
package ch11.five;
import android.app.Activity;
import android.os.Bundle;
import android.widget.*;
import android.view.*;
import android.graphics.drawable.Drawable;
import java.io.*;
public class Example11_5 extends Activity {
    TextView text;
    ImageButton image;
    public void onCreate(Bundle savedInstanceState) {
        super.onCreate(savedInstanceState);
        setContentView(R.layout.ch11_5);
        text = (TextView)findViewById(R.id.text);
        image = (ImageButton)findViewById(R.id.image);
    }
    public void showText(View view) {
        text.setText(null);
        try {
            InputStream in = getResources().openRawResource(R.raw.ok);
            int n = -1;
            byte a[] = new byte[1024];
            while((n = in.read(a))!= -1)
                text.append(new String(a,0,n));
        }
        catch(IOException exp){
            text.setText("" + exp);
        }
    }
    public void showImage(View view) {
```

```
        try {
            InputStream in = getResources().openRawResource(R.raw.flower);
            Drawable drawable = Drawable.createFromStream (in,"");
            image.setImageDrawable(drawable);
        }
        catch(Exception exp){
            text.setText("" + exp);
        }
    }
}
```

(5) 启动 AVD,进入工程的根目录,用快捷方式编译工程、安装应用程序到 AVD(有关知识点参见 1.5 节)。对于本例子,用命令行进入 D:\2000\ch11_5,执行如下命令:

```
D:\2000 > ch11_5 > ant debug install
```

11.6 解析 XML 文件

当我们创建工程后,工程的根目录下就有一个名字为 res(资源)的目录,编译器(Debug)会为 res 的子目录中的文件在系统的 R.java 文件中生成一个资源 ID(参见 2.8 节)。系统允许在 res 下再建立名字为 xml 的子目录,该子目录下可以存放 XML 文件。

1. XmlResourceParser 解析器

存放在\res\xml 目录中的 XML 文件只要符合 XML 语法规则即可,对标记没有限制,不同于 Android 系统规定的某些 XML 文件(对标记有严格的限制),比如\res\layout 下和视图相关的 XML 文件,因此人们习惯地称\res\xml 目录中的 XML 文件为任意的 XML 文件。当编译器(Debug)发现程序解析\res\xml 目录中的 XML 文件时,编译器要首先检查 XML 文件是否是规范的(有关 XML 的知识可参见有关的教科书,比如作者在清华大学出版社出版的《XML 程序设计》),如果 XML 文件是规范的,就会将程序要解析的 XML 文件打包在应用程序中(apk 文件中),否则编译器将停止编译,要求用户修改不规范的 XML 文件。因此程序实际解析的是打包在应用程序中的 XML 文件。

假设\res\xml 目录中有文件 word.xml,获得解析器的步骤如下。

1) 得到资源对象

android.content 包中的 Context 类以及 android.view 包中的 View 类提供了获取资源的 public Resources getResources()方法(参见 2.8 节),该方法返回一个 Resources 类的实例(Resource 类在 android.content.res 包中)。例如,Activity 类是 Context 类的子类,返回资源对象的代码如下:

```
Resources resource = getResources();
```

2) 得到解析器

Resources 对象调用 XmlResourceParser getXml(int id)方法返回一个解析器。假设保存在\res\xml 下的文件的文件名为 word.xml,那么可以用 R 类得到资源 ID:R.xml.word。下面是返回解析器 parse 的代码:

```
XmlResourceParser parser = resource.getXml(R.xml.word);
```

XmlResourceParser解析器的工作原理类似于Java API中的SAX解析器的工作原理，即是基于事件的解析器（建议参看有关XML语言的教材）。当XmlResourceParser解析器调用next()方法解析一个XML文件时，可能导致的事件有"文档开始"、"标记开始"、"标记结束"、"文本"或"文档结束"事件，程序可以根据导致的事件进行必要的操作。解析器用next()方法解析器XML文件的过程中只能导致一次"文档开始"事件和"文档结束"事件"。

2．示例

例子11-6中，程序使用解析器解析XML文件（见例子11-6中的student.xml)中的数学和英语成绩，然后计算出数学和英语的平均成绩。运行效果如图11.6所示。

例子11-6

（1）创建名字为ch11_6的工程，主要Activity子类的名字为Example11_6，使用的包名为ch11.six。用命令行进入D:\2000，创建工程 D:\2000＞android create project -t 3 -n ch11_6 -p ./ch11_6 -a Example11_6 -k ch11.six。

（2）将名字为stdudent.xml的文件保存在工程的/res/xml目录中，保存时必须将编码选择为"UTF-8"、保存类型选择为"所有文件"。

图11.6 解析XML文件

student.xml

```
<?xml version = "1.0" encoding = "UTF-8" ?>
<成绩单>
    <学生>
        <name>张三</name>
        <math> 86 </math>
        <english> 99 </english>
    </学生>
    <学生>
        <name>李四</name>
        <math> 87 </math>
        <english> 96 </english>
    </学生>
</成绩单>
```

（3）将下列和视图相关的XML文件保存到工程的\res\layout目录中。

ch11_6.xml

```
<?xml version = "1.0" encoding = "utf-8"?>
<LinearLayout xmlns:android = "http://schemas.android.com/apk/res/android"
    android:orientation = "vertical"
    android:layout_width = "match_parent"
    android:layout_height = "match_parent">
    <Button
        android:layout_width = "match_parent"
        android:layout_height = "wrap_content"
```

```xml
            android:text = "显示考试情况"
            android:onClick = "show" />
    <ScrollView
        android:layout_width = "match_parent"
        android:layout_height = "match_parent"
        android:scrollbarStyle = "outsideOverlay">
        <TextView
            android:id = "@+id/text"
            android:background = "#dddddd"
            android:textColor = "#0000FF"
            android:layout_width = "match_parent"
            android:layout_height = "wrap_content"
            android:textSize = "16sp" />
    </ScrollView>
</LinearLayout>
```

(4) 修改工程\src\ch11\six 目录下的 Example11_6.java 文件，修改后的内容如下：

Example11_6.java

```java
package ch11.six;
import android.app.Activity;
import android.os.Bundle;
import android.widget.*;
import android.view.*;
import android.content.res.XmlResourceParser;
import org.xmlpull.v1.XmlPullParser;
public class Example11_6 extends Activity {
    TextView text;
    public void onCreate(Bundle savedInstanceState) {
        super.onCreate(savedInstanceState);
        setContentView(R.layout.ch11_6);
        text = (TextView)findViewById(R.id.text);
    }
    public void show(View view) {
        text.setText(null);
        String s = "";
        double math = 0, english = 0;
        int mathCount = 0, englishCount = 0;
        try {
            XmlResourceParser xpp = getResources().getXml(R.xml.student);
            int eventType = xpp.getEventType();
            text.append("*****************\n");
            while (eventType != XmlPullParser.END_DOCUMENT) {
                if(eventType == XmlPullParser.START_DOCUMENT) {}
                else if(eventType == XmlPullParser.START_TAG) {
                    s = xpp.getName();
                    if(s.equals("math")) {
                        text.append("\n" + s + ":");
                        mathCount++;
                    }
                    if(s.equals("english")) {
```

```java
                    text.append("\n" + s + ":");
                    englishCount++;
                }
            }
            else if(eventType == XmlPullParser.END_TAG) {}
            else if(eventType == XmlPullParser.TEXT) {
                if(s.equals("name")) {
                    text.append("\n");
                    text.append(xpp.getText());
                }
                if(s.equals("math")) {
                    text.append(xpp.getText());
                    math = math + Double.parseDouble(xpp.getText());
                }
                if(s.equals("english")) {
                    text.append(xpp.getText() + "\n");
                    english = english + Double.parseDouble(xpp.getText());
                }
            }
            eventType = xpp.next();
        }
        text.append("\n*****************\n");
        math = math/mathCount;
        english = english/englishCount;
        text.append("\nMath aver:" + String.format("%5.2f",math));
        text.append("\nenglish aver:" + String.format("%5.2f",english));
    }
    catch(Exception exp){
        text.setText("" + exp);
    }
}
```

（5）启动 AVD，进入工程的根目录，用快捷方式编译工程、安装应用程序到 AVD（有关知识点参见 1.5 节）。对于本例子，用命令行进入 D:\2000\ch11_6，执行如下命令：

D:\2000 > ch11_6 > ant debug install

11.7　基于文本文件的电话簿

本节给出一个略微复杂的综合例子，除了需要本章的知识外，还涉及 3.10 节，8.6 节和 8.12 节中的一些主要知识点。

1. 用多个 Activity 对象体现电话簿的功能

一个应用程序可以包含若干个 Activity 对象，当包含多个 Activity 对象时，需要在配置文件中为每个 Activity 对象配置一个＜activity＞标记，即向程序注册 Activity 对象，只有这样程序才可以使用这个 Activity 对象。当程序略微复杂后，就需要使用多个 Activity 对象协调工作来体现整个程序的功能。比如电话簿有 3 个模块，即 3 个 Activity 对象，一个负责显示电话列表的 Activity 对象,一个负责增加联系人的 Activity 对象,一个负责删除联系人的 Activity 对象。

2. 示例

以下例子 11-7 使用文件实现一个电话簿,应用程序中有 3 个 Activity 对象,一个主要的 Activity 对象,由 Example11_7 类负责创建,一个负责向文件写入联系人信息的 Activity 对象,由 WriteActivity 类负责创建,一个负责删除联系人的 Activity 对象,由 DelActivity 类负责创建。用户在程序的主要 Activity 对象上可以看见联系人列表,选择某个联系人,将启动系统内置的负责打电话的 Activity 对象。用户可以单击增加联系人按钮启动增加联系人的 Activity 对象,单击删除联系人按钮启动删除联系人的 Activity 对象。运行的部分效果如图 11.7(a)、11.7(b)所示。

(a) 联系人列表　　　　(b) 拨打电话

图 11.7　运行的部分效果

例子 11-7

(1) 创建名字为 ch11_7 的工程,主要 Activity 子类的名字为 Example11_7,使用的包名为 ch11.seven。用命令行进入 D:\2000,创建工程 D:\2000＞android create project -t 3 -n ch11_7 -p ./ch11_7 -a Example11_7 -k ch11.seven。

(2) 将下列和视图相关的 XML 文件 ch11_7.xml(Example11_7 创建的 Activity 对象使用的视图)、people.xml(WriteActivity 创建的 Activity 对象使用的视图)、delete.xml(DelActivity 创建的 Activity 对象使用的视图)保存到工程的\res\layout 目录中。

ch11_7.xml

```
<?xml version = "1.0" encoding = "utf-8"?>
<LinearLayout xmlns:android = "http://schemas.android.com/apk/res/android"
    android:orientation = "vertical"
    android:layout_width = "match_parent"
    android:layout_height = "match_parent">
    <TextView
```

```xml
        android:background = "#dddddd"
        android:textColor = "#000AFF"
        android:layout_width = "match_parent"
        android:layout_height = "wrap_content"
        android:textSize = "15sp"
        android:text = "联系人列表(单击联系人可拨打电话): " />
    <ListView
        android:layout_width = "wrap_content"
        android:layout_height = "wrap_content"
        android:divider = "#0000FF"
        android:dividerHeight = "3dp"
        android:background = "#A82CA9"
        android:id = "@+id/phone_list" />
    <Button
        android:layout_width = "match_parent"
        android:layout_height = "wrap_content"
        android:text = "添加联系人"
        android:onClick = "goToWriteActivity" />
    <Button
        android:layout_width = "match_parent"
        android:layout_height = "wrap_content"
        android:text = "删除联系人"
        android:onClick = "goToDelActivity" />
</LinearLayout>
```

people.xml

```xml
<?xml version = "1.0" encoding = "utf-8"?>
<LinearLayout xmlns:android = "http://schemas.android.com/apk/res/android"
    android:orientation = "vertical"
    android:layout_width = "match_parent"
    android:layout_height = "match_parent"
    android:background = "#87CEEB">
    <TextView
        android:background = "#555555"
        android:layout_width = "wrap_content"
        android:layout_height = "wrap_content"
        android:text = "输入姓名:" />
    <EditText
        android:id = "@+id/name"
        android:layout_width = "match_parent"
        android:layout_height = "wrap_content" />
    <TextView
        android:background = "#555555"
        android:layout_width = "wrap_content"
        android:layout_height = "wrap_content"
        android:text = "输入电话:" />
    <EditText
        android:id = "@+id/phone"
        android:layout_width = "match_parent"
        android:layout_height = "wrap_content"
```

```xml
            android:phoneNumber = "true" />
        <Button
            android:layout_width = "match_parent"
            android:layout_height = "wrap_content"
            android:text = "保存"
            android:onClick = "writeFile"  />
        <Button
            android:layout_width = "match_parent"
            android:layout_height = "wrap_content"
            android:text = "返回到电话簿"
            android:onClick = "back"   />
</LinearLayout>
```

delete.xml

```xml
<?xml version = "1.0" encoding = "utf-8"?>
<LinearLayout xmlns:android = "http://schemas.android.com/apk/res/android"
    android:orientation = "vertical"
    android:layout_width = "match_parent"
    android:layout_height = "match_parent"
    android:background = "#87CEEB">
    <TextView
        android:background = "#555555"
        android:layout_width = "wrap_content"
        android:layout_height = "wrap_content"
        android:text = "单击要删除的联系人:" />
    <ListView
        android:layout_width = "wrap_content"
        android:layout_height = "wrap_content"
        android:divider = "#0000FF"
        android:dividerHeight = "3dp"
        android:background = "#A82CA9"
        android:id = "@+id/phone_list" />
    <Button
        android:layout_width = "match_parent"
        android:layout_height = "wrap_content"
        android:text = "返回到电话簿"
        android:onClick = "back"   />
</LinearLayout>
```

（3）将下列 WriteActivity.java 和 DelActivity.java 保存到工程\src\ch11\seven 目录下，并修改程序\src\ch11\seven 目录下 Example11_7.java 文件。WriteActivity.java，DelActivity.java 和修改后的 Example11_7.java 文件如下：

Example11_7.java

```java
package ch11.seven;
import android.app.*;
import android.os.*;
import android.widget.*;
import android.view.*;
import android.content.*;
```

```java
import android.net.*;
import android.graphics.Color;
import java.util.*;
public class Example11_7 extends Activity implements AdapterView.OnItemClickListener{
    ListView listView;
    public void onCreate(Bundle savedInstanceState) {
        super.onCreate(savedInstanceState);
        setContentView(R.layout.ch11_7);
        listView = (ListView)findViewById(R.id.phone_list);
        listView.setOnItemClickListener(this);
    }
    public void onItemClick(AdapterView parent,View view,int pos,long id) {
        String nameAndPhone = listView.getItemAtPosition(pos).toString();
        Uri uri = Uri.parse("tel:" + nameAndPhone);
        Intent intent = new Intent(Intent.ACTION_DIAL,uri);
        startActivity(intent);
    }
    public void goToWriteActivity(View v) {
        Intent intent = new Intent(this,WriteActivity.class);
        startActivity(intent);
    }
    public void goToDelActivity(View v) {
        Intent intent = new Intent(this,DelActivity.class);
        startActivity(intent);
    }
    public void onStart() {
        super.onStart();
        listPeople();
    }
    public void onResume() {
        super.onResume();
        listPeople();
    }
    public void listPeople() {
        ArrayList<String> listItem = new ArrayList<String>();
        MyAdapter adapter = new MyAdapter() ;   //适配器
        SharedPreferences people = getSharedPreferences("phone",Context.MODE_PRIVATE) ;
        HashMap map = (HashMap)people.getAll();
        Collection coll = map.values();
        Iterator iterator = coll.iterator();
        while(iterator.hasNext()) {
            String s = (String)iterator.next();
            listItem.add(s);
        }
        adapter.setContext(this);
        adapter.setArrayList(listItem);
        listView.setAdapter(adapter);
    }
}
class MyAdapter extends BaseAdapter {
    ArrayList<String> list;
```

```java
        Context context;
        public void setContext(Context context) {
            this.context = context;
        }
        public void setArrayList(ArrayList<String> list) {
            this.list = list;
        }
        public int getCount() {
            return list.size();
        }
        public String getItem (int position) {
            String item = list.get(position);
            return item;
        }
        public long getItemId (int position) {
            return 0;
        }
        public View getView (int position, View convertView, ViewGroup parent) {
            TextView nameAndPhone = new TextView(context);
            nameAndPhone.setTextSize(1,16);
            nameAndPhone.setTextColor(Color.BLACK);
            nameAndPhone.setText(getItem (position));
            return nameAndPhone;
        }
    }
```

WriteActivity.java

```java
package ch11.seven;
import android.app.Activity;
import android.os.Bundle;
import android.widget.*;
import android.view.*;
import android.content.*;
public class WriteActivity extends Activity {
    EditText nameEdit, phoneEdit;
    public void onCreate(Bundle savedInstanceState) {
        super.onCreate(savedInstanceState);
        setContentView(R.layout.people);
        nameEdit = (EditText)findViewById(R.id.name);
        phoneEdit = (EditText)findViewById(R.id.phone);
    }
    public void writeFile(View view) {
        String name = nameEdit.getText().toString();
        String phone = phoneEdit.getText().toString();
        String mess = name + ":" + phone;
        SharedPreferences people = getSharedPreferences("phone",Context.MODE_PRIVATE);
        SharedPreferences.Editor editFile = people.edit();
        editFile.putString(name,mess);
        editFile.commit();
        Toast
        toast = Toast.makeText (this,"success",Toast.LENGTH_SHORT);
        toast.setGravity(Gravity.TOP,10,30);
```

```java
            toast.show();
        }
        public void back(View view) {
            Intent intent = new Intent(this,Example11_7.class);
            startActivity(intent);
        }
    }
```

DelActivity.java

```java
package ch11.seven;
import android.app.*;
import android.os.Bundle;
import android.widget.*;
import android.view.*;
import android.content.*;
import java.util.*;
public class DelActivity extends Activity implements
AdapterView.OnItemClickListener,DialogInterface.OnClickListener{
    ListView listView;
    String delete_name;
    public void onCreate(Bundle savedInstanceState) {
        super.onCreate(savedInstanceState);
        setContentView(R.layout.delete);
        listView = (ListView)findViewById(R.id.phone_list);
        listView.setOnItemClickListener(this);
        listPeople();
    }
    public void onItemClick(AdapterView parent,View view,int pos,long id) {
        String nameAndPhone = listView.getItemAtPosition(pos).toString();
        delete_name = nameAndPhone.substring(0,nameAndPhone.indexOf(":"));
        getAlertDialogWithButton(delete_name).show();
    }
    AlertDialog getAlertDialogWithButton(String name) {
        AlertDialog.Builder builder = new AlertDialog.Builder(this);
        builder.setTitle("delete");
        builder.setMessage(name);
        builder.setIcon(R.drawable.ic_launcher);
        builder.setPositiveButton("Yes",this);
        builder.setNegativeButton("No",this);
        builder.setCancelable(false);
        AlertDialog dialog = builder.create();
        return dialog;
    }
    public void onClick(DialogInterface dialog,int which) {
        if(which == dialog.BUTTON_POSITIVE) {
            delete(delete_name);
        }
    }
    public void delete(String s) {
        SharedPreferences people = getSharedPreferences("phone",Context.MODE_PRIVATE);
        SharedPreferences.Editor editFile = people.edit();
        editFile.remove(s);
        editFile.commit();
```

```
            listPeople();
        }
        public void listPeople() {
            ArrayList<String> listItem = new ArrayList<String>();
            MyAdapter adapter = new MyAdapter() ;   //使用了和 Example11_7 中同样的适配器
            SharedPreferences people = getSharedPreferences("phone",Context.MODE_PRIVATE) ;
            HashMap map = (HashMap)people.getAll();
            Collection coll = map.values();
            Iterator iterator = coll.iterator();
            while(iterator.hasNext()) {
                String s = (String)iterator.next();
                listItem.add(s);
            }
            adapter.setContext(this);
            adapter.setArrayList(listItem);
            listView.setAdapter(adapter);
        }
        public void back(View view) {
            Intent intent = new Intent(this,Example11_7.class);
            startActivity(intent);
        }
    }
}
```

（4）修改工程下的配置文件，注册程序中需要的 Activity 对象，即增加＜activity＞标记，修改后的配置文件的内容如下：

AndroidManifest.xml

```
<?xml version = "1.0" encoding = "utf-8"?>
<manifest
    <!-- 略去此处原有内容,以下是新增的 activity 标记 -->
        <activity android:name = "WriteActivity"
            android:label = "写联系人到通讯录">
        </activity>
        <activity android:name = "DelActivity"
            android:label = "删除联系人">
        </activity>
    <!-- 略去此处原有内容, -->
</manifest>
```

（5）启动 AVD，进入工程的根目录，用快捷方式编译工程、安装应用程序到 AVD（有关知识点参见 1.5 节）。对于本例子，用命令行进入 D:\2000\ch11_7，执行如下命令：

```
D:\2000>ch11_7>ant debug install
```

11.8　基于 XML 数据库的英-汉字典

本节使用 XML 文件和 XPath 语言实现一个能查询英语单词的英-汉字典程序，除了需要本章的知识外，还涉及 XPath 语言的主要知识点，讲解 XPath 语言不在本书范畴内，建议读者参看作者在清华大学出版社出版的《XML 程序设计》一书中的第 8 章的内容。

1. XML 文件与 Xpath 语言

当 XML 文件的结构是侧重于程序对其进行查询时,人们习惯地将这样的 XML 文件称作一个 XML 数据库。W3C 在 1999 年推出 XML Path Language(XPath) Version 1.0,简称 XPath 1.0 语言规范,并在 2007 年对 XPath 1.0 进行了补充,正式公布了 XML Path Language(XPath) Version 2.0,简称 XPath 2.0 语言规范。XPath 1.0 是 Xpath 的一个子集,XPath 2.0 中有大约 80% 和 XPath 1.0 相同。

使用 XPath 可以很容易地编写查询 XML 中数据的 XPath 路径表达式,和 DOM 和 SAX 解析器的侧重点不同,XPath 语言为应用程序从 XML 文件中获得所需要的特殊数据提供了更加方便、快捷的语法,XPath 的作用非常类似 SQL 语言在关系数据库中的地位。

Android SDK 包含有 Java SDK 为 XPath 语言提供的 API(在 javax.xml.xpath 包中)。

2. 将 XML 数据库打包在应用程序中

对于 Android 系统,当希望编写一个基于 XML 数据库的应用程序时,应当将 XML 数据库打包在应用程序中。可以将 XML 数据库存放到工程的\res\raw 目录中。当编译器(Debug)发现程序引用了\res\raw 目录中的 XML 数据库时,就会将程序读取的 XML 数据库打包在应用程序中(apk 文件中),因此程序实际读取的是打包在应用程序中的 XML 数据库。需要注意的是,由于编译器不对\res\raw 目录中的 XML 数据库进行文法检查,程序设计者需要用 XML 编辑器或浏览器检查 XML 数据库是否有文法错误,否则程序运行后,Xpath 语言无法查询 XML 数据库(会发生 SAXParseException 等异常)。

3. 示例

以下例子 11-8 使用 XML 文件和 XPath 语言实现一个能查询英语单词的英-汉字典程序。运行的部分效果如图 11.8(a),11.8(b)所示。

(a) 精确查询   (b) 模糊查询

图 11.8 运行的部分效果

例子 11-8

(1) 创建名字为 ch11_8 的工程,主要 Activity 子类的名字为 Example11_8,使用的包名为 ch11.eight。用命令行进入 D:\2000,创建工程 D:\2000>android create project -t 3 -n ch11_8 -p ./ch11_8 -a Example11_8 -k ch11.eight。

(2) 将下列 XML 文件,习惯称作 XML 数据库,保存到工程的\res\raw 目录中。在实际项目中最好对 XML 数据库进行约束(使用内部 DTD),并在程序代码中检查 XML 数据库是否满足 DTD 给出的约束,对 XML 数据库进行约束便于软件的维护和升级。本例子为简化代码,没有对 XML 数据库进行约束,所以需要在这里事先强调。用户在调试代码时,XML 数据库的根标记必须是<worlkist>,<worlkist>标记可以有多个<word>子标记,每个<word>子标记必须有 name 属性,该属性的值是一个英文单词。如果准备开发一个有 1 万条单词的英-汉字典,就需要有 1 万个<word>标记,本例子使用简单的文本编辑器编写了下列 XML 文件(有许多快速编辑 XML 文件的软件,可在网络上查找有关的编辑工具)。需要再次强调的是 XML 文件默认的是 UTF-8 编码,因此在保存 XML 文件时必须将编码选择为"UTF-8"、保存类型选择为"所有文件"。

word.xml

```
<?xml version = "1.0" encoding = "utf-8"?>
<wordlist>
    <word name = "appearance">
        英[ə'pɪər(ə)ns] 美[ə'pɪrəns]
        n. 外貌,外观;出现,露面
        The car is not bad in appearance
    </word>
    <word name = "append">
        英[ə'pend] 美[ə'pɛnd]
        vt. 附加;贴上;盖章
        n. 设置数据文件的搜索路
        If it's not, append it to the root element
    </word>
    <word name = "advertising">
        英['ædvətaɪzɪŋ] 美['ædvətaɪzɪŋ]
        n. 广告;广告业;登广告
        adj. 广告的;广告业的
        v. 公告;为…做广告(advertise 的 ing 形式)
        The shopping centre agreed to desist from false advertising
    </word>
</wordlist>
```

(3) 将下列和视图相关的 XML 文件 ch11_8.xml 保存到工程的\res\layout 目录中。

ch11_8.xml

```
<?xml version = "1.0" encoding = "utf-8"?>
<LinearLayout xmlns:android = "http://schemas.android.com/apk/res/android"
    android:orientation = "vertical"
    android:layout_width = "match_parent"
    android:layout_height = "match_parent"
    android:background = "#87CEEB">
    <LinearLayout
        android:layout_width = "match_parent"
        android:layout_height = "wrap_content">
        <TextView
            android:background = "#dddddd"
            android:textColor = "#000AFF"
            android:layout_width = "wrap_content"
            android:layout_height = "wrap_content"
            android:layout_weight = "1"
```

```xml
            android:text = "输入单词:"/>
        <EditText
            android:id = "@ + id/edit"
            android:layout_width = "wrap_content"
            android:layout_weight = "5"
            android:layout_height = "wrap_content" />
    </LinearLayout>
    <LinearLayout
        android:layout_width = "match_parent"
        android:layout_height = "wrap_content">
        <Button
            android:layout_width = "wrap_content"
            android:layout_height = "wrap_content"
            android:text = "精确查询"
            android:layout_weight = "1"
            android:onClick = "findExactly" />
        <Button
            android:layout_width = "wrap_content"
            android:layout_height = "wrap_content"
            android:text = "模糊查询"
            android:layout_weight = "1"
            android:onClick = "findFuzzy" />
    </LinearLayout>
    <ScrollView
        android:layout_width = "match_parent"
        android:layout_height = "match_parent"
        android:scrollbarStyle = "outsideOverlay"
        android:background = "#87CEEB">
        <TextView
            android:id = "@ + id/show"
            android:background = "#dddddd"
            android:textColor = "#000AFF"
            android:layout_width = "match_parent"
            android:layout_height = "match_parent" />
    </ScrollView>
</LinearLayout>
```

（4）修改 Example11_8.java 文件，修改后的 Example11_8.java 文件如下：

Example11_8.java

```java
package ch11.eight;
import android.app.*;
import android.os.*;
import android.widget.*;
import android.view.*;
import javax.xml.xpath.*;
import org.w3c.dom.*;
import org.xml.sax.*;
import java.io.*;
public class Example11_8 extends Activity {
    XPathFactory xPathFactory;
```

```java
        XPath xPath;
        InputSource source;
        TextView showResult;
        EditText editWord;
        InputStream in;
        public void onCreate(Bundle savedInstanceState) {
            super.onCreate(savedInstanceState);
            setContentView(R.layout.ch11_8);
            showResult = (TextView)findViewById(R.id.show);
            editWord = (EditText)findViewById(R.id.edit);
            xPathFactory = XPathFactory.newInstance();
        }
        public void findExactly(View v) {
            showResult.setText(null);
            String wantFindWord = editWord.getText().toString().trim();
            String path = "/wordlist/word[(@name = '" + wantFindWord + "')]";
            findWord(path);
        }
        public void findFuzzy(View v) {
            showResult.setText(null);
            String wantFindWord = editWord.getText().toString().trim();
            String path = "/wordlist/word[contains(@name,'" + wantFindWord + "')]";
            findWord(path);
        }
        void findWord(String path) {
            xPath = xPathFactory.newXPath();
            in = getResources().openRawResource(R.raw.word);
            source = new InputSource(in);
            try{
                NodeList nodelist =
                (NodeList)xPath.evaluate(path, source, XPathConstants.NODESET);
                int size = nodelist.getLength();
                for( int k = 0;k < size;k++){
                    Node node = nodelist.item(k);
                    NamedNodeMap map = node.getAttributes();
                    Attr attrNode = (Attr)map.item(0);
                    String attValue = attrNode.getValue();
                    showResult.append(attValue + ":");
                    String value = node.getTextContent();;
                    showResult.append("" + value + "\n");
                }
                if(size == 0)
                    showResult.setText("can not find");
            }
            catch(Exception exp){
                showResult.setText("" + exp.toString());
            }
        }
    }
}
```

(5) 启动 AVD,进入工程的根目录,用快捷方式编译工程、安装应用程序到 AVD(有关知识点参见 1.5 节)。对于本例子,用命令行进入 D:\2000\ch11_8,执行如下命令:

D:\2000>ch11_8>ant debug install

习　题　11

1. 编写一个程序,向系统的数据区/data/data/应用程序包名/files 写入一个名字为 student 的文本文件,向该文件写入的内容是"We are students",然后读取该文件,将文件的内容显示在一个 TextView 视图中(参考例子 11-1)。

2. 编写一个程序,用户可以把"商品"和"商品的价格"作为"键/值"数据保存到文件中,例如"TV/8976","Phone/1298"等数据保存到文件中,用户可以查询文件中的商品(参考例子 11-2)。

3. 将一个文本文件保存到工程的\assets\myfile 目录中,编写一个程序,单击一个 Button 视图,程序读取\assets\myfile 目录中的文本文件,并将读取的内容显示在一个 TextView 视图中。

第 12 章　使用 SQLite 数据库

主要内容：
- 连接 SQLite 数据库；
- 外挂 SQLite 数据库；
- SQLiteDatabase 类的两个重要方法；
- 事务；
- 基于数据库的消费记载。

SQLite 数据库属于关联式数据库，它占用资源非常低、内存非常少（大概只占用几百 KB 的内存），而且处理数据的速度也很快（比 Mysql 数据库要快）。SQLite 数据库特别适合用于嵌入式设备中，因此 Android 系统选择 SQLite 数据库作为它的内置数据库，方便用户使用数据库开发有关的应用程序。SQLite 第一个 Alpha 版本诞生于 2000 年 5 月，2005 年推出 SQLite 3 版本，Android 4.1 系统使用的版本是 SQLite version 3.7.11（2012 年 3 月 20 推出的版本）。

本章不能重点讲解 SQLite 数据库本身的知识内容，而是侧重讲解怎样在 Android 应用程序中使用 SQLite 数据库。

12.1　连接 SQLite 数据库

1. SQLite 数据库的数据类型

当用户在数据库中建立表时，用户需要知道数据库支持的数据类型。SQLite 数据库支持常见的数据类型有 NULL、INTEGER、FLOAT、TEXT、BOOLEAN、BLOB（二进制对象）、VARCHAR(n) 和 NVARCHAR(n)。

使用 SQLite 数据库的方便之处之一就是，在建立表的时候，如果该字段（列）的类型不是 PRIMARY KEY（主键），那么该字段的值可以是数据库支持的任何数据类型的数据，即忽略字段的数据类型，例如，可以在 INTEGER 类型的字段中存放日期、浮点数、字符串等数据，在布尔型字段中存放整数、浮点数、日期、字符串，或者在字符型字段中存放日期、整数、浮点数等数据。但对于主键字段，比如，类型是 INTEGER PRIMARY KEY 的字段，该字段的值只能是整数，否则将会产生错误。

在 SQLite 数据库建立表时，比如 student 表，下列两个写法是等价的：

```
CREATE TABLE student (student_id INTEGER PRIMARY KEY,student_name TEXT);
CREATE TABLE student (student_id INTEGER PRIMARY KEY,student_name);
```

SQLite 数据库在处理 SQL 语句时,会忽略 SQL 语句中给字段(列)定义的数据类型(除 PRIMARY KEY 字段)。

2. SQLiteOpenHelper 类与数据库的连接

当应用程序需要创建、连接数据库时,需要使用 android.database.sqlite 包中的 SQLiteOpenHelper 类,SQLiteOpenHelper 类的实例能完成创建、连接数据库,管理数据库的版本号(用户可自定义一个版本号)以及更新数据库等工作。

1) 指定数据库的名称与版本号

SQLiteOpenHelper 类的构造方法负责创建指定数据库的名称与版本号,即用 SQLiteOpenHelper 类构造方法创建一个数据库,构造方法如下:

SQLiteOpenHelper(Context context, String name, SQLiteDatabase.CursorFactory factory, int version);

参数意义如下。

- context:上下文对象,取值可以是 Context 类的任何子类的实例的引用,例如取值是 Activity 对象的引用。
- name:数据库的名字,例如,name 取值是"stdent",那么数据库就是 stdent.db 数据库。如果 name 取值是 null,表示不指定任何数据库。
- factory:一个游标工厂类的实例,一般取值 null 即可,表示使用系统默认的游标工厂类的实例。
- version:数据库的版本号,取正整数(其值主要用于数据版本的更新)。

SQLiteOpenHelper 类是 abstract 类,当需要创建一个数据库时,需要扩展 SQLiteOpenHelper 类,即使用 SQLiteOpenHelper 类的子类来创建一个数据库。

用 SQLiteOpenHelper 类的子类创建对象本质上就是创建一个数据库,如果数据库不存在,系统将按着子类的构造方法中给出的数据库名称和版本号创建一个数据库,并调用子类重写的 void onCreate(SQLiteDatabase db)方法,以便决定是否在数据库中创建表或更新表(onCreate()方法是 SQLiteOpenHelper 类的抽象方法,子类必须要重写)。当执行 onCreate(SQLiteDatabase db)方法时,参数 db 就是构造方法创建的当前数据库,而且系统会自动打开 db 数据库,onCreate 方法执行完毕后,系统自动关闭 db 数据库。

当 SQLiteOpenHelper 类的子类创建一个数据库时,如果数据库已经存在,但版本号与构造方法中给出的版本号不同(但构造方法给出的版本号必须要大于当前数据库的版本号,否则发生运行异常),那么将按着子类的构造方法中给出的数据库版本号更新已有数据库版本号(不更改数据库的其他结构),但是,不再调用子类重写的 void onCreate(SQLiteDatabase db)方法,而是调用子类重写的 onUpgrade(SQLiteDatabase db, int oldVersion, int newVersion)方法(onUpgrade()方法是 SQLiteOpenHelper 类的抽象方法,子类必须要重写)。当执行 onUpgrade(SQLiteDatabase db, int oldVersion, int newVersion)方法时,参数 db 就是构造方法创建的当前数据库,而且系统会自动打开 db 数据库,onUpgrade 方法执行完毕后,系统自动关闭 db 数据库。用户可以在 onUpgrade 方法中决定是否需要更新数据库的表。

当 SQLiteOpenHelper 类的子类创建一个数据库时,如果数据库已经存在,版本号与构造方法中给出的版本号相同,那么子类的构造方法将不做任何事情,即保持原有数据库,也不调用子类重写的 void onCreate()方法和 onUpgrade()方法。

2）打开数据库

程序想和数据库交互数据，就必须要连接到数据库，即和数据建立连接。和数据库建立连接也称为打开数据库。SQLiteOpenHelper 类提供的打开数据库的方法如下。

- SQLiteDatabase getReadableDatabase()：用只读方法打开数据库，并返回数据库的引用。该方法在打开数据库时，只依赖 SQLiteDatabase 实例给出的数据库名，忽略版本号。
- SQLiteDatabase getWritableDatabase()：用读写方式打开数据库，并返回数据库的引用。该方法在打开数据库时，只依赖 SQLiteDatabase 实例给出的数据库名，忽略版本号。

编写一个名字为 CreateSQLiteDatabase 子类的代码如下：

```
import android.database.sqlite.*;
public class CreateSQLiteDatabase extends SQLiteOpenHelper {
    public CreateSQLiteDatabase
        (Context context,String name,SQLiteDatabase.CursorFactory factory,int version){
        super(context,name,factory,version);
    }
    public void onCreate(SQLiteDatabase db) {
        //选择是否在数据库 db 中创建一些新表
    }
    void onUpgrade(SQLiteDatabase db,int oldVersion,int newVersion) {
        //选择是否更新数据库 db 中已有的表
    }
}
```

子类重写的 onUpgrade(SQLiteDatabase db,int oldVersion,int newVersion)方法主要目的是当数据库版本号增大时，完成数据库的更新工作。

3. 关闭连接

SQLiteOpenHelper 类的子类的实例调用 close()方法可以关闭它曾打开的数据库。

4. 数据库所在目录

系统将应用程序创建的数据库存放在系统提供的数据区内，目录是\data\data\应用程序的包名\databases\。

例如，假设数据库的名字为 student.db，应用程序的包名为 ton.jiafei，那么数据库的路径是\data\data\tom\jiafei\databases\stdent.db。

5. 示例

例子 12-1 中，单击"打开数据库"按钮，系统打开数据库，并向数据库添加一个表，向表中插入一条记录，然后显示表中的全部记录。运行效果如图 12.1(a)，12.1(b)所示。

例子 12-1

（1）创建名字为 ch12_1 的工程，主要 Activity 子类的名字为 Example12_1，使用的包名为 ch12.one。用命令行进入 D:\2000，创建工程 D:\2000＞android create project -t 3 -n ch12_1 -p ./ch12_1 -a Example12_1 -k ch12.one。

（2）将下列和视图相关的 XML 文件保存到工程的\res\layout 目录中。

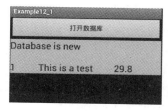
(a) 新建数据库的效果

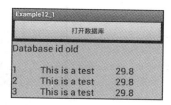
(b) 打开已有数据库的效果

图 12.1　运行效果

ch12_1.xml

```xml
<?xml version = "1.0" encoding = "utf-8"?>
<LinearLayout xmlns:android = "http://schemas.android.com/apk/res/android"
    android:orientation = "vertical"
    android:layout_width = "match_parent"
    android:layout_height = "match_parent">
    <Button
        android:layout_width = "match_parent"
        android:layout_height = "wrap_content"
        android:text = "打开数据库"
        android:onClick = "openDatababse" />
    <TextView
        android:id = "@ + id/text"
        android:background = "#AAEE00"
        android:textColor = "#000000"
        android:layout_width = "match_parent"
        android:layout_height = "wrap_content"
        android:textSize = "20sp" />
</LinearLayout>
```

（3）修改工程\src\ch12\one 目录下的 Example12_1.java 文件，修改后的内容如下：

Example12_1.java

```java
package ch12.one;
import android.app.Activity;
import android.os.Bundle;
import android.widget.*;
import android.view.*;
import android.database.sqlite.*;
import android.database.Cursor;
import android.content.Context;
public class Example12_1 extends Activity {
    TextView show;
    SQLiteDatabase database;
    int version = 100;
    public void onCreate(Bundle savedInstanceState) {
        super.onCreate(savedInstanceState);
        setContentView(R.layout.ch12_1);
        show = (TextView)findViewById(R.id.text);
```

```java
        }
        public void openDatababse(View view) {
            show.setText(null);
            try {
                String dabaseName = "firstSQlite";version++;
                CreateSQLiteDatabase dabaseHelper =
                new CreateSQLiteDatabase(this,dabaseName,null,version);
                                                                //dabaseHelper 负责打开数据库
                database = dabaseHelper.getWritableDatabase() ;   //打开数据库
                show.append("" + dabaseHelper.mess + "\n");
                String sql = "CREATE TABLE IF NOT EXISTS booklist" +
                        "(book_id INTEGER PRIMARY KEY,book_name TEXT,book_price FLOAT)";
                database.execSQL(sql);                          //创建表
                sql = "select * from booklist";
                Cursor cursor = database.rawQuery(sql,null);     //查询表
                int n = cursor.getCount() ;
                n++;
                sql = "insert into booklist(book_id,book_name,book_price) values(" + n + ",'This is a test',29.8)";
                database.execSQL(sql);
                sql = "select * from booklist";
                cursor = database.rawQuery(sql,null);
                while (cursor.moveToNext()) {
                    String ISBN = cursor.getString(0);           //列的索引从 0 开始
                    String name = cursor.getString(1);
                    float price = cursor.getFloat(2);
                    show.append("\n" + ISBN + "\t\t\t" + name + "\t\t\t" + price);
                }
                cursor.close();
                database.close();
            }
            catch(Exception exp){
                show.append("" + exp);
            }
        }
    }
    class CreateSQLiteDatabase extends SQLiteOpenHelper {
        String mess = "Database id old";
        public CreateSQLiteDatabase
        (Context context,String name,SQLiteDatabase.CursorFactory factory, int version){
            super(context,name,factory,version);
        }
        public void onCreate (SQLiteDatabase db) {
            mess = "Database is new";
        }
        public void onUpgrade(SQLiteDatabase db, int oldVersion, int newVersion) {}
    }
```

（4）启动 AVD，进入工程的根目录，用快捷方式编译工程、安装应用程序到 AVD（有关知识点参见 1.5 节）。对于本例子，用命令行进入 D:\2000\ch12_1，执行如下命令：

D:\2000 > ch12_1 > ant debug install

12.2 外挂 SQLite 数据库

1. 使用输入流外挂数据库

在某些设计中,可能不希望应用程序在初次运行时自己创建数据库、打开数据库,并在数据库中建立表。比如,开发一个字典查询的手机程序,那么数据库应当是事先设计完毕的,包括其中的表以及表中的记录。应用程序只需打开数据库查询记录即可。所谓外挂 SQLLite 数据库,就是将其他 SQLite 开发工具开发设计好的 SQLite 数据库打包在应用程序中,程序首次运行时将打包在程序中的数据库存放到系统的数据区内,然后程序就可以打开数据库,查询其中的记录了。

可以将已有的数据库保存到项目的\res\raw 文件夹中(有关知识点参见 11.5 节),程序使用输入流读取数据库文件,并将读取的数据库文件写入到系统的数据区,即写入到如下的目录中:

/data/data/程序包名/databases

例如,假设程序的包名为 dalian.ok,\res\raw 目录存放的数据库为 tom.db,代码如下:

```
InputStream in = getResources().openRawResource(R.raw.tom);
File file = new File("/data/data/dalian.ok/databases/","tom");
FileOutputStream out = new FileOutputStream(file);
int n = -1;
byte a[] = new byte[1024];
while((n = in.read(a))!= -1)
    out.write(a,0,n);
```

当编译器(Debug)发现程序引用了/res/raw 目录中的数据库文件时,就会将程序引用的数据库文件打包在应用程序中(apk 文件中),因此程序实际读取的是打包在应用程序中的数据库文件,然后将其写入到系统的数据区中。

2. 获得一个 SQLite 数据库

目前有很多用于开发 SQLite 数据库的开发工具,可以登录 SQLite 官方网站 http://www.sqlite.org/下载相应的开发工具。本书在网上搜索了一个开发 SQLite 数据库的小软件,使用它设计了名字为 tom.db 的 SQLite 数据库。tom.db 中有一个名字为 student 的表,该表的结构是(number INTEGER,name TEXT,age INTEGER)。在该表中输入了 4 条记录(见例子 12-2 运行效果图 12.2 所示)。

3. 示例

例子 12-2 中,单击"打开外挂数据库"按钮,系统打开应用程序外挂的 tom.db 数据库,然后显示数据库中 student 表中的全部记录。运行效果如图 12.2 所示。

例子 12-2

(1) 创建名字为 ch12_2 的工程,主要 Activity 子类的名字为 Example12_2,使用的包名为 ch12.two。用命令行进入

图 12.2 使用外挂数据库

D:\2000,创建工程 D:\2000＞android create project -t 3 -n ch12_2 -p ./ch12_2 -a Example12_2 -k ch12.two。

(2) 将下列和视图相关的 XML 文件保存到工程的\res\layout 目录中。

ch12_2.xml

```xml
<?xml version="1.0" encoding="utf-8"?>
<LinearLayout xmlns:android="http://schemas.android.com/apk/res/android"
    android:orientation="vertical"
    android:layout_width="match_parent"
    android:layout_height="match_parent">
    <Button
        android:layout_width="match_parent"
        android:layout_height="wrap_content"
        android:text="打开外挂数据库"
        android:onClick="openDatababse" />
    <TextView
        android:id="@+id/text"
        android:background="#AAEE80"
        android:textColor="#000000"
        android:layout_width="match_parent"
        android:layout_height="wrap_content"
        android:textSize="20sp" />
</LinearLayout>
```

(3) 将其他开发工具设计的 SQLite 数据库保存到工程的\res\raw 目录中,本例子保存到工程的\res\raw 目录中的数据库是 tom.db。

(4) 修改工程\src\ch12\two 目录下的 Example12_2.java 文件,修改后的内容如下:

Example12_2.java

```java
package ch12.two;
import android.app.Activity;
import android.os.Bundle;
import android.widget.*;
import android.view.*;
import android.database.sqlite.*;
import android.database.Cursor;
import android.content.Context;
import java.io.*;
public class Example12_2 extends Activity {
    TextView show;
    SQLiteDatabase database;
    public void onCreate(Bundle savedInstanceState) {
        super.onCreate(savedInstanceState);
        setContentView(R.layout.ch12_2);
        show = (TextView)findViewById(R.id.text);
        File file = new File("/data/data/ch12.two/databases/","tom");
        if(!file.exists()) {
            try {
                InputStream in = getResources().openRawResource(R.raw.tom);
                FileOutputStream out = new FileOutputStream(file);
```

```java
            int n = -1;
            byte a[] = new byte[1024];
            while((n = in.read(a))!= -1)
                out.write(a,0,n);
            out.close();
            in.close();
        }
        catch(IOException exp){ }
    }
}
public void openDatababse(View view) {
    show.setText(null);
    try {
        String dabaseName = "tom";
        CreateSQLiteDatabase dabaseHelper =
        new CreateSQLiteDatabase(this,dabaseName,null,1001);
        database = dabaseHelper.getWritableDatabase();
        String sql = "select * from student";
        Cursor cursor = database.rawQuery(sql,null);
        while (cursor.moveToNext()) {
            int number = cursor.getInt(0);
            String name = cursor.getString(1);
            int age = cursor.getInt(2);
            show.append("\n" + number + "\t\t\t" + name + "\t\t\t" + age);
        }
        cursor.close();
        database.close();
        dabaseHelper.close();
    }
    catch(Exception exp){
        show.append("" + exp);
    }
}
}
class CreateSQLiteDatabase extends SQLiteOpenHelper {
    public CreateSQLiteDatabase
    (Context context, String name, SQLiteDatabase.CursorFactory factory, int version){
        super(context,name,factory,version);
    }
    public void onCreate (SQLiteDatabase db) {}
    public void onUpgrade(SQLiteDatabase db, int oldVersion, int newVersion) {}
}
```

（5）启动 AVD，进入工程的根目录，用快捷方式编译工程、安装应用程序到 AVD（有关知识点参见 1.5 节）。对于本例子，用命令行进入 D:\2000\ch12_2，执行如下命令：

D:\2000>ch12_2> ant debug install

12.3 SQLiteDatabase 类的两个重要方法

应用程序打开数据库后，最主要的工作就是和数据库进行数据交互，比如对数据库中的表进行查询、更新、删除等操作。SQLiteOpenHelper 类的实例在打开数据库的同时返回一

个 SQLiteDatabase 类的实例:

```
SQLiteDatabase getReadableDatabase();
SQLiteDatabase getWritableDatabase();
```

SQLiteDatabase 提供了许多操作数据库的方法,本节介绍两个常用的、重要的方法 execSQL()和 rawQuery()方法,对于熟悉 SQL 语法的程序员而言,直接使用这两个方法执行 SQL 语句就能完成和数据库相关的添加、删除、更新、查询等操作。

1. execSQL()方法

1) void execSQL(String sql)

void execSQL(String sql)可以让数据库执行参数指定的 SQL 语句,但 SQL 语句不能是 SELECT 语句,可以是 INSERT、DELETE、UPDATE 和 CREATE 等更改数据库中记录或表的 SQL 语句,即不能是有返回结果的 SQL 语句。

例如:

```
SQLiteDatabase db = getWritableDatabase();
String sql = "INSERT INTO book(id,name,price) values('879 - 23','java',25.7)";
db.execSQL(sql);
```

2) void execSQL(String sql,Object[] bindArgs)

void execSQL(String sql,Object[] bindArgs)可以让数据库执行参数指定的 SQL 语句,但 SQL 语句不能是 SELECT 语句,可以是 INSERT、DELETE、UPDATE 和 CREATE 等更改数据库中记录或表的 SQL 语句,即不能是有返回结果的 SQL 语句。该方法较 execSQL(String sql)方法的优势就是,SQL 语句中不必指定字段(列)的值,而是使用通配符号"?"代表字段的值,这些通配符号所代表的值由方法中的第 2 个参数 bindArgs 来指定,因此要求 SQL 语句中通配符号"?"的个数必须和数组 bindArgs 的长度一致。SQL 语句中通配符号"?"按从左到右的顺序,依次由数组 bindArgs 各个元素(索引从 0 开始)的值来指定。例如下列代码等价于前面的代码(但编写代码更加方便、灵活,而且不必担心单引号问题了):

```
SQLiteDatabase db = getWritableDatabase();
String sql = "INSERT INTO book(id,name,price) values(?,?,?)";
Object[] bindArgs = {"879 - 23","java",25.7};
db.execSQL(sql,bindArgs);
```

2. rawQuery 方法

public Cursor rawQuery (String sql,String[] selectionArgs)方法的第一个参数 sql 应当是一个 SELECT 语句,如果 SELECT 语句中没有使用通配符号"?"代表字段的值,方法的第 2 个参数可以是 null。例如:

```
SQLiteDatabase db = getWritableDatabase();
String sql = "SELECT FROM * book WHRER id = '879 - 23' AND name = 'java'";
Cursor cursor = db.rawQuery (sql,null);
```

如果 SELECT 语句中没有使用通配符号"?"代表字段的值(这些值必须是字符型数据),这些通配符号所代表的值由方法中的第 2 个参数 selectionArgs 来指定,因此要求 SQL 语句中通配符号"?"的个数必须和数组 selectionArgs 的长度一致。SQL 语句中通配符号"?"按从左到右的顺序,依次由数组 selectionArgs 各个元素(索引从 0 开始)的值来指定。例如,下列代码等价于前面的代码(但编写代码更加方便、灵活,而且不必担心单引号问题了):

```
SQLiteDatabase db = getWritableDatabase();
String sql = "SELECT FROM * book WHRER id = ? AND name = ?";
String[] selectionArgs = {"879 - 23","java"};
Cursor cursor = db.rawQuery(sql,selectionArgs);
```

3. Cursor 类

rawQuery 方法将查询结果返回到一个 Cursor 类的实例中,Cursor 类类似 Java API 中的 ResultSet 类。也就是说 SQL 查询语句对数据库的查询操作将返回一个 Cursor 对象,Cursor 对象是由统一形式的列组织的数据行组成,习惯上称 Cursor 对象为结果集,例如,对于 Cursor cursor=db.rawQuery("SELECT * FROM book")结果集 cursor 的列数和表 goods 的列数相同,而对于 Cursor cursor=db.rawQuery("SELECT name,price FROM book")结果集 cursor 只有两列,第一列是 name 列,第二列是 price 列。

结果集 cursor 最初的查询游标在第 1 行之前,结果集 cursor 使用 moveToNext()方法将查询游标移到下一数据行,获得一行数据后,结果集 cursor 可以使用诸如 getXxx 方法获得字段值(列值),将位置索引(列索引从 0 开始,第一列使用 0,第二列使用 1 等)传递给 getXxx 方法的参数即可。

如果 moveToNext()方法已经到了结果集 cursor 的最后一行的后面,moveToNext()方法返回 false,否则返回 true。另外 Cursor 还有常用的 moveToPrevious()方法(用于将查询游标从当前行移动到上一行,如果已经移到了结果集的第一行的前面,该返回 false,否则返回 true)、moveToFirst()方法(用于将查询游标移动到结果集的第一行,如果结果集为空,该返回 false,否则返回 true)和 moveToLast()方法(用于将查询游标移动到结果集的最后一行,如果结果集为空,该方法返回 false,否则返回 true)。

4. 示例

例子 12-3

例子 12-3 中,程序在创建数据库的同时向数据库中的一个表中添加了 3 条记录,用户单击"显示记录"按钮,程序显示插入的 3 条记录。运行效果如图 12.3 所示。

(1) 创建名字为 ch12_3 的工程,主要 Activity 子类的名字为 Example12_3,使用的包名为 ch12.three。用命令行进入 D:\2000,创建工程 D:\2000＞android create project -t 3 -n ch12_3 -p ./ch12_3 -a Example12_3 -k ch12.three。

图 12.3 使用 execSQL 方法

(2) 将下列视图文件 ch12_3.xml 保存到工程的\res\layout 目录中。

ch12_3.xml

```
<?xml version = "1.0" encoding = "utf - 8"?>
<LinearLayout xmlns:android = "http://schemas.android.com/apk/res/android"
    android:orientation = "vertical"
    android:layout_width = "match_parent"
    android:layout_height = "match_parent"
    android:background = "#87CEEB">
```

```xml
<Button
    android:layout_width = "match_parent"
    android:layout_height = "wrap_content"
    android:text = "显示记录"
    android:onClick = "show" />
<ScrollView
    android:layout_width = "match_parent"
    android:layout_height = "match_parent"
    android:scrollbarStyle = "outsideOverlay"
    android:background = "#87CEEB">
    <TextView
        android:id = "@+id/text"
        android:background = "#AAEE00"
        android:textColor = "#000000"
        android:layout_width = "match_parent"
        android:layout_height = "wrap_content"
        android:textSize = "20sp" />
</ScrollView>
</LinearLayout>
```

(3) 修改工程\src\ch12\three 目录下的 Example12_3.java 文件,修改后的内容如下。

Example12_3.java

```java
package ch12.three;
import android.app.Activity;
import android.os.Bundle;
import android.widget.*;
import android.view.*;
import android.database.sqlite.*;
import android.database.Cursor;
import android.content.Context;
public class Example12_3 extends Activity {
    TextView show;
    String dabaseName;
    SQLiteDatabase db;
    CreateSQLiteDatabase dabaseHelper;
    public void onCreate(Bundle savedInstanceState) {
        super.onCreate(savedInstanceState);
        setContentView(R.layout.ch12_3);
        show = (TextView)findViewById(R.id.text);
        dabaseName = "Students";
        dabaseHelper = new CreateSQLiteDatabase(this,dabaseName,null,6);
    }
    public void show(View view) {
        show.setText(null);
        try {
            db = dabaseHelper.getWritableDatabase();
            String sql = "select * from people_list";
            Cursor cursor = db.rawQuery(sql,null);
            while (cursor.moveToNext()) {
                int number = cursor.getInt(0);
```

```
                String name = cursor.getString(1);
                int age = cursor.getInt(2);
                show.append("\n" + number + "\t\t\t" + name + "\t\t\t" + age);
            }
            cursor.close();
            db.close();
        }
        catch(Exception exp){
            show.append("" + exp);
        }
    }
}
class CreateSQLiteDatabase extends SQLiteOpenHelper {
    public CreateSQLiteDatabase
    (Context context, String name, SQLiteDatabase.CursorFactory factory, int version){
        super(context, name, factory, version);
    }
    public void onCreate (SQLiteDatabase db) {
        String sql = "CREATE TABLE IF NOT EXISTS people_list(number INTEGER, name, age)";
        db.execSQL(sql);
        sql = "INSERT INTO people_list(number, name, age) values(?,?,?)";
        Object[] a = {1,"GengLing",25};
        Object[] b = {2,"ZengYing",26};
        Object[] c = {3,"WengXing",20};
        db.execSQL(sql,a);
        db.execSQL(sql,b);
        db.execSQL(sql,c);
    }
    public void onUpgrade(SQLiteDatabase db, int oldVersion, int newVersion) {}
}
```

（4）启动 AVD，进入工程的根目录，用快捷方式编译工程、安装应用程序到 AVD（有关知识点参见 1.5 节）。对于本例子，用命令行进入 D:\2000\ch12_3，执行如下命令：

D:\2000 > ch12_3 > ant debug install

12.4 事 务

事务由一组 SQL 语句组成，所谓事务处理是指应用程序保证事务中的 SQL 语句，要么全部都执行，要么一个都不执行。事务处理是保证数据库中数据完整性与一致性的重要机制。应用程序和数据库建立连接之后，可能使用多条 SQL 语句操作数据库中的一个表或多个表，比如，一个管理资金转账的应用程序为了完成一个简单的转账业务可能需要两条 SQL 语句，即需要将数据库 user 表中 id 号是 0001 的记录的 userMoney 字段的值由原来的 100 更改为 50，然后将 id 号是 0002 的记录的 userMoney 字段的值由原来的 20 更新为 70。应用程序必须保证这两条 SQL 语句要么全都执行，要么全都不执行。

1. 事务处理步骤

1) beginTransaction()

数据库对象，比如 db，调用 beginTransaction()方法开始一个事务，例如：

```
db.beginTransaction();
```

2) 执行事务中的 SQL 语句

数据库对象，比如 db，调用方法执行事务中的 SQL 语句，但这些 SQL 语句不会立刻生效。

3) 让事务中的 SQL 语句生效

数据库对象，比如 db，调用 setTransactionSuccessful()方法让事务中的 SQL 语句生效：

```
db.setTransactionSuccessful();
```

4) 结束事务

数据库对象，比如 db，调用 endTransaction()方法结束事务：

```
db.endTransaction();
```

当调用 endTransaction()方法时，如果发现 setTransactionSuccessful()方法没有让事务中的所有 SQL 语句都生效，该方法将取消事务中曾生效的 SQL 语句。

上述步骤的代码如下：

```
db.beginTransaction();
try {
    ...//事务中的 SQL 语句
    db.setTransactionSuccessful();
}catch(Exception exp){}
finally {
    db.endTransaction();
}
```

2. 示例

例子 12-4

下面的例子 12-4 使用了事务处理，将图书库存表 stock_list 中的 id 字段是"987-765"的 amount 的值减少 100，并将减少的 100 增加到销量表 sell_list 字段是"987-765"的 amount 上。运行效果如图 12.4 所示。

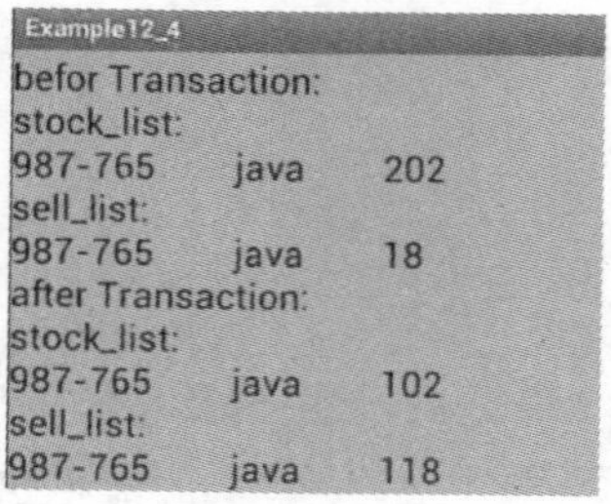

(1) 创建名字为 ch12_4 的工程，主要 Activity 子类的名字为 Example12_4，使用的包名为 ch12.four。用命令行进入 D:\2000，创建工程 D:\2000＞android create project -t 3 -n ch12_4 -p ./ch12_4 -a Example12_4 -k ch12.four。

(2) 将下列视图文件保存到工程的\res\layout 目录中。

ch12_4.xml

```
<?xml version = "1.0" encoding = "utf-8"?>
<LinearLayout xmlns:android = "http://schemas.android.com/apk/res/android"
```

图 12.4 事务

```
        android:orientation = "vertical"
        android:layout_width = "match_parent"
        android:layout_height = "match_parent"
        android:background = "#87CEEB">
    <ScrollView
            android:layout_width = "match_parent"
            android:layout_height = "match_parent"
            android:scrollbarStyle = "outsideOverlay"
            android:background = "#87CEEB">
        <TextView
            android:id = "@+id/text"
            android:background = "#AAEE00"
            android:textColor = "#000000"
            android:layout_width = "match_parent"
            android:layout_height = "wrap_content"
            android:textSize = "20sp" />
    </ScrollView>
</LinearLayout>
```

（3）修改工程\src\ch12\four 目录下的 Example12_4.java 文件，修改后的内容如下：

Example12_4.java

```
package ch12.four;
import android.app.Activity;
import android.os.Bundle;
import android.widget.*;
import android.view.*;
import android.database.sqlite.*;
import android.database.Cursor;
import android.content.Context;
public class Example12_4 extends Activity {
    TextView show;
    String dabaseName;
    SQLiteDatabase db;
    CreateSQLiteDatabase dabaseHelper;
    public void onCreate(Bundle savedInstanceState) {
        super.onCreate(savedInstanceState);
        setContentView(R.layout.ch12_4);
        show = (TextView)findViewById(R.id.text);
        dabaseName = "ComputerBook";
        dabaseHelper = new CreateSQLiteDatabase(this,dabaseName,null,1);
        show.append("befor Transaction:\n");
        showMess();
        doTransaction();
        show.append("\nafter Transaction:\n");
        showMess();
    }
    void doTransaction() {
        db = dabaseHelper.getWritableDatabase();
        int n = 100;
        String sql = "select amount from stock_list where id = '987-765' ";
```

```java
            Cursor cursor = db.rawQuery(sql,null);
            boolean boo = cursor.moveToNext();
            if(boo == false) return;
            int amount_1 = cursor.getInt(0);
            int stockAmount = amount_1 - n > 0?amount_1 - n:0;
            sql = "select amount from sell_list where id = '987 - 765'";
            cursor = db.rawQuery(sql,null);
            boo = cursor.moveToNext();
            if(boo == false) return;
            int amount_2 = cursor.getInt(0);
            int sellAmount = amount_2 + (amount_1 - n > 0?n:amount_1);
            db.beginTransaction();
                try {
                sql = "update stock_list SET amount = " + stockAmount + " where id = '987 - 765'";
                db.execSQL(sql);
                sql = "update sell_list SET amount = " + sellAmount + " where id = '987 - 765'";
                db.execSQL(sql);
                db.setTransactionSuccessful();
            }catch(Exception exp){}
            finally {
                db.endTransaction();
            }
            cursor.close();
            db.close();
        }
        public void showMess() {

            show.append("stock_list:\n");
            try {
                db = dabaseHelper.getWritableDatabase() ;
                String sql = "select * from stock_list";
                Cursor cursor = db.rawQuery(sql,null);
                while (cursor.moveToNext()) {
                    String id = cursor.getString(0);
                    String name = cursor.getString(1);
                    int amount = cursor.getInt(2);
                    show.append("" + id + "\t\t\t" + name + "\t\t\t" + amount);
                }
                cursor.close();
                db.close();
            }
            catch(Exception exp){
                show.append("" + exp);
            }
            show.append("\nsell_list:\n");
            try {
                db = dabaseHelper.getWritableDatabase() ;
                String sql = "select * from sell_list";
                Cursor cursor = db.rawQuery(sql,null);
                while (cursor.moveToNext()) {
                    String id = cursor.getString(0);
```

```
                    String name = cursor.getString(1);
                    int amount = cursor.getInt(2);
                    show.append("" + id + "\t\t\t" + name + "\t\t\t" + amount);
                }
                cursor.close();
                db.close();
            }
            catch(Exception exp){
                show.append("" + exp);
            }
        }
    }
    class CreateSQLiteDatabase extends SQLiteOpenHelper {
        public CreateSQLiteDatabase
        (Context context,String name,SQLiteDatabase.CursorFactory factory,int version){
            super(context,name,factory,version);
        }
        public void onCreate (SQLiteDatabase db) {
            String sql = "CREATE TABLE IF NOT EXISTS stock_list(id TEXT,name,amount)";
            db.execSQL(sql);
            sql = "INSERT INTO stock_list(id,name,amount) values('987 - 765','java',202)";
            db.execSQL(sql);
            sql = "CREATE TABLE IF NOT EXISTS sell_list(id TEXT,name,amount)";
            db.execSQL(sql);
            sql = "INSERT INTO sell_list(id,name,amount) values('987 - 765','java',18)";
            db.execSQL(sql);
        }
        public void onUpgrade(SQLiteDatabase db, int oldVersion, int newVersion) {}
    }
}
```

（4）启动 AVD，进入工程的根目录，用快捷方式编译工程、安装应用程序到 AVD（有关知识点参见 1.5 节）。对于本例子，用命令行进入 D:\2000\ch12_4，执行如下命令：

D:\2000 > ch12_4 > ant debug install

12.5　基于数据库的消费记载

如果读者希望用手机记载自己的消费情况，可以在学习例子 12-5 的基础上编写符合自己需要的手机程序。在例子 12-5 中，读者可以把自己消费的商品名称、价格作为一条记录保存到数据库的表中，读者也可以查询（支持模糊查询）自己的消费情况。运行效果如图 12.5(a)，12.5(b)所示。

例子 12-5

例子 12-5 给出的应用程序中有两个 Activity 对象，一个由 Example12_5 类创建的主要的 Activity 对象，负责录入商品的名称和价格，一个由 QueryGoods 类创建的 Activity 对象，负责查询已购买的商品以及列出全部的商品。

（1）创建名字为 ch12_5 的工程，主要 Activity 子类的名字为 Example12_5，使用的包名为 ch12.five。用命令行进入 D:\2000，创建工程 D:\2000>android create project -t 3 -n

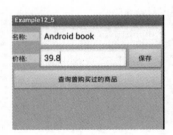
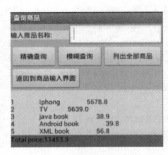

(a) 输入并保存商品　　　　(b) 查询或列出全部商品

图 12.5　运行效果

ch12_5 -p ./ch12_5 -a Example12_5 -k ch12.five。

（2）将下列主要的 Activity 对象以及 QueryGoods 类创建的 Activity 对象使用的视图文件 ch12_5.xml 和 find_goods.xml 保存到工程的\res\layout 目录中。

ch12_5.xml

```xml
<?xml version = "1.0" encoding = "utf-8"?>
<LinearLayout xmlns:android = "http://schemas.android.com/apk/res/android"
    android:orientation = "vertical"
    android:layout_width = "match_parent"
    android:layout_height = "match_parent"
    android:background = "#87CEEB">
    <LinearLayout
        android:layout_width = "match_parent"
        android:layout_height = "wrap_content">
        <TextView
            android:background = "#dddddd"
            android:textColor = "#000AFF"
            android:layout_width = "wrap_content"
            android:layout_height = "wrap_content"
            android:layout_weight = "1"
            android:text = "名称:"/>
        <EditText
            android:id = "@+id/edit_name"
            android:layout_width = "wrap_content"
            android:layout_weight = "5"
            android:layout_height = "wrap_content" />
    </LinearLayout>
    <LinearLayout
        android:layout_width = "match_parent"
        android:layout_height = "wrap_content">
        <TextView
            android:background = "#dddddd"
            android:textColor = "#000AFF"
            android:layout_width = "wrap_content"
            android:layout_height = "wrap_content"
            android:layout_weight = "1"
            android:text = "价格:"/>
```

```xml
<EditText
    android:id = "@+id/edit_price"
    android:layout_width = "wrap_content"
    android:layout_weight = "5"
    android:layout_height = "wrap_content" />
<Button
    android:layout_width = "wrap_content"
    android:layout_height = "wrap_content"
    android:layout_weight = "1"
    android:text = "保存"
    android:onClick = "saveGoods" />
    </LinearLayout>
        <Button
            android:layout_width = "match_parent"
            android:layout_height = "wrap_content"
            android:text = "查询曾购买过的商品"
            android:onClick = "findGoods" />
</LinearLayout>
```

find_goods.xml

```xml
<?xml version = "1.0" encoding = "utf-8"?>
<LinearLayout xmlns:android = "http://schemas.android.com/apk/res/android"
    android:orientation = "vertical"
    android:layout_width = "match_parent"
    android:layout_height = "match_parent"
    android:background = "#87CEEB">
    <LinearLayout
        android:layout_width = "match_parent"
        android:layout_height = "wrap_content">
        <TextView
            android:background = "#dddddd"
            android:textColor = "#000AFF"
            android:layout_width = "wrap_content"
            android:layout_height = "wrap_content"
            android:layout_weight = "1"
            android:text = "输入商品名称:"/>
        <EditText
            android:id = "@+id/edit_name"
            android:layout_width = "wrap_content"
            android:layout_weight = "5"
            android:layout_height = "wrap_content" />
    </LinearLayout>
    <LinearLayout
        android:layout_width = "match_parent"
        android:layout_height = "wrap_content">
        <Button
            android:layout_width = "wrap_content"
            android:layout_height = "wrap_content"
            android:text = "精确查询"
            android:layout_weight = "1"
```

```xml
            android:onClick = "queryExactly" />
    < Button
            android:layout_width = "wrap_content"
            android:layout_height = "wrap_content"
            android:text = "模糊查询"
            android:layout_weight = "1"
            android:onClick = "queryFuzzy" />
    < Button
            android:layout_width = "wrap_content"
            android:layout_height = "wrap_content"
            android:text = "列出全部商品"
            android:layout_weight = "1"
            android:onClick = "listAllGoods" />
</LinearLayout >
< Button
        android:layout_width = "wrap_content"
        android:layout_height = "wrap_content"
        android:text = "返回到商品输入界面"
        android:onClick = "back" />
< ScrollView
        android:layout_width = "match_parent"
        android:layout_height = "wrap_content"
        android:scrollbarStyle = "outsideOverlay"
        android:background = " #87CEEB">
    < TextView
            android:id = "@ + id/show_goods"
            android:background = " #dddddd"
            android:textColor = " #000AFF"
            android:layout_width = "match_parent"
            android:layout_height = "match_parent" />
</ScrollView >
< TextView
        android:id = "@ + id/show_total_price"
        android:background = " #FF0850"
        android:textColor = " #000000"
        android:layout_width = "match_parent"
        android:layout_height = "100dp" />
</LinearLayout >
```

(3)将负责创建查询的 Activity 对象的 QueryGoods.java 以及负责建立和打开数据库的 CreateSQLiteDatabase.java 保存到工程的\src\ch12\five 目录下,并修改工程\src\ch12\five 目录下的 Example12_5.java 文件。CreateSQLiteDatabase.java,QueryGoods.java 和修改后的 Example12_5.java 内容如下:

CreateSQLiteDatabase.java

```java
package ch12.five;
import android.content.*;
import android.database.sqlite.*;
public class CreateSQLiteDatabase extends SQLiteOpenHelper {
    public CreateSQLiteDatabase
```

```java
    (Context context, String name, SQLiteDatabase.CursorFactory factory, int version){
        super(context, name, factory, version);
    }
    public void onCreate(SQLiteDatabase db) {
        String sql = "CREATE TABLE IF NOT EXISTS goods_list" +
        "(id INTEGER primary key autoincrement, name, price)";
        db.execSQL(sql);
    }
    public void onUpgrade(SQLiteDatabase db, int oldVersion, int newVersion) {}
}
```

QueryGoods.java

```java
package ch12.five;
import android.app.*;
import android.os.Bundle;
import android.widget.*;
import android.view.*;
import android.content.*;
import android.database.sqlite.*;
import android.database.Cursor;
public class QueryGoods extends Activity {
    EditText edit_name;
    TextView show_goods;
    TextView show_total_price;
    String dabaseName ;
    SQLiteDatabase db;
    float sum = 0;
    CreateSQLiteDatabase dabaseHelper;
    public void onCreate(Bundle savedInstanceState) {
        super.onCreate(savedInstanceState);
        setContentView(R.layout.find_goods);
        dabaseName = "GoodsMess";
        dabaseHelper = new CreateSQLiteDatabase(this, dabaseName, null, 1);
        edit_name = (EditText)findViewById(R.id.edit_name);
        show_goods = (TextView)findViewById(R.id.show_goods);
        show_total_price = (TextView)findViewById(R.id.show_total_price);
    }
    public void queryExactly(View view) {
        String name = edit_name.getText().toString().trim();
        String sql = "select * from goods_list where name = '" + name + "'";
        if(name == null) return;
        show_goods.setText(null);
        query(sql);
    }
    public void queryFuzzy(View view) {
        String name = edit_name.getText().toString().trim();
        String sql = "select * from goods_list where name like '%" + name + "%'";
        if(name == null) return;
        show_goods.setText(null);
        query(sql);
```

```java
    }
    public void listAllGoods(View view) {
        String name = edit_name.getText().toString().trim();
        String sql = "select * from goods_list ";
        if(name == null) return;
        show_goods.setText(null);
        query(sql);
        show_total_price.setText("Total price:" + sum);
    }
    public void query(String sql) {
        sum = 0;
        try {
            db = dabaseHelper.getWritableDatabase();
            Cursor cursor = db.rawQuery(sql,null);
            while (cursor.moveToNext()) {
                int id = cursor.getInt(0);
                String name = cursor.getString(1);
                float price = cursor.getFloat(2);
                sum = sum + price;
                show_goods.append("\n" + id + "\t\t\t" + name + "\t\t\t" + price);
            }
            cursor.close();
            db.close();
            show_total_price.setText(null);
        }
        catch(Exception exp){}
    }
    public void back(View view) {
        Intent intent = new Intent(this,Example12_5.class);
        startActivity(intent);
    }
}
```

Example12_5.java

```java
package ch12.five;
import android.app.*;
import android.os.Bundle;
import android.widget.*;
import android.view.*;
import android.content.*;
import android.database.sqlite.*;
public class Example12_5 extends Activity {
    EditText edit_name,edit_price;
    String dabaseName ;
    SQLiteDatabase db;
    CreateSQLiteDatabase dabaseHelper;
    public void onCreate(Bundle savedInstanceState) {
        super.onCreate(savedInstanceState);
        setContentView(R.layout.ch12_5);
        dabaseName = "GoodsMess";
```

```java
        dabaseHelper = new CreateSQLiteDatabase(this,dabaseName,null,1);
        edit_name = (EditText)findViewById(R.id.edit_name);
        edit_price = (EditText)findViewById(R.id.edit_price);
    }
    public void saveGoods(View view) {
        String name = edit_name.getText().toString().trim();
        String price = edit_price.getText().toString().trim();
        Toast toast;
        if(name == null||price == null) return;
        try {
            db = dabaseHelper.getWritableDatabase();
            String sql = "INSERT INTO goods_list(id,name,price) values(?,?,?)";
            Object[] a = {null,name,price};
            db.execSQL(sql,a);
            toast = Toast.makeText (this,"success!",Toast.LENGTH_SHORT);
            toast.setGravity(Gravity.TOP,60,60);
            toast.show();
        }
        catch(Exception exp){
            toast = Toast.makeText (this,"fail!",Toast.LENGTH_SHORT);
            toast.setGravity(Gravity.TOP,20,60);
            toast.show();
        }
    }
    public void findGoods(View view) {
        Intent intent = new Intent(this,QueryGoods.class);
        startActivity(intent);
    }
}
```

（4）程序包含了两个 Activity 对象，需要修改工程根目录下的配置文件 AndroidManifest.xml，加入一个 Activity 标记，该标记对应着 QueryGoods 类创建的 Activity 对象，修改后的配置文件的内容如下：

AndroidManifest.xml

```xml
<?xml version = "1.0" encoding = "utf-8"?>
<manifest
        <!-- 此处省略了原有内容,以下是新增的activity标记 -->
        <activity android:name = "QueryGoods"
                android:label = "查询商品">
        </activity>
    <!-- 此处省略了原有内容 -->
</manifest>;
```

（5）启动 AVD，进入工程的根目录，用快捷方式编译工程、安装应用程序到 AVD（有关知识点参见 1.5 节）。对于本例子，用命令行进入 D:\2000\ch12_5，执行如下命令：

```
D:\2000>ch12_5> ant debug install
```

习 题 12

1. 编写一个程序,该程序首次运行时创建一个名字为 employee 的数据库,并创建一个名字为 popel_list 的表。该表的结构是(number INTEGER,name TEXT,age INTEGER)。程序提供视图,用户可以输入一条记录到数据库的表中。

2. 下载一个 SQLite 数据库的开发工具,设计好一个 SQL 数据库,并将该数据库外挂在读者的程序中(参考例子 12-2)。

3. 编写一个基于数据库的、积累型的英-汉字典程序(参考例子 12-5)。

教学资源支持

敬爱的教师：

感谢您一直以来对清华版计算机教材的支持和爱护。为了配合本课程的教学需要，本教材配有配套的电子教案（素材），有需求的教师请到清华大学出版社主页（http://www.tup.com.cn）上查询和下载，也可以拨打电话或发送电子邮件咨询。

如果您在使用本教材的过程中遇到了什么问题，或者有相关教材出版计划，也请您发邮件告诉我们，以便我们更好地为您服务。

我们的联系方式：

地　　　址：北京海淀区双清路学研大厦 A 座 707

邮　　　编：100084

电　　　话：010-62770175-4604

课件下载：http://www.tup.com.cn

电子邮件：weijj@tup.tsinghua.edu.cn

教师交流 QQ 群：136490705

教师服务微信：itbook8

教师服务 QQ：883604

（申请加入时，请写明您的学校名称和姓名）

用微信扫一扫右边的二维码，即可关注计算机教材公众号。

扫一扫
课件下载、样书申请
教材推荐、技术交流